Power-Transistor and TTL Integrated-Circuit Applications

Edited by

BRYAN NORRIS

Manager, Applications Laboratory
Texas Instruments Limited

McGRAW-HILL BOOK COMPANY

New York St. Louis San Francisco Auckland Bogotá
Düsseldorf Johannesburg London Madrid Mexico
Montreal New Delhi Panama Paris São Paulo
Singapore Sydney Tokyo Toronto

Library of Congress Cataloging in Publication Data
Main entry under title:

Power-transistor and TTL integrated-circuit applications.

(Texas Instruments electronics series)
Includes bibliographical references and index.
1. Power transistors. 2. Integrated circuits.
3. Semiconductor switches. 4. Logic curcuits.
1. Norris, Bryan. II. Texas Instruments Limited.
Applications Laboratory.
TK7871.92.P68 621.381'73 77-6258
ISBN 0-07-063754-7

1234567890 HDHD 786543210987

Contents

Preface

The chapters in this book have been chosen with a broad industrial bias to make it readily acceptable and of interest to a wide spectrum of engineers using semi-conductor devices. Each chapter has been written for engineers by a professional specialist engineer in the Texas Instruments Limited, Applications Laboratory in Bedford.

Even if the reader is not in the industrial field, he is likely sometime to be interested, for example, in Inverter design and Chapter X covers this extensively. Also I have yet to meet an engineer not interested in audio amplifier design—extensively covered for both the amateur and professional.

The chapters of the book group together readily to form four sections. The first covers power control with Triacs, Thyristors and Unijunction circuits. The first chapters of this section serve to give fundamentals, with constructions, characteristics and comparisons made between Triacs and Thyristors, followed by their applications in basic circuits. Chapters IV and V describe more specific applications, i.e. burst firing techniques and circuits and a reversible d.c. supply. Chapter VI gives extensive coverage of the theory, operation and circuits of Unijunction Transistors.

Section 2, Power Transistor Applications, concentrates on power supply designs, particularly of the switching mode type, and inverters. Although Chapter VIII talks specifically about a TV application, the regulator design procedure is applicable to all applications.

Section 3 is devoted to the design of audio amplifiers from 1 Watt to over 100 Watts, plus a high fidelity preamplifier.

The final section aims to show how the most widely used family of integrated circuits, i.e. Transistor-Transistor-Logic (TTL), can be applied. Basic descriptions and examples of how counters and registers may be used and connected are given (Chapter XIV), followed by assorted, fairly simple, circuit applications (Chapter XV). The final chapter (XVI) describes how TTL may be safely employed should it proposed environment be an electrically noisy one.

BRYAN NORRIS

Applications and Contracts Manager
Texas Instruments Limited

TEXAS INSTRUMENTS ELECTRONICS SERIES

Applications Laboratory Staff of Texas Instruments Incorporated ▪ DIGITAL INTEGRATED CIRCUITS AND OPERATIONAL AMPLIFIER AND OPTO-ELECTRONIC CIRCUIT DESIGN

Applications Laboratory Staff of Texas Instruments Incorporated ▪ ELECTRONIC POWER CONTROL AND DIGITAL TECHNIQUES

Applications Laboratory Staff of Texas Instruments Incorporated ▪ MOS AND SPECIAL-PURPOSE BIPOLAR INTEGRATED CIRCUITS AND R-F POWER TRANSISTOR CIRCUIT DESIGN

Applications Laboratory Staff of Texas Instruments Incorporated ▪ POWER-TRANSISTOR AND TTL INTEGRATED-CIRCUIT APPLICATIONS

Carr and Mize ▪ MOS/LSI DESIGN AND APPLICATION

Crawford ▪ MOSFET IN CIRCUIT DESIGN

Delhom ▪ DESIGN AND APPLICATION OF TRANSISTOR SWITCHING CIRCUITS

The Engineering Staff of Texas Instruments Incorporated ▪ CIRCUIT DESIGN FOR AUDIO, AM/FM, AND TV

The Engineering Staff of Texas Instruments Incorporated ▪ SOLID-STATE COMMUNICATIONS

The Engineering Staff of Texas Instruments Incorporated ▪ TRANSISTOR CIRCUIT DESIGN

The IC Applications Staff of Texas Instruments Incorporated ▪ DESIGNING WITH TTL INTEGRATED CIRCUITS

Hibberd ▪ INTEGRATED CIRCUITS

Hibberd ▪ SOLID-STATE ELECTRONICS

Kane and Larrabee ▪ CHARACTERIZATION OF SEMICONDUCTOR MATERIALS

Luecke, Mize, and Carr ▪ SEMICONDUCTOR MEASUREMENTS AND INSTRUMENTATION

Runyan ▪ SEMICONDUCTOR MEASUREMENTS AND INSTRUMENTATION

Runyan ▪ SILICON SEMICONDUCTOR TECHNOLOGY

Sevin ▪ FIELD-EFFECT TRANSISTORS

SECTION 1

POWER CONTROL WITH TRIACS, THYRISTORS AND UNIJUNCTION TRANSISTORS

I TRIACS - THEORY AND GENERAL APPLICATIONS

by Jurek Budek

The name of Thyristor is derived from the Greek word "η θυρα" meaning a door. Thus, strictly, all four layer (PNPN) devices with gates, such as silicon controlled rectifiers (SCR), bidirectional switches (Triacs) and gate controlled switches should be called Thyristors, However, the word "Thyristor" has become accepted in the past few years to mean a silicon controlled rectifier (SCR). In this book , therefore, it will have the same meaning.

Thyristor: A semiconductor switch. It is most commonly a three terminal semiconductor device with electrical characteristics similar to those of a thyratron, except that it has a very low internal dissipation. The unilateral Thyristor, commonly called SCR, and the bilateral switch commonly called Triac, are often mounted in a stud package as shown in Figure 1. The Thyristor has three connections, where the stud is the anode, the largest of two isolated leads on the case is the cathode, and the smaller lead is the gate. The Triac is identical with the Thyristor, except that the large lead is normally referred to as Main Terminal 1 (MT1) and the case as Main Terminal 2 (MT2).

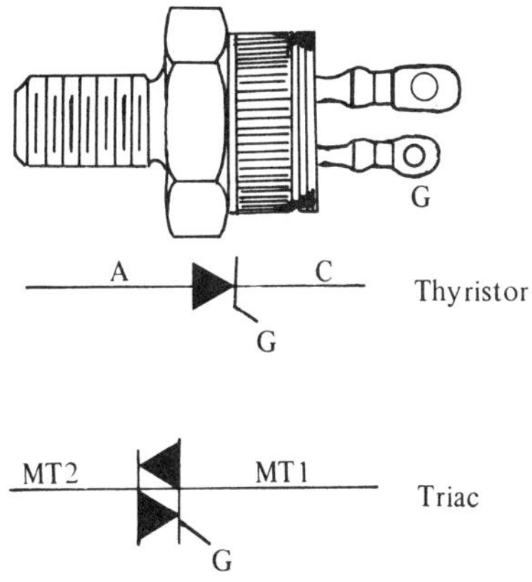

FIGURE 1

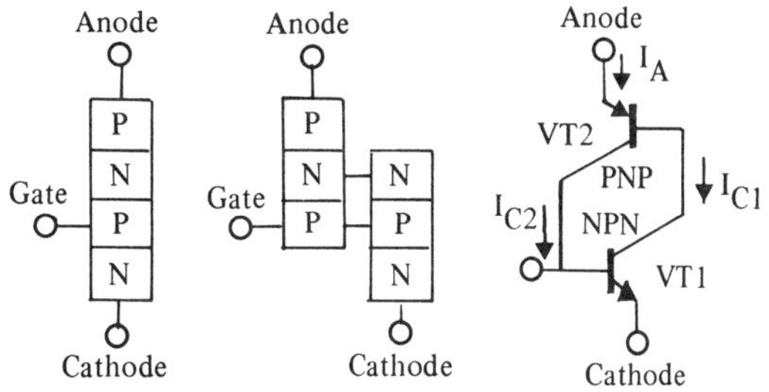

FIGURE 2

How a Four Layer (PNPN) Device Works

Four layer (PNPN) devices in use are either in one of two stable states; that is, they are either 'on' or 'off'. They are like switches in that they can be either open ('off') or closed ('on').

A two transistor analogy can best describe their action:–

$$I_A \text{ (total anode current)} = I_{C1} + I_{C2}$$

$$I_A = \frac{(1 + h_{FE1})\,(1 + h_{FE2})\,(I_{CBO1} + I_{CBO2})}{1 - (h_{FE1}\,h_{FE2})} \qquad (1)$$

With positive bias applied to the transistor pair (+ anode to cathode), h_{FE1} and h_{FE2} are both low and their product is less than unity. This is true since only small leakage currents are involved. Since h_{FE} is directly proportional to the collector current, the current gains are also small. Thus, the equation develops a value of I_A that is only slightly higher than the sum of the collector-base leakage currents $I_{CBO1} + I_{CBO2}$.

Regeneration takes place if the product of h_{FE1} and h_{FE2} is made to approach unity. In equation (1), I_A approaches infinity. Here the current increases and drives the transistor pair into saturation, thus turning it into the 'on' state. Thereafter, no further gate signal is required to maintain the 'on' state as long as the anode current is above the holding current.

Regeneration to cause an 'on' condition may be made to occur by:–

(i) applying a gate to cathode voltage (external base current to VT1) which causes the 'on' state to be reached by transistor action.

(ii) applying a fast rising voltage (dv/dt) to initiate turn 'on'.

(iii) exceeding the avalanche breakover voltage (V_{BO}) of the device.

Thyristors require a positive gate to cathode voltage and will turn 'on' only when the anode voltage is positive with respect to the cathode.

Triacs, on the other hand, can be designed to turn 'on' with either polarity of gate to MT1 and with either polarity of applied voltage between MT1 and MT2. For example, the Triac can be turned 'on' when MT2 is positive with respect to MT1 and the gate is negative with respect to MT1.

FEATURES

Thyristors

1. Three terminal devices similar to the thyratron except that it has a lower internal dissipation loss.
2. Conducts current in one direction only. (Figure 3). Basic construction of the Thyristor is shown in Figure 4.
3. Case is normally anode.
4. Triggered into conduction with a positive voltage on the gate of the device with respect to the cathode, with the main positive voltage applied to the anode.
5. Gate loses control when device is triggered on.
6. Turn on time is typically 1μs.
7. Conduction ceases when current drops below the holding current.
8. Turn off time is typically 5 - 60μs.
9. Technology is planar up to 2A, 250V and mesa up to 1000A, 2000V.

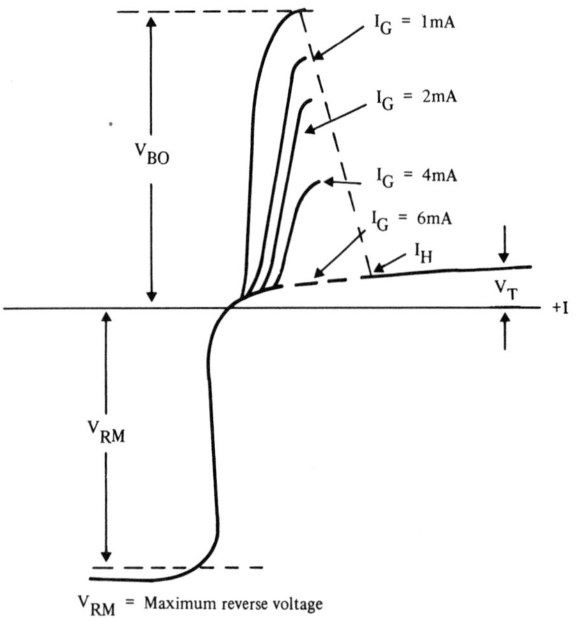

FIGURE 3 Typical Thyristor Characteristics

In a Thyristor, increasing the forward voltage does tend to increase the leakage current very slightly until the point at which avalanche begins. On exceeding this value, the leakage current will change rapidly into full conduction and the Thyristor will stay 'on' until the anode current drops below the holding current. On increasing the gate current, the forward break over voltage is reduced as shown in Figure 3. If the gate current is sufficiently large the forward blocking voltage is removed and Thyristor behaves as a diode.

Triacs

1. Conducts current (Figure 5) in either direction, similar to Thyristors operating in inverse parallel connection. Figure 6 shows basic Triac structure.
2. May trigger into conduction either by positive or by negative gate signals with either polarity of main terminal voltage.
3. Case is normally main terminal 2 (MT2).
4. Gate loses control when device is triggered on.
5. Turn on time is typically 1μs.
6. Conduction ceases when current drops below minimum holding current (I_H).
7. Turn off time is typically 50μs.
8. It offers simplified circuitry over two Thyristors, eliminating additional components needed in the triggering.
9. Present technology is limited to mesa, up to 200A, 1000V.

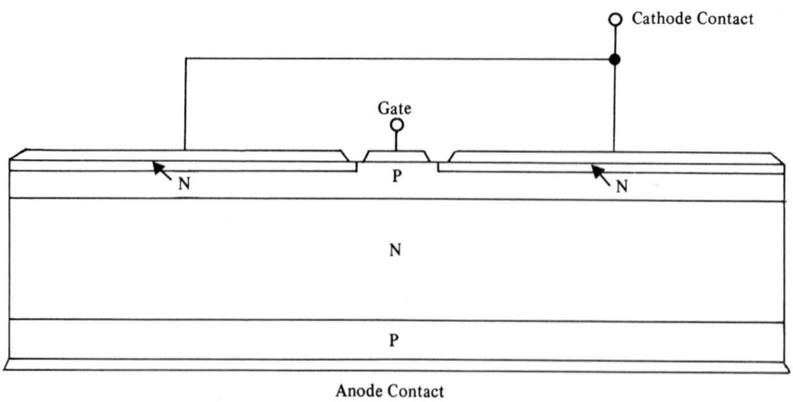

FIGURE 4 *Basic Thyristor Structures*

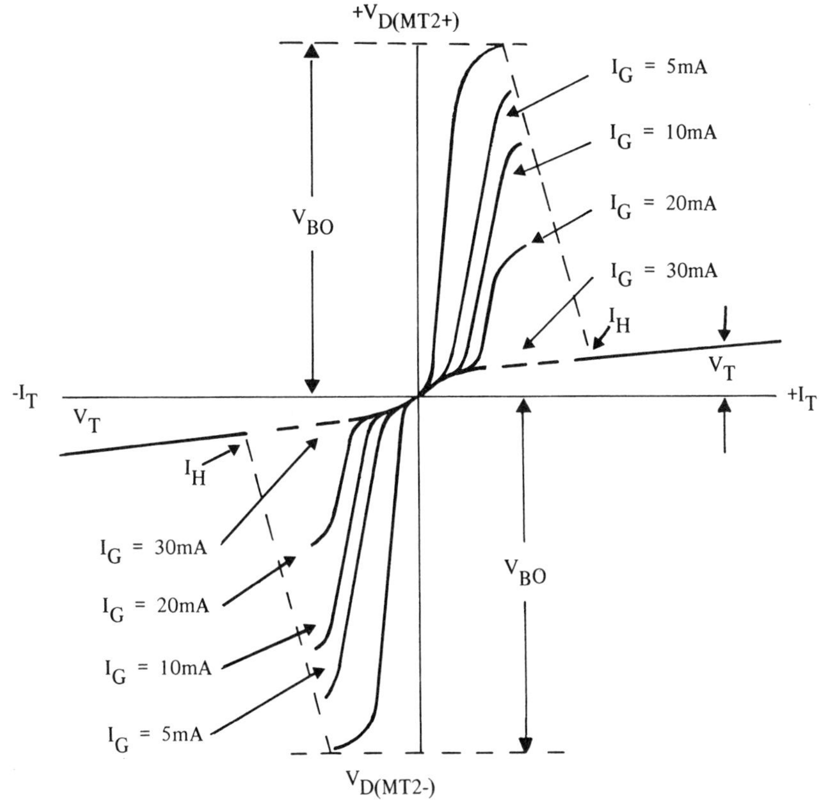

FIGURE 5 Typical Triac Characteristics

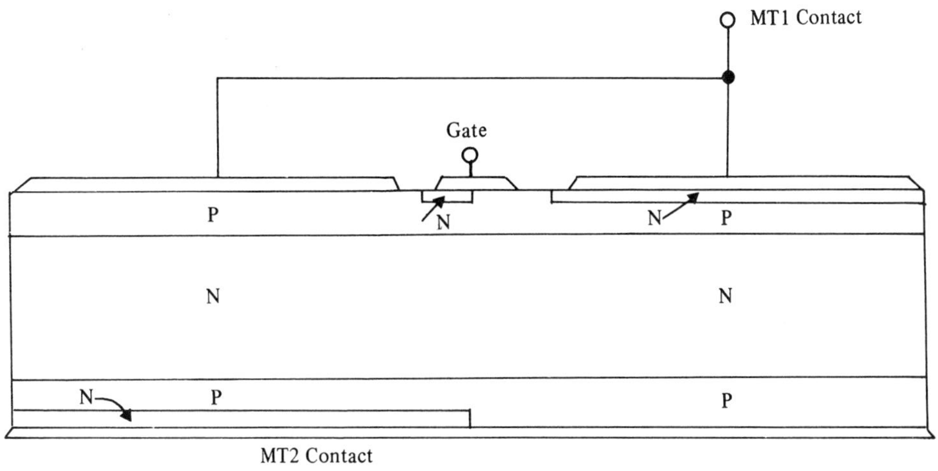

FIGURE 6 Basic Triac Structure

TRIAC ELECTRICAL CHARACTERISTICS

When the Triac is turned 'on', the current which flows between MT1 and MT2 is referred to as the principal current. The principal voltage-current characteristics of a Triac are shown in Figure 7; the voltage at MT1 is taken as a reference potential. Curve A is the V-I trace for a zero level gate current, and shows that for either polarity of applied voltage, the principal current is small and reasonably constant until the breakover voltage V_{BO} is reached. Curve B is the V-I trace for a gate current to the Triac; this current is measured with respect to MT1. Either positive or negative values of gate current (and hence, gate voltage) may be used to trigger the device. A hysteresis effect is noted along the current axis; two current levels, the latching current, I_L, and the holding current, I_H, are illustrated. The gate-drive signal must increase the principal current above the level of I_L in order for the Triac to latch into conduction.

Once the device is on at a high current level, the principal current must drop below the I_H level before the device will turn 'off'.

Figure 7 shows that the Triac operates in quadrants 1 and 3 of the V-I curve. As a rule, the Triac is difficult to trigger when device operation is in the third quadrant and the gate current is positive. Consequently, the Triac is generally not operated in this mode.

The repetitive peak off state current, I_{DRM}, illustrated in Figure 7 is measured at a value of repetitive peak reverse voltage, V_{DRM}, applied to MT2. The value of this reverse voltage is selected to be below the breakdown voltage, V_{BO}, of the device. A worst case condition for this measurement is the application of a DC reverse voltage equal to the magnitude of V_{DRM}.

As the Triac is controlled by a charge at the gate terminal, the peak gate trigger current required to turn 'on' a device is dependent upon the width of gate-drive pulse. The peak gate trigger current, I_{GTM}, is also determined by the polarity of voltage V_{supply} across the main terminals, in addition to the polarity of gate voltage. Values of V_{supply}, load resistance and pulse width are specified for the measurement of I_{GTM} by the Triac manufacturer.

The peak gate trigger voltage, V_{GTM}, is measured under the same circuit conditions as used for the measurement of I_{GTM}. This voltage parameter is simply the voltage drop from gate to MT1.

Triacs, like Thyristors, have a critical rate of rise of 'off' state voltage designated dv/dt. This is a measure of the maximum rate at which 'off' state voltage can be applied to the device, and is generally given in V/μs. False firing may occur if the dv/dt rating is exceeded.

Figure 8 shows the effect of temperature on the peak gate-trigger current, I_{GTM}, required to trigger typical Triacs, for various polarities of V_{supply}. (The gate pulse width $t_{p(g)}$ is 50μs). These curves indicate that less trigger current is required for negative levels of gate drive signals.

'On' state voltage drop across the main terminals is another important parameter of the Triac. This voltage is specified as V_{TM}, the peak 'on' state voltage. It is measured at a peak 'on' state current, I_{TM}.

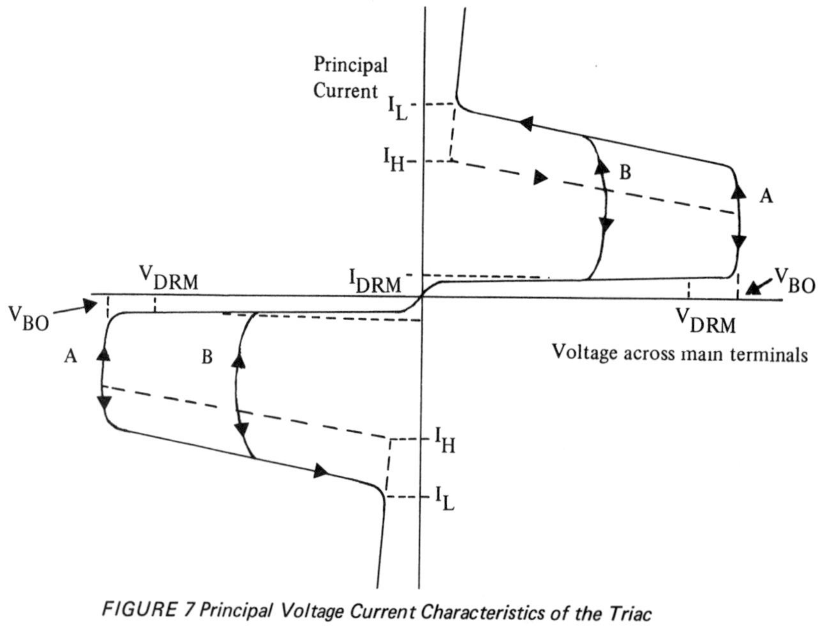

FIGURE 7 Principal Voltage Current Characteristics of the Triac

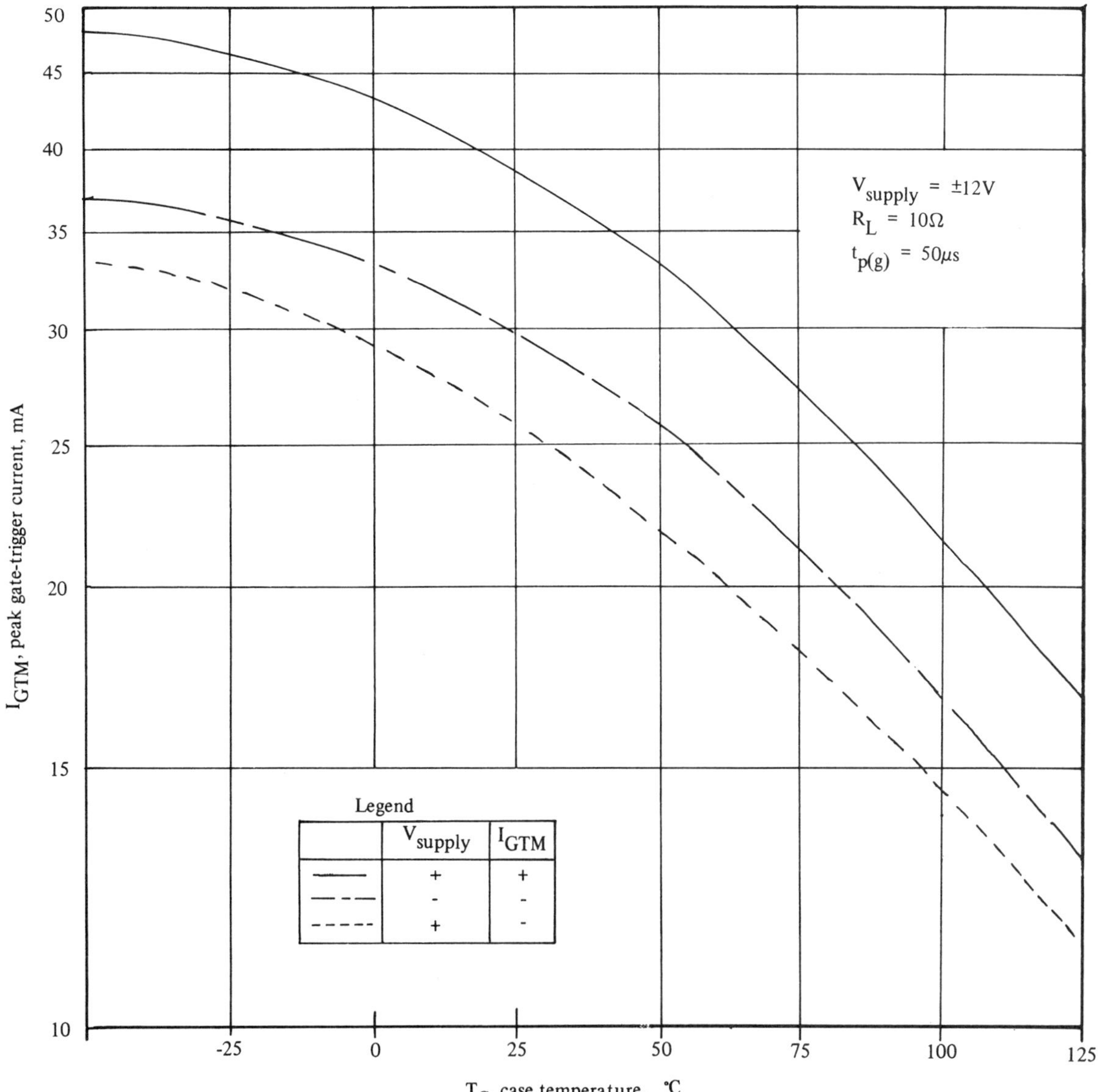

FIGURE 8

Typical curves of peak gate-trigger current versus case temperature for various polarities of V_{supply} and I_{GTM}.

APPLICATIONS

Comparison of Thyristor and Triac Circuits

A conventional method of providing full-wave phase control utilises two Thyristors connected as shown in Figure 9 (a). Each device conducts load current for up to a half cycle period, one during the positive and the other during the negative cycle of input AC waveform. This is accomplished by applying gate turn 'on' current during the interval that the anode voltage of the Thyristor becomes positive with respect to its cathode. Figure 9(b) shows typical voltage and current waveforms in the resistive circuit: gate current I_G, anode voltage, V_{anode}, and anode current, I_{anode}, are shown for one Thyristor. The gate-drive current is delayed by Θ electrical degrees for this particular case. By advancing or retarding Θ, proportional phase control is obtained.

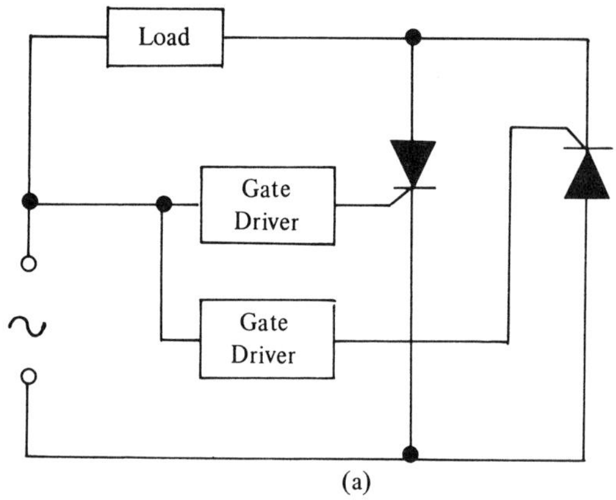

Thyristor full-wave phase-control circuit (a)

This circuit is highly efficient, as the Thyristor has a relatively low power dissipation in both the 'on' and 'off' state. In addition, the Thyristor switches between these two states in only a few microseconds. An undesirable feature of using two Thyristors for full wave control is the requirement that two separate gate drive signals must be provided. This is not necessary by using the Triac in place of two Thyristors. Furthermore, only one power component is needed instead of two, hence simplicity in mounting.

Figure 10(a) shows full-wave phase control in resistive circuit utilising a Triac. Waveforms of alternating supply voltage, gate current, I_G, voltage, V_D, across the main terminals and principal current, I_T, are shown in Figure 10(b). A comparison of Figures 9 and 10 show that the Triac is electrically similar to an inverse parallel connection of two Thyristors, with a common gate for the two devices. Consequently, the gate-drive circuit is less complex for the Triac than for the two Thyristors. The ability to conduct either polarity of load current makes the Triac well suited for full-wave phase control.

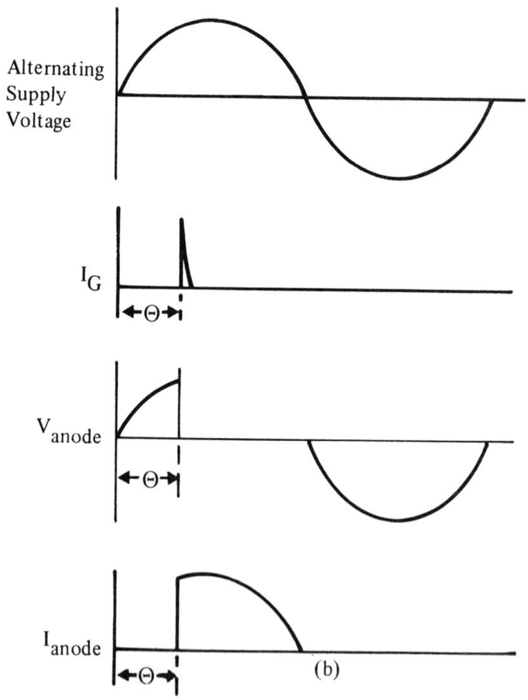

Voltage and current waveforms (b)
(Resistive loads)

FIGURE 9

The manner in which the required gate-trigger current varies as a function of current through the main terminals is shown in Figure 11 for various widths of gate drive pulses. Data is plotted for device operation in the first quadrant and for positive gate signals; gate current requirements are largest for these conditions. These curves show that the gate-trigger current for a given width of gate drive signal is relatively constant over a wide range of 'on' state currents. It is found that the gate-drive current is independent of V_{supply}.

In order for a Triac to be triggered to the 'on' state, the gate drive signal must not be removed until the principal current, I_T, rises above the latch current I_L. Figure 12 shows I_L as a function of gate pulse width for two different values of V_{supply}. These curves show that the value of I_L decreases as the gate drive pulse is made wider.

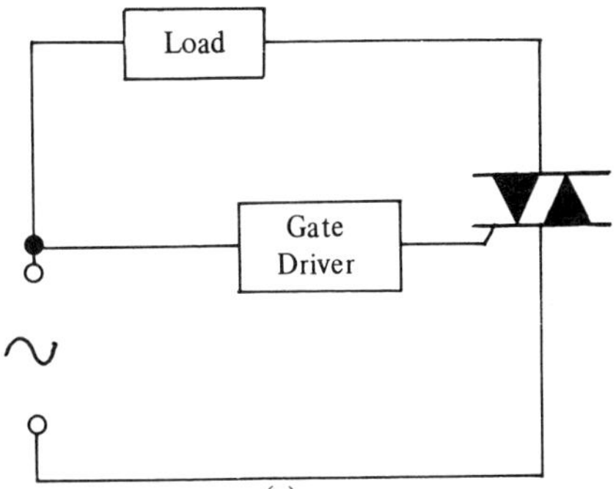

(a)

Triac full-wave phase-control circuit

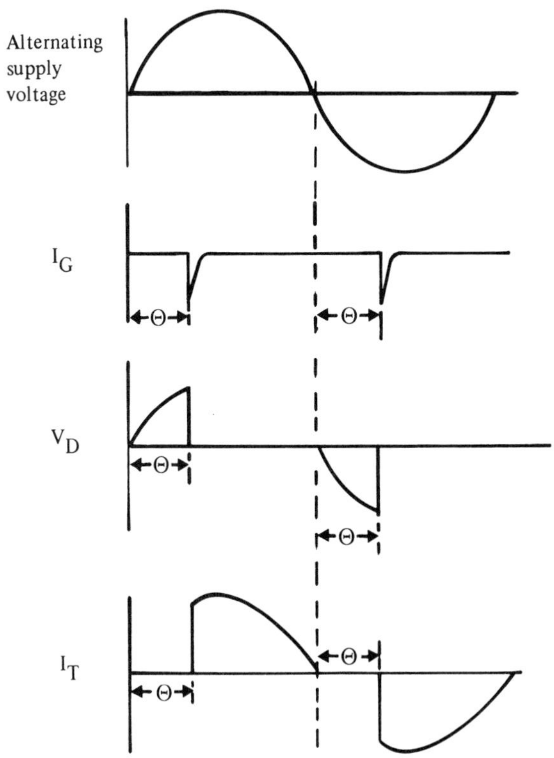

(b)

Voltage and current waveforms (Resistive load)

FIGURE 10

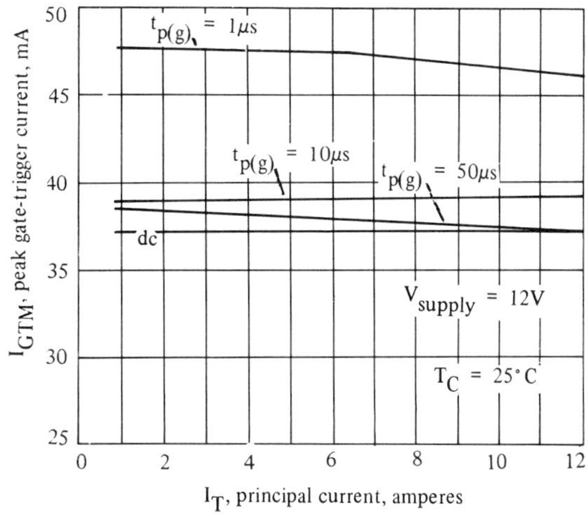

FIGURE 11 Typical curves of peak gate-trigger current versus principal current for various widths of gate-drive pulse.

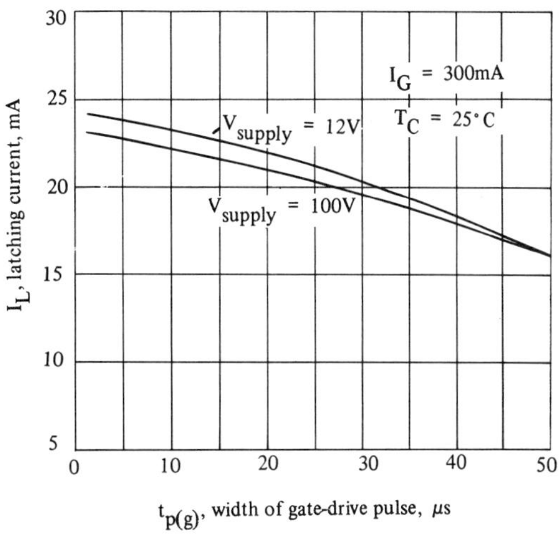

FIGURE 12 Typical curves of latching current versus width of gate-drive signal, for two values of V_{supply}

Triac Applications[1]

There are several circuit configurations for triggering the gate of a Triac. To a large extent, the trigger circuit depends upon the particular applications. Regardless of which trigger circuit is used, the gate turn-on current and the latching current must have at least minimum levels.

The gate drive circuit may consist simply of a resistor, R_G and a switch, S1, connected as shown in Figure 13. Gate current is positive when MT2 is positive, and negative when MT2 is negative. The undersirable operating mode of a positive gate current and a negative MT2 will not occur. The Triac will turn 'off' each time the principal current drops to zero. The recommended value for R_G is about 400 ohms for 240V A.C. supply. This circuit can be used as a solid state contactor.

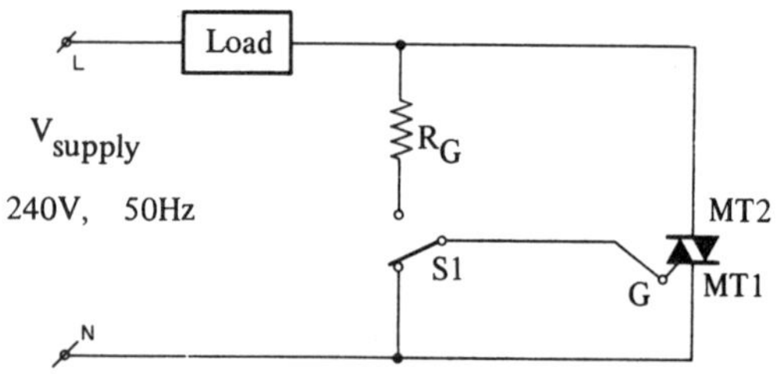

FIGURE 13 OFF - ON Power Control Circuit

As a spread in I_{GTM} among various Triacs exists, each device would conduct at a different value of V_{supply}. This problem of variation in the gate-trigger point can be overcome by using a voltage-threshold detector as shown in Figure 14. A bidirectional trigger diode D1 may be used to detect both polarities of V_{supply}. Similar results can be obtained by using a single phase bridge, connected in series with the gate of a Triac and shorting out the DC output of the bridge by a unidirectional trigger diode.

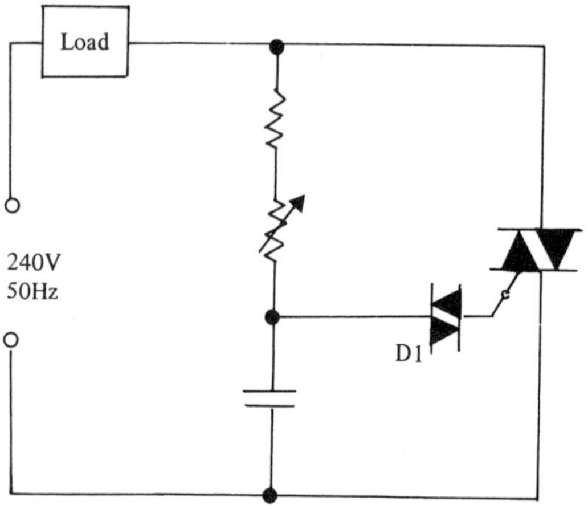

FIGURE 14 Single phase control using bidirectional trigger diode

In order to achieve an extended range for gate triggering a Triac, a double CR network may be used as shown in Figure 15. This circuit offers a smooth control of a DC motor and lamp dimming without a flicker. Voltage across trigger diode D1 is shifted in Phase and attenuated by two CR networks. When this voltage rises to the avalanche breakdown voltage of D1, a pulse of turn 'on' current will flow through the gate of the Triac.

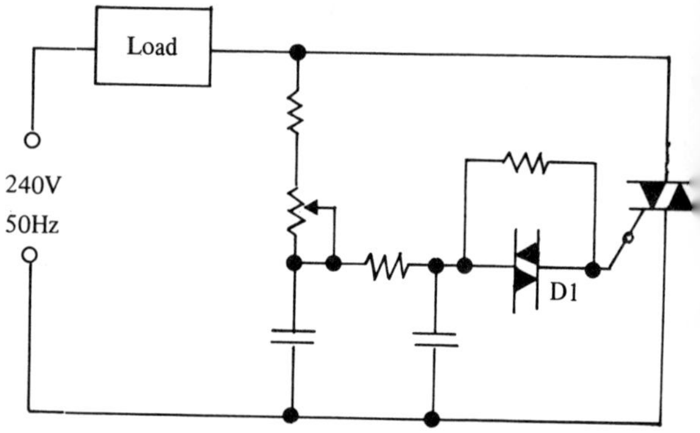

FIGURE 15 Phase Control with Bidirectional Trigger Diode

A light sensitive device can be utilised in controlling the operation of a Triac. Such a circuit could be used for applications as burglar alarms, automatic opening of doors, etc. This circuit is shown in Figure 16.

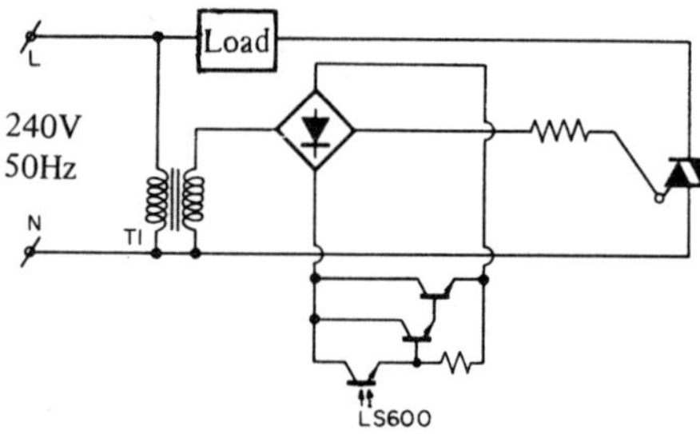

FIGURE 16 Triggering of Triac by Light Sensing Device

Integrated logic circuits may be used to control the Triac. Figure 17 shows an example of this.

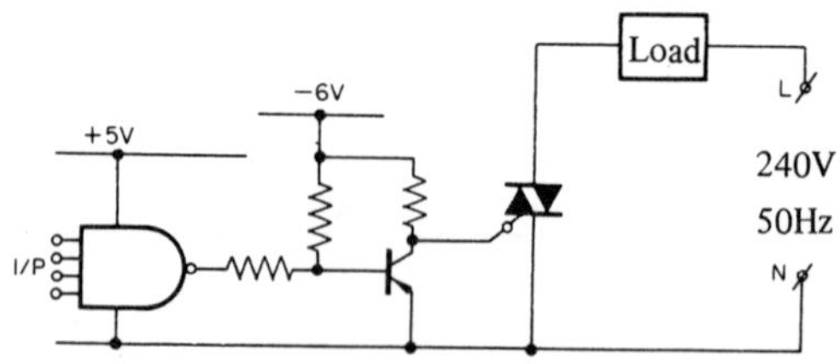

FIGURE 17 Integrated logic circuit for Triac trigger control

All these circuits are described in detail in later chapters[2] where component values are given.

PROTECTION[3]

To ensure the satisfactory operation of equipment using semiconductor devices, the problem of protection must be solved. There are several points to which a designer must pay attention, i.e. the rate of rise of voltage, (dv/dt), the rate of rise of current, (di/dt), voltage surges and overloads.

'Commutation' and 'Critical dv/dt'

When there are inductive loads, Triacs and Thyristors are subjected to a sudden rise of voltage, see Figure 18. If a device is not capable of withstanding such fast rises in voltages, it will lose control. The problem of dv/dt becomes important especially when dealing with 'commutation dv/dt' as the capability of a device is somewhat lower in the commutation case.

From Figure 18 it can be seen that two Thyristors connected in inverse parallel are subjected to 'critical dv/dt' as opposed to a Triac in a similar circuit being exposed to 'commutation dv/dt'. The 'critical dv/dt' can be defined as a blocking dv/dt applied to a device after the device has fully recovered from a principal current conduction. 'Commutation dv/dt' occurs when the blocking voltage is being stressed across the device during the recovery time from the principal current. Malfunctioning can even occur due to the fast voltage transients initiated by switching, etc. This characteristic phenomenon of Triacs and Thyristors is caused by the CR coupling between gate and MT2 terminal for Triacs and gate and anode for Thyristors. (R is the impedance between MT1 and gate). Unwanted turn 'on' at full voltage could bring some problems, especially when the equipment is feeding low impedance loads such as stationary motors or a bank of filament lamps. Excessive currents flowing, even for half a cycle, could blow the protective fuses or damage some of the semiconductors. The gates of the Triacs will resume control after half a cycle. The values of dv/dt are published in the manufacturers' data sheets. One way of slowing down the voltage rise of the switching supply voltage or transients is to bring the circuit into oscillation by adding a CR network and inductance if required. A suitable place to connect R and C is across the Triac.

Solving and analysing the following equation:–

$$L\,(di/dt) + Ri + (1/C).\int i.dt = V_{max} \quad \ldots\ldots \quad (2)$$

gives

$$(dv/dt)_{max} = V_{max} \Big/ \sqrt{LC}$$

hence

$$C = V_m{}^2/L.\ (dv/dt)^2 \quad \ldots\ldots\ldots\ldots \quad (3)$$

where V_{max} = peak supply voltage
and dv/dt = values obtained from the manufacturers data sheets (V/μs)

If the supply reactance = X%, then

$$(\omega L\ .\ I\ .\ 100)/V = X \quad \ldots\ldots\ldots\ldots \quad (4)$$

where I = supply current, rms
V = supply voltage, rms

Substituting L from equation (4) into equation (3), the value of the capacitor becomes:–

$$C = \frac{2.V.\pi.I}{(dv/dt)^2\ .\ 10^2\ .\ X}\ \mu F \quad \ldots\ldots\ldots\ldots \quad (5)$$

assuming a frequency of 50Hz and (dv/dt) in volts/μs.

The discharge of the capacitor through a Triac should be limited by adding a series resistor. This resistor should also be able to damp the ringing of the capacitance with the load inductance.

A suggested value for R is:–

$$R = 2\sqrt{\frac{L}{C}} \quad \ldots\ldots\ldots\ldots \quad (6)$$

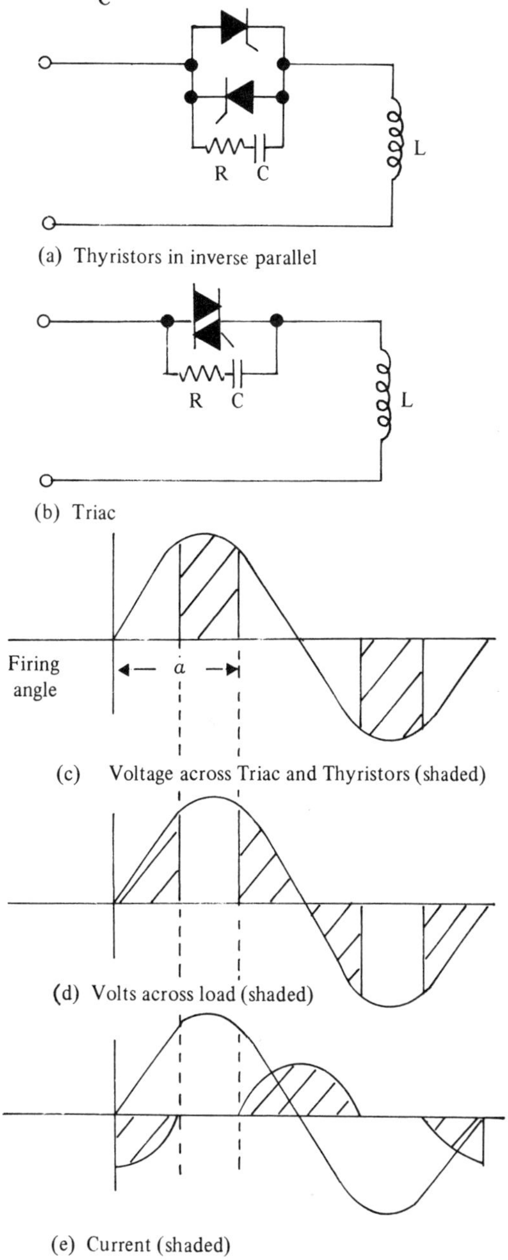

FIGURE 18 dv/dt in the inductive loads

Rate of Rise of Current (di/dt)

If the rate of rise of current is very high when compared with the speed with which the current turning 'on' can spread across the junction of the Triac or Thyristor, a local "hot spot" may develop causing a device to fail.

In most cases, there is no cause for concern as sufficient inductance exists in the circuit. However, where there is a fast rise of current, the value of di/dt should be examined and a choke added in the circuit to slow down the rise if necessary.

The value of inductance can be established from the expression:–

$$e = L.\frac{di}{dt} \qquad (7)$$

where e is a supply voltage and di/dt is the maximum permissible value of the rate of rise of current.

Voltage Transients

Diodes and Thyristors can be destroyed when subjected to excessive voltage transients, unless they incorporate avalanche characteristics. The Triac, being a bidirectional switch, will simply break over in one or other direction, turning 'on' into conduction. In spite of these self protective capabilities, the turning 'on' of a Triac, even for one isolated pulse, is not acceptable as explained on the previous page. An effective surge absorbing device, therefore, is necessary for reliable operation. Voltage transients can be initiated by various means such as switching transformers, inductors, from Thyristor circuits as commutation spikes, etc.

The suppression of transient voltages can easily be achieved if the sources and causes of transients are known. Voltage transients generated by switching 'off' transformers are known to a certain extent. The energy stored in the magnetic field can be evaluated from the equation $E_{(energy)} = Li^2/2$ where 'i' is peak magnetising current. By using a CR network as a surge absorbing device, a simple comparison of the magnetic energy with energy to be absorbed by the capacitor used ($CV^2/2$) will give its required value. There are many surge absorbing devices on the market in the form of CR networks, selenium plates connected back to back, or silicon carbide units. If the magnitude of the transient is unknown, a trial and error approach to the problem is often unavoidable.

Overloads and Short Circuits

The decision on what form the protection of an equipment should take rests with the designer. Depending on the application, the cost of protective devices must obviously be related to the price of the equipment. There are numerous problems to be borne in mind at the designing stage. The protective devices should be chosen and arranged in such a way as to obtain the desired discrimination.

AC and DC circuit breakers, AC and DC contactors, fuses and current limiting circuits are normally used to protect semiconductor equipments against overloads. There are some basic points which it may be helpful to remember. A circuit breaker is designed to protect against overloads and to open and isolate a healthy circuit during the short circuit. Contactors are, as a rule, cheaper than circuit breakers and are used for frequent switching and operation at overloads but must be backed up by fuses as they do not have the clearing capability during short circuits. Fuses, however, are the cheapest and most commonly used in medium and small sized equipments for protection against damage to wiring and components in heavy overloads and short circuits. When frequent faults are expected, the loss of fuses during each failure may present an economic problem. A circuit breaker or contactor could then become a more attractive proposition.

In order to withstand an inrush current to a transformer, the fuse in the input side of the transformer is usually rated at twice or three times the nominal current, unless other provision is made to reduce the inrush current.

When fuses and breakers are adapted for the protection of Triacs, correct discrimination can be achieved by studying the overload characteristics of fuses, Triacs and circuit breakers. Normally Triacs or Thyristors will be protected against a short circuit and a heavy overload by fuses, and against other conditions of overload by circuit breakers. Care must be taken to keep below the Triac overload curve at every point, by either fuse or breaker characteristics. The transformer reactance is a useful tool to control a fault current. The peak fault current can be evaluated from a transformer secondary rms current by using the following formula:–

$$I_{s/c\ peak} = (I_{rms} \cdot \sqrt{2} \cdot 100)/X$$

where X is a percentage reactance of a system (transformer plus supply). As an example, in three phase bridge connection, each device will carry a current approaching sinusoidal shape for 180° during short circuit and not 120° as in the case of a nominal load. Then the peak value, related to the nominal DC current, will be:–

$$I_{s/c\ peak} = \frac{I_{rms} \cdot \sqrt{2} . 100}{X} = \frac{2.\sqrt{2}.\ I_{DC} . 100}{\sqrt{3}.X}$$

Hence

$$I_{s/c\ peak} = \frac{116\ .\ I_{DC}}{X} \qquad (8)$$

From equation (8), it can be seen that the peak fault current can vary with the variation of transformer reactance. Furthermore, the first pulse of a fault current can be much higher than a steady state short circuit. This first pulse is called an "asymmetric peak" and its magnitude is a function of the "R/X" of the circuit. For practical purposes, this value can be assumed to be 1.6 times the symmetric short circuit. The asymmetric peak current is very important when dealing with isolated fault pulses and for obvious reasons can be neglected in long term overloads.

The knowledge of i^2t (let through energy) is very useful in cases where a fault current is expected to be interrupted between 2 and 8 milliseconds. In this time interval, the i^2t can be assumed to be approximately constant. A simple comparison of the Triac 'i^2t' with total fuse 'i^2t' will give the rating of the fuse needed. The total 'i^2t' of the fuse consists of pre-arcing and arcing 'i^2t. The technique of using 'i^2t' is dicussed in an article, reference 2.

COOLING

As the Triac can be used in various circuit configurations, it will be necessary to know the power dissipation of the device in terms of rms or mean current needed. Most of the Triac manufacturers provide information on power dissipation with reference to the rms current.

Figure 19 shows the ratio of current $\frac{\text{rms}}{\text{mean}} = f(\Theta)$, where Θ is the delay angle. It should be noted that the rms value and mean value are calculated for a Thyristor with one pulse per one cycle, for the same peak current.

$$I_{rms} = \sqrt{\frac{1}{2\pi}\int_a^{\pi} (I_m \sin X)^2 \, dx} = \frac{I_m}{2}\sqrt{\frac{1}{\pi}\left(\pi - a + \frac{\sin 2a}{2}\right)} \quad \ldots(9)$$

$$I_{mean} = \frac{1}{2\pi}\int_a^{\pi} I_m \sin X \, dx = \frac{I_m}{2\pi}(1 + \cos a) \quad \ldots(10)$$

The results of the above formulae are shown in Table 1.

The minimum heat sink requirements may be calculated for any 'on' state current, heat sink combination by the following procedure:–

(a) Determine the worst case power dissipation from Figure 20.

(b) Calculate the maximum allowable case to free air thermal resistance, $\Theta_{C\text{-}A}$, by use of the equation:–

$$\Theta_{C\text{-}A} = \frac{T_J - T_A}{P(av)} - \Theta_{J\text{-}C} \quad \ldots\ldots\ldots\ldots\ldots\ldots(11)$$

where T_J = junction temperature

T_A = ambient temperature

$P(av)$ = average power dissipation (Figure 20)

$\Theta_{J\text{-}C}$ = junction to case thermal resistance = 1.75°/W

(c) If possible, determine the area of heat sink from Figure 21. If Figure 21 cannot be used, obtain further information from the heat sink manufacturer.

The Triac (Figure 21) is mounted in the centre of a square heat sink vertically positioned in still free air with both sides exposed. The heat sink area is twice the area of one side. $\Theta_{C\text{-}A}$ includes the case to heat sink thermal resistance $\Theta_{C\text{-}HS}$ in addition to the heat sink to air thermal resistance $\Theta_{HS\text{-}A}$ and is defined by the equation:–

$$\Theta_{C\text{-}A} = \Theta_{C\text{-}HS} + \Theta_{HS\text{-}A} \quad \ldots\ldots\ldots\ldots\ldots\ldots(12)$$

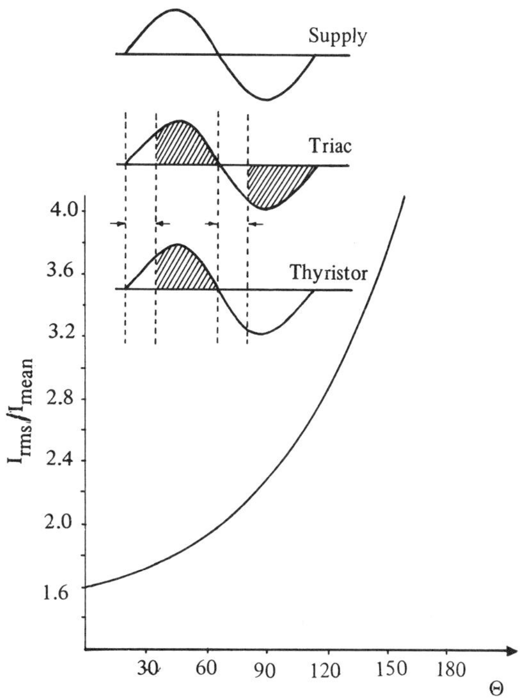

FIGURE 19 Ratio between rms and mean versus delay angle Θ

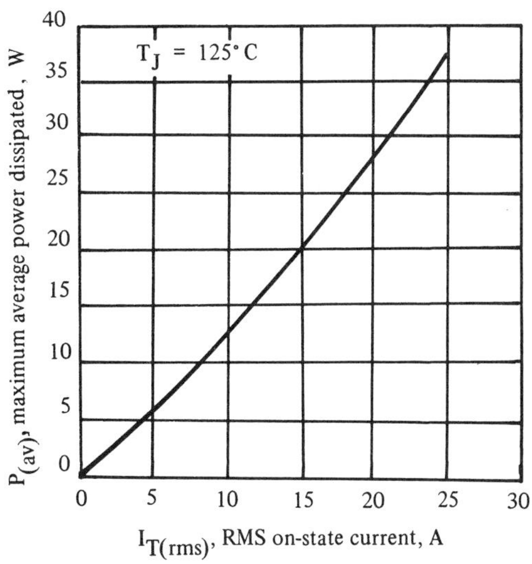

FIGURE 20

Maximum average power dissipation versus rms on-state current

Table 1.

Θ°	0	30	60	90	120	150	180
I_m X (rms)	0.5	0.492	0.447	0.353	0.22	0.084	0
I_m X (mean)	0.319	0.298	0.24	0.159	0.08	0.0215	0
$\frac{I_{rms}}{I_{mean}}$	1.57	1.65	1.86	2.22	2.75	3.9	∞

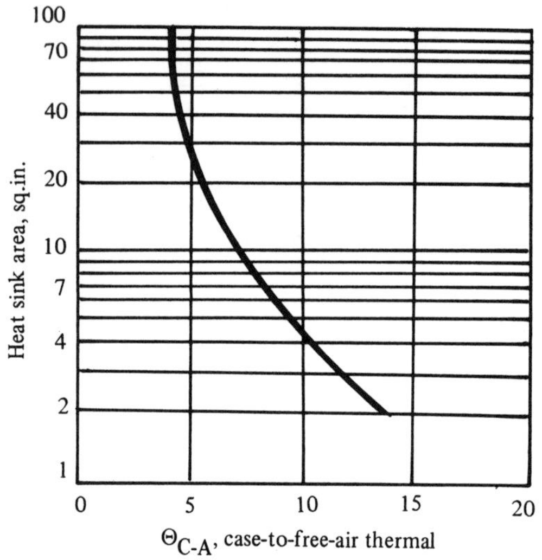

FIGURE 21

Case to free air thermal resistance (1/32" thick aluminium heat sink)

Example

Determine the minimum size of 1/32" thick aluminium heat sink for safe operation of the Triac at an rms current of 10A, with the device mounted directly on the heat sink, given that the maximum:–

$T_J = 125°C.$

$T_A = 35°C$

$\Theta_{J-C} = 1.75°/W$

From Figure 20, P (av) – 13W for 10A rms. Using equation (11),

$$\Theta_{C-A} = \frac{125°C - 35°C}{13W} - 1.75°/W = 5.1°/W.$$

For $\Theta_{C-A} = 5.1$ (Figure 21), the area is 24 square inches (total), the minimum dimension then is $\sqrt{\frac{24}{2}} = 3.5"$. The dimension of the heat sink will be 3.5" x 3.5".

TERMS AND SYMBOLS

dv/dt	Critical rate of 'off' state voltage
I_{DRM}	Repetitive peak 'off' state current
I_{GM}	Peak gate current
I_{GTM}	Peak gate trigger current
I_H	Holding current
I_L	Latching current
I_T	'On' state current (rms)
I_{TSM}	Peak 'on' state surge current
$P_{G(av)}$	Average gate power dissipation
P_{GM}	Peak gate power dissipation
T_C	Operating case temperature
V_{BO}	Breakover voltage
V_D	Continuous 'off' state voltage
V_{DRM}	Repetitive peak 'off' state voltage
V_{GM}	Peak gate voltage
V_{GTM}	Peak gate trigger voltage
V_{TM}	Peak 'on' state voltage

REFERENCES

1. *SCR Manual,* Fifth Edition, General Electric, Syracuse N.Y., 13201, 1972.
2. Chapter 2, pp 13-16, and Chapter 3, pp 17-
3. J. A. Budek and A. H. Marchant, 'Protective Methods for Silicon Rectifier Equipment', *Direct Current,* 1964.

II TRIACS WITH RESISTIVE AND INDUCTIVE LOADS

by Jurek Budek

The performance and reliability of semiconductors have proved to be acceptable in a variety of applications ranging from space vehicles to high voltage light current or low voltage heavy current equipments.

The thyristor has established its position in the industrial and domestic field with the circuits for switching or controlling a.c. or D.C. powers. The Triac, replacing 'back to back' connected Thyristors, offers definite advantages in certain applications over the Thyristor.

The most common application of the Triac is that of controlling resistive or inductive loads. The triggering, however, in the two cases is different. This report will attempt to explain certain phenomena occurring in controlling these loads by Triacs together with recommended triggering for inductive or resistive loads.

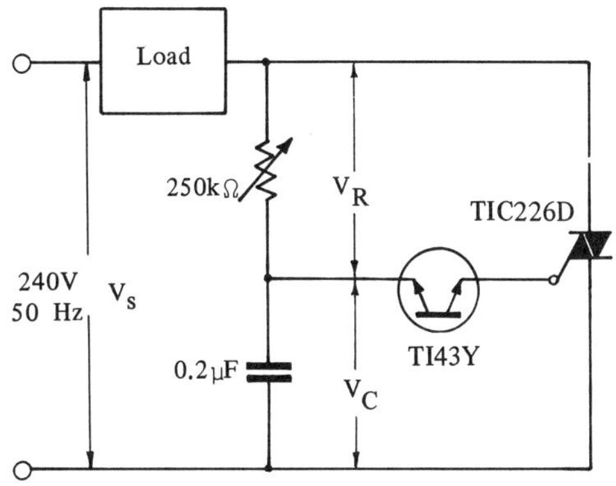

FIGURE 1

TRIAC OPERATED BY A TRIGGER DIODE (FOR RESISTIVE LOADS)

A trigger diode is the main device in the Triac gating circuit. This is usually a diffused silicon trigger bidirectional device with forward and reverse breakover voltages of, say, 32 ± 6V. The breakdown differential voltage, however, is typically guaranteed to be within 2 volts which is very well suited for a symmetrical firing of the positive and negative half cycles. Furthermore, the breakback voltage 'V' is of the order of 8 volts and this, together with a triggering capacitor, provides sufficient energy to turn-on the Triac effectively.

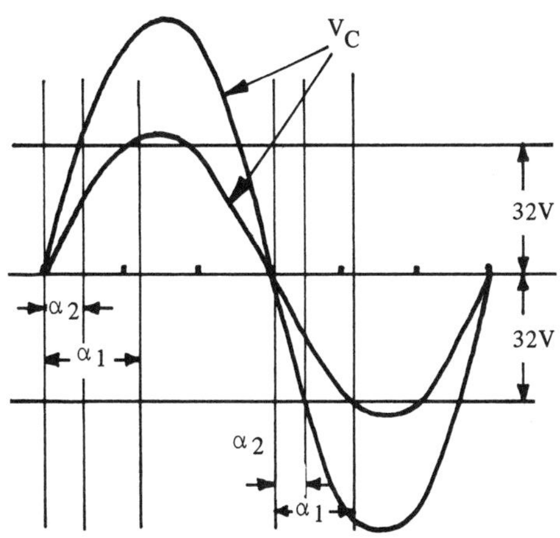

FIGURE 2

A Simple Trigger Circuit

Two factors contribute to the phase shift in the circuit shown in Figure 1.

When the voltage across the capacitor, V_C, is increased by reducing the value of the potentiometer resistor, the trigger diode will break down earlier at 'α2' and not 'α1' as is shown in Figure 2 – hence the phase shift.

On the other hand, by varying the ratio of potentiometer 'R' to '1/ωC', the phase shift is also obtained between the capacitor voltage 'V_C' and the input voltage 'V_S'. This phase shift is shown in Figure 3.

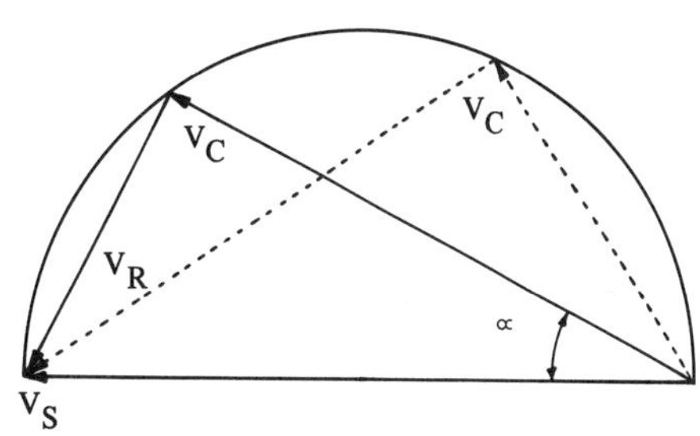

FIGURE 3

The actual phase shift as applied to the trigger diode will be, therefore, a combination of the two phase shifts discussed above. Figure 4 shows the phase shift caused by the variation of 'R/ωC' only, (1/ωC is assumed to be constant) and, also, the actual phase shift. The curves have

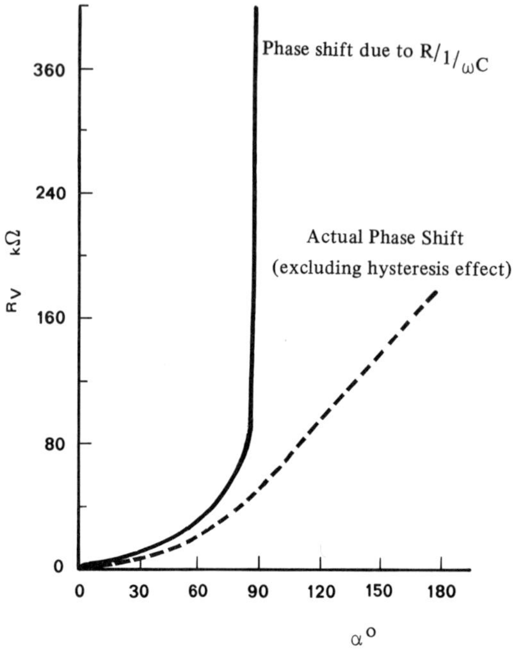

FIGURE 4 Relation Between Delay Angle 'α' and Potentiometer Resistor

been plotted for 240V a.c. input, 50Hz and 0.2μF capacitor. Figure 4 shows that control is possible from approximately 0° to 180° by using the circuit shown in Figure 1.

Hysteresis Effect

It was noticed, when using the circuit shown in Figure 1, that once the Triac has been turned 'on' at the minimum volts, the volts could be further reduced by increasing the potentiometer resistance. In other words, the striking voltage seemed to be higher than the turning 'off' voltage. This hysteresis type phenomenon is attributed to the trigger diode and it can be explained by examining Figure 5. The peak voltage across the capacitor, V_C, must be slightly higher than that of the breakdown voltage of the trigger diode or the Triac may never turn 'on' there being insufficient volts between terminals MT1 and MT2.

As soon as the trigger diode breaks down, part of the energy from the capacitor is transferred to the gate of the Triac. Figure 5 shows that the first breakdown occurs when the voltage across the capacitor reaches the forward breakdown voltage, giving a phase shift β_1. The triac will then turn 'on' and the capacitor, partly discharged, will charge again. Thus when the second and succeeding breakdowns occur at the reverse and forward breakdown voltages alternately, they will give a phase shift β_2. The steady state condition, therefore, will be established at β_2. Once the firing point is set at β_2, it is then possible to reduce the load voltage by increasing the resistance of the potentiometer until the peak of the shifted, V_C, can be reached at the trigger diode breakdown line.

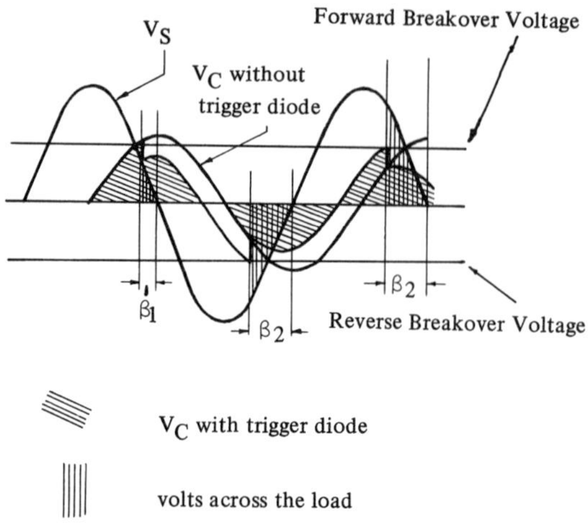

FIGURE 5

Double C/R Circuit

It is not always desirable to turn on the Triac with the voltage, V_C, around the peak value since any small variation in V_C could cause a large change in the delay angle 'α'. This problem can be eliminated by adding a second C/R network to that of the circuit shown in Figure 1. It is clear from Figure 6 that a further fixed delay is achieved at the gate of a Triac by adding resistor R2 and capacitor C2. This delay is shown in Figure 7.

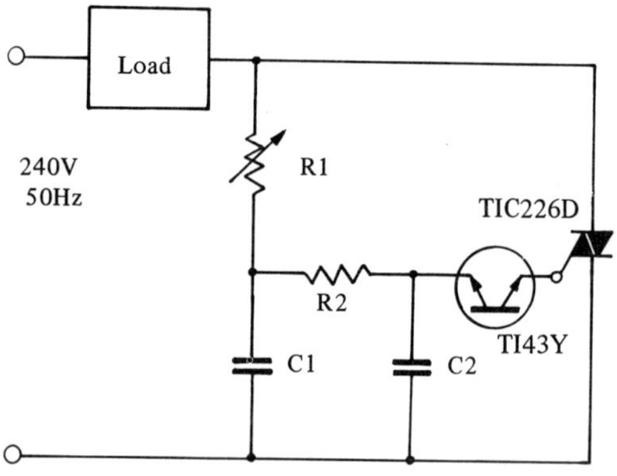

FIGURE 6

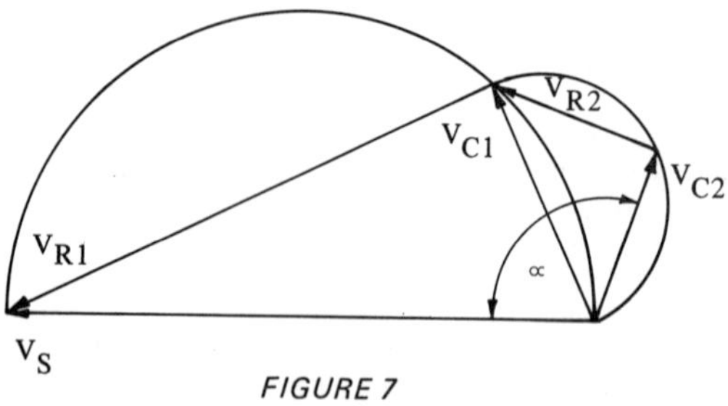

FIGURE 7

It is possible for the trigger diode to breakdown during the fast rising section of the, V_{C2}, sine wave. The reduction of the hysteresis effect is a further advantage of using a double C/R circuit. After the trigger diode has turned 'on', the partly discharged capacitor, C2, is recharged by some of the energy from capacitor C1. This occurs during the time the Triac is conducting, hence the smaller phase shift and hysteresis.

Choice of Circuit for Resistive Loads

Figure 8 shows the circuit recommended for use with resistive loads. It is a cheap and simple circuit which gives adequate control in lamp dimming without flicker and with minimized hysteresis effect. The resistor, connected in series with the potentiometer, limits the maximum voltage applied to the capacitors. The resistor across the trigger diode ensures smooth control at all angles of delay.

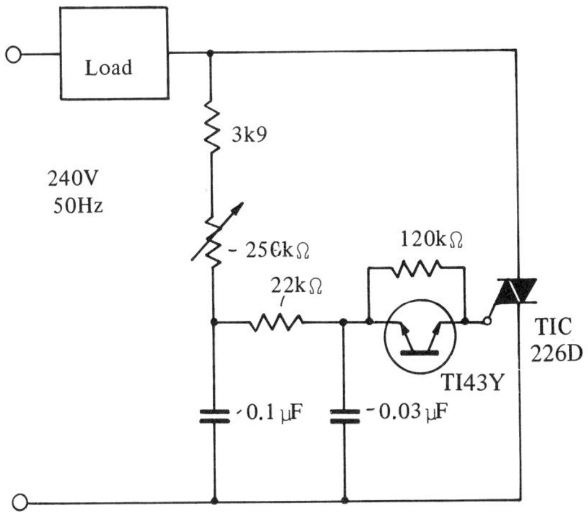

FIGURE 8

The 'load' in Figure 8 can be replaced by a single phase bridge with the D.C. outputs connected to a fractional H.P. D.C. motor. By adding a 50µF capacitor in series with a 4.7Ω(5W) resistor across the armature, a smooth speed control of the D.C. motor can be obtained.

TRIACS WITH INDUCTIVE LOADS

The circuit discussed in the previous paragraph cannot be used for controlling inductive loads. With the resistive loads (Figure 1) the current and voltage are in phase, hence the capacitor starts charging again at the time when the voltage reverses. Thus, the positive and negative half cycles will reach the trigger diode's breakdown voltage at the same time, resulting in symmetrical firing of the Triac. In the case of an inductive load, however, for a set value of the potentiometer, the current will still flow when the a.c. voltage reverses, causing the second half of the cycle to be delayed. This is because the capacitor will begin to recharge from the time when the current ceases to flow through the Triac. Figure 9 illustrates the positive and negative half

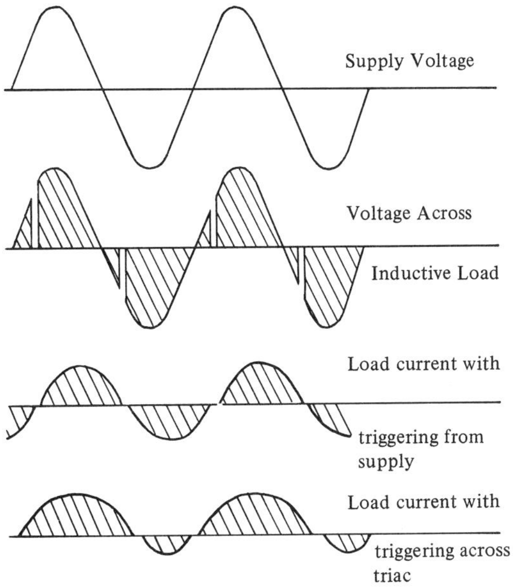

FIGURE 9

cycles with inductive loads where triggering is directly from the supply and across the Triac.

When the triggering is directly from the supply it provides one remedy for correcting the unbalance. Figure 10 shows the recommended circuit for controlling inductive loads of up to 6kW.

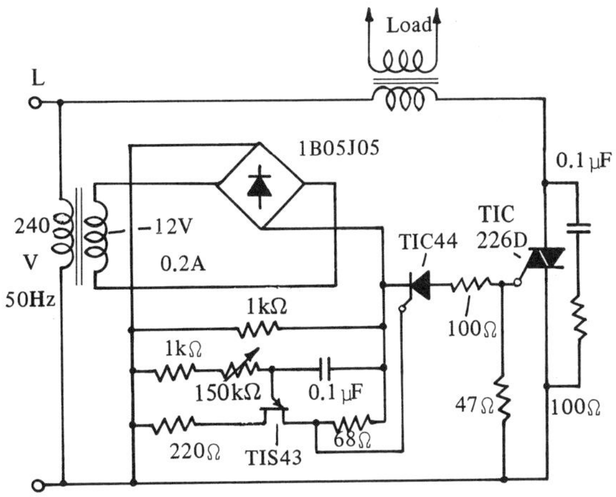

FIGURE 10

The low supply voltage to the triggering circuit (12V) means that it is necessary to replace the trigger diode by a unijunction transistor. The pulses generated by a basic unijunction transistor circuit cannot be used for direct gating of a Triac as they are of a very short duration and may disappear just before the Triac reaches the latching current. One way of extending these pulses is to use a 'slave Thyristor' through which the resistive current will flow. The thyristor will then turn 'off' each time the full wave rectifier voltage reaches zero as shown in Figure 9. The normal precautions against dv/dt and di/dt are arranged as in Chapter I. The unbalanced current flowing through the transformer will become even more unbalanced due to the presence of D.C. component.

A very large unbalance in one direction would certainly destroy the Triac. The circuit shown in Figure 10 gives a reasonable balance between the positive and negative half cycles. A further improvement, if required, can be accomplished by connecting a resistor across the primary winding of the transformer.

III SOLID-STATE SWITCHING USING TRIACS AND THYRISTORS

by Jurek Budek

Mechanical contactors and relays, having moving parts, are subject to wear and tear and require periodic maintenance. There are cases when frequent switching operations are needed; for instance, controlling a fresh water supply by switching on and off electric pumps, controlling electric heaters, solenoids, etc. In addition to the cost of the maintenance of contactors, the inconvenience caused by the interruption of work, especially in nonattended stations, is another factor that has to be considered.

A triac, used as a semiconductor switch, has features which overcome these difficulties and is ideally suited for frequent operation. It is small and reliable. It turns on when a gate pulse is applied and in a.c. circuits switches off each time the current passes through zero. It requires a very small gate power to control the main (principal) current. It must be remembered, however, that by using a solid-state contactor, complete isolation is not possible. In fact a triac inserts in the circuit a very high impedance when in off state, turning this impedance to a very low value in the on position. For complete electrical isolation an off load switch is recommended.

This chapter is concerned mainly with single-phase switching circuits, but some 3-phase and d.c. static contactors are discussed.

SIMPLE TRIAC SWITCH

The gate drive circuit may consist simply of a resistor and a switch, connected as shown in Fig 1. Gate current is positive when MT_2 is positive, and negative when MT_2 is negative. The triac will turn off each time the principal current drops to zero.

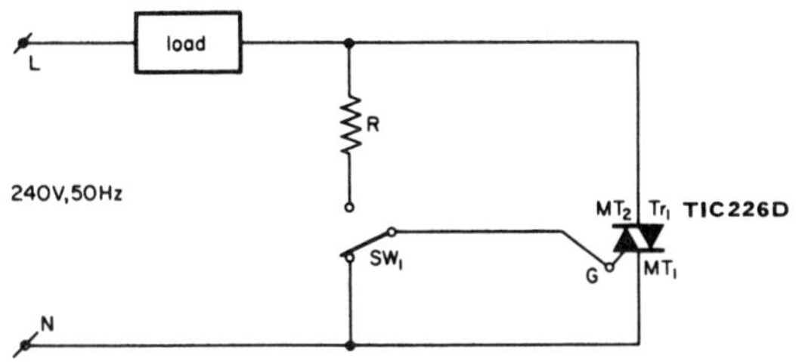

FIGURE 1

The resistor R limits the gate current to the desired value. Assuming the worst condition when the triac is turned on at the peak supply voltage V, the resistor R, must be

$$R > \sqrt{2}/I_{GM}$$

where I_{GM} is the permissible maximum peak gate current. If R is made too large, the triac will not trigger at the beginning of each cycle, resulting in the loss of some of the output voltage. The recommended resistor R is 120Ω. The toggle switch can be replaced by a reed relay or pilot thermostat contacts, thus eliminating the wear of contacts associated with breaking high currents.

TRIAC SWITCH OPERATED BY A LIGHT SOURCE

A light-sensing device can be utilised in controlling the operation of a triac. A circuit, shown in Fig 2, could be used for applications such as burglar alarms, automatic opening of doors, switching motors, etc.

The light-sensor amplifier comprises a high-gain Darlington pair with collector and emitter terminals connected across a single-phase bridge. When the light falls on and activates the LS600 light sensor, a voltage is developed across the 100kΩ resistor which turns on the Darlington pair. This in turn shorts out the bridge and fires the Triac. By changing over the positions of the light sensor and 100kΩ resistor, the operation can be reversed.

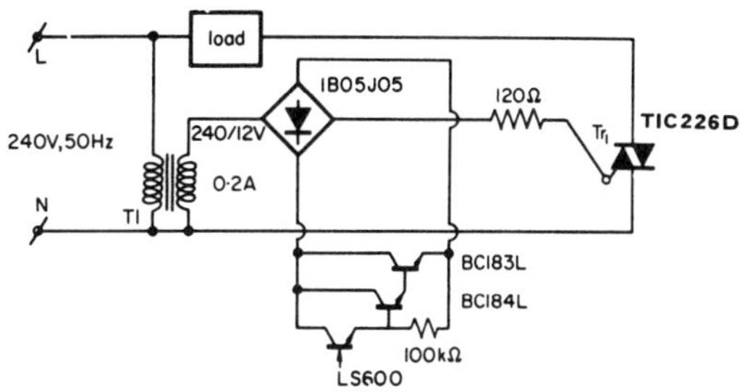

FIGURE 2

TRIAC SWITCH OPERATED BY A LOGIC NAND CIRCUIT

An integrated logic circuit may be used to control the triac; Fig. 3 shows an example of this.

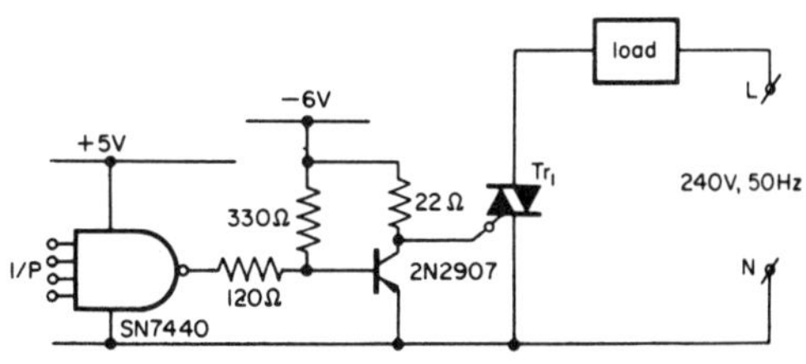

FIGURE 3

The NAND gate controls the state of a pnp transistor which, in turn, controls the gate current to the triac. When all inputs to the NAND gate are at a positive level, the pnp transistor is turned on; the gate is now brought down to almost neutral potential of the supply through the saturated pnp device. If any input to the NAND gate decreases to a low level, the pnp device turns off, and turn on gate current flows from the triac. In this circuit the components have been chosen to give sufficiently large gate power to turn on a 25A r.m.s. Triac. It should be noted that the neutral of supply is not at the earth potential.

SWITCHING 3-PHASE LOADS WITH A TRIAC

Switching on and off 3-phase inductive or resistive loads can be achieved by controlling power to the gates of triacs as shown in Fig. 4. Three resistive loads R_L are connected in star via three triacs Tr_1, Tr_2, and Tr_3. The supply to the gates is arranged through 100Ω resistors $R_{1/3}$ and blocking diodes $D_{1/3}$. As shown in Fig. 4, the load can be controlled either by a toggle switch, a pushbutton, a relay contact or light sensing. For complete isolation, a switch S_1 has been added and fuses F_2 and F_1 are also included to protect the main circuit and auxiliaries against a fault.

TRIGGERING TRIACS FROM A VERY-HIGH-IMPEDANCE SOURCE

It is sometimes necessary to control loads from a very-high-impedance source. As an example, a circuit is given in Fig. 5 which could be used in plastic welding equipments as a protection against flashover. The impedance of the plastic material, during normal welding, may be of many thousand megohms. As the plastic is not always perfect, a pinhole or unevenness could reduce this impedance to a very low level, possibly 1 or 2 MΩ. As a result there could be flashover and arcing. This, of course, in undesirable. The flashover may damage the welding electrodes and, if the welding was on plastic bags containing liquid, puncturing and thus some splashing could occur.

In order to prevent the flashover, a triac could be used as a solid-state contactor, being turned off when the impedance between electrodes is reduced to a few megohms and thus cutting off the supply to the h.v. transformer just before the breakdown voltage is reached.

Referring to Fig. 5, the sensing arm of the balanced bridge is connected to the welding electrodes; E to the earthed electrode and S to the live one. The high-voltage high-frequency across electrodes has been filtered out externally

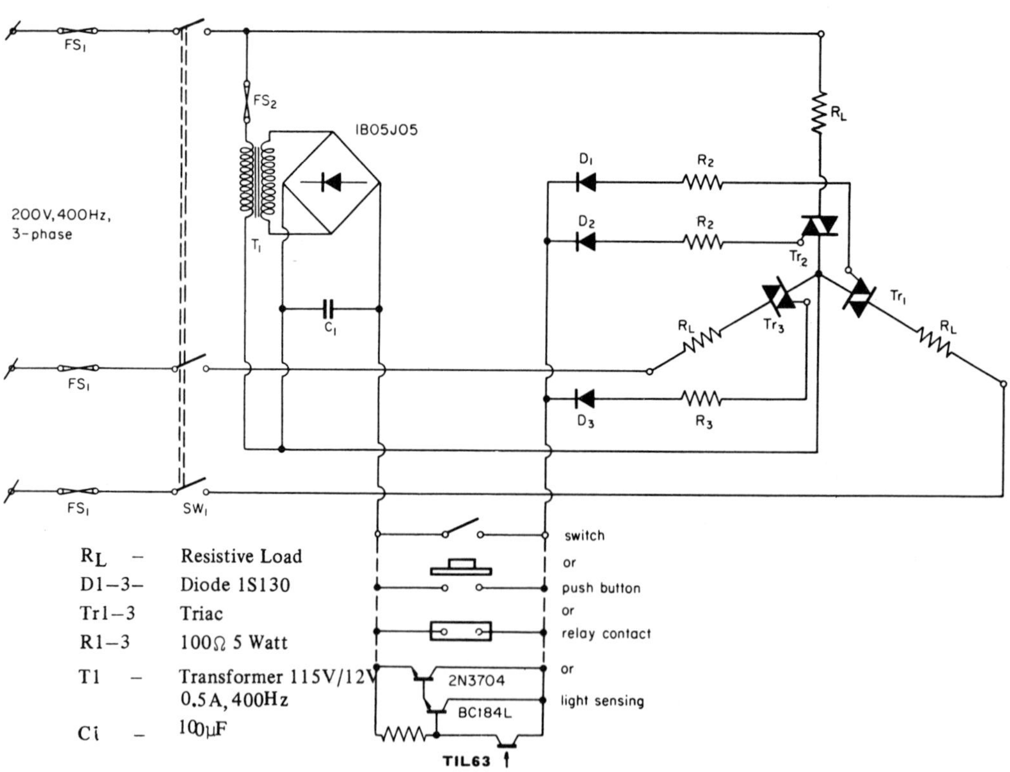

FIGURE 4

and looking from the input side E-S represents impedance only. The sensing bridge, consisting of resistors R_3, R_4, R_5 and potentiometer RV_1 is balanced for E-S higher than 1 to 8MΩ adjusted by RV_1. The output of the sensing bridge is injected into the f.e.t. differential amplifier (VT_1 and VT_2) and then to the modified Schmitt-trigger circuit consisting of VT_4 and VT_5 transistors. When the transistor VT_6 is on the voltage developed across R_{19} will turn on transistor VT_7 and hence the single-phase bridge SPB_2 will be shorted out, completing primary connection to the transformer T_3.

The d.c. current, limited by resistor R_{22}, will now flow through the gate and terminal MT_1 of triac Tr_1.

Capacitor C_2 ensures that the triac is either in on or off state and eliminates unbalanced phase control which is undesirable in inductive loads[1]. The purpose of transformer T_3 is to isolate the earthed electrode E from the neutral of the supply N. An f.e.t. differential amplifier is being used because of the high impedances involved in the sensing bridge. Back to back zener diodes ZD_1 and ZD_2 protect the f.e.t differential amplifier against excessive voltage transients generated during arcing while adjusting potentiometer RV_1.

The protection against excessive dV/dt is arranged by connecting resistor R_{25} in series with capacitor C_4 across the triac[2].

The user normally provides an inrush current limiting resistor R_{24} which is short time rated and is shorted out as soon as the steady-state conditions are obtained. The triac protective fuse F_2 can be then rated closely to the nominal current.

The circuit shown in Fig. 5 suggests a means of triggering triac from a very high impedance source. By redesigning the sensing bridge, the usage of this circuit can be extended.

USING A TRIAC TO CHANGE TRANSFORMER TAPS

Triacs can be used in place of mechanical contactors to change the tappings of a transformer. Care must be taken, however, to eliminate the possibility of triggering both triacs at the same time, as, in such cases, a short circuit current will flow in the winding shorted out by both triacs.

Fig. 6 shows a circuit consisting of two triacs controlling two tappings of a transformer. Both triacs are triggered by a logic NAND circuit. The operation of the circuit is similar to that described earlier. When all inputs to the NAND gate are at a positive level, triac Tr_2 is on giving a lower voltage from transformer T_1.

If any input to the NAND gate decreases to a low level, triac Tr_1 is turned on giving higher voltage output from the transformer. Both i.c. circuits G_1 and G_2 are mounted on the same chip with G_2 inverting the output signal from G_1 so that at any time when triac Tr_1 is on triac Tr_2 is off, or vice versa. By switching off −6V supply, both triacs will turn off.

There is no danger here that the triacs Tr_1 and Tr_2 may short out part of the winding when changing the state from on to off, as the switching over is made with triacs conducting the current in the same direction. Similar protection will be needed against transformer in-rush current as discussed on this page. When the current and voltage permit the above circuit can be adopted for the secondary windings of the transformer. An advantage would be that there are no transformer in-rush current problems.

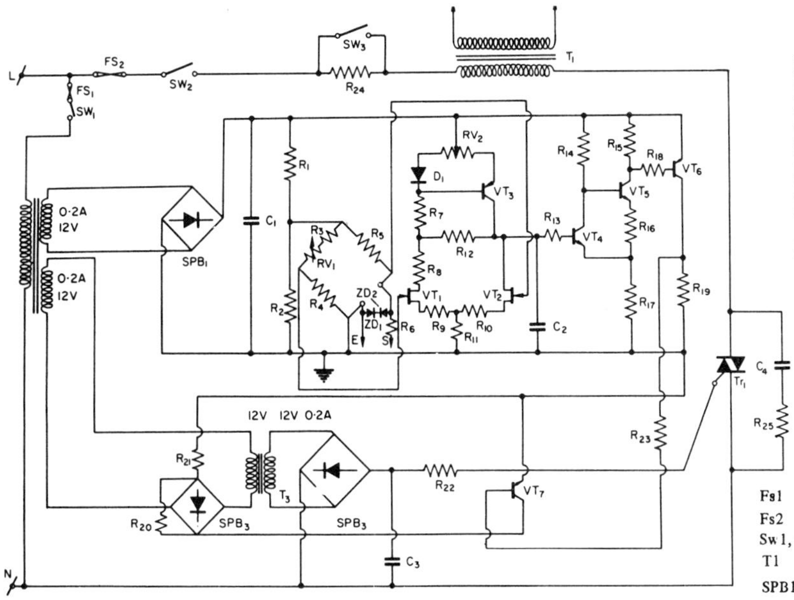

C_1, C_2	100μF, 25V
C_3	4μF, 25V
C_4	0.1μF, 240V r.m.s.
VT_1, VT_2	f.e.t. type 2N3819
VT_3, VT_6	2N3702
VT_4, VT_5 VT_7	BC183L
D_1	1S130
Tr_1	triac,
R_1, R_2, R_{12}	10kΩ, ½W
R_3	0.33MΩ, ½W
R_4	1.2MΩ, ½W
R_5	2.2MΩ, ½W
R_6	100kΩ, ½W
R_7	560Ω, ½W
R_8, R_9, R_{10}, R_{16}	1kΩ, ½W
R_{11}	2.2kΩ, ½W
R_{13}	39kΩ, ½W
R_{14}	5.6kΩ, ½W
R_{15}	1.2kΩ, ½W
R_{17}	6.8kΩ, ½W
R_{18}	4.7kΩ, ½W
R_{19}, R_{20}, R_{23}	12kΩ, ½W
R_{21}	15Ω, ½W
R_{22}	100Ω, ½W
R_{24}	3Ω, 20W
R_{25}	100Ω, ½W
RV_1	2MΩ potentiometer
RV_2	1kΩ potentiometer
ZD_1, ZD_2	Zener diode 1S2180A
Fs1	25 ARMS, type GS450/25
Fs2	2A
Sw1, Sw2, Sw3	Off load switches
T1	HV Trans 6kVA, 240V, 50Hz.
SPB1/2/3	1 φ bridge type 1B05J05

FIGURE 5

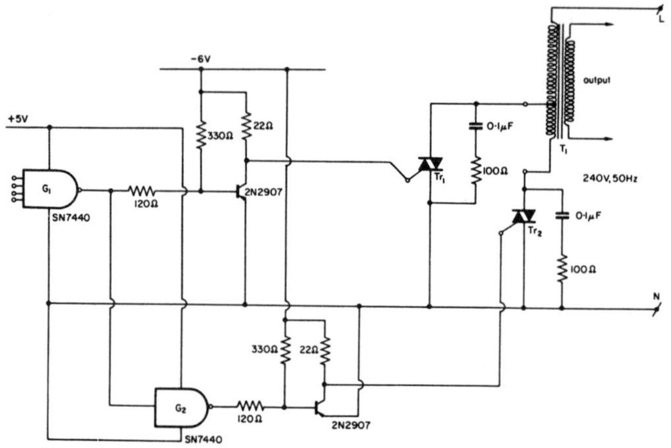

FIGURE 6

LIGHT SENSOR/THYRISTOR CONTROL OF D.C. TOY MOTOR

The gate has no control over its thyristor once anode to cathode current is flowing. In D.C. circuits, therefore, the current through the thyristor can be interrupted only by a mechanical switch connected either in series or across the thyristor, or, by forced commutation as used in inverter circuits. However, small thyristors can be switched 'off' by reverse biassing the gate. This method should be applied with caution as far as the gate to cathode reverse voltage is concerned. With increasing areas of the thyristor junction, i.e. for larger devices, negative gate bias has less effect.

Figure 7 gives a circuit suitable for switching 'on' and 'off' a small D.C. Toy Motor by means of a thyristor.

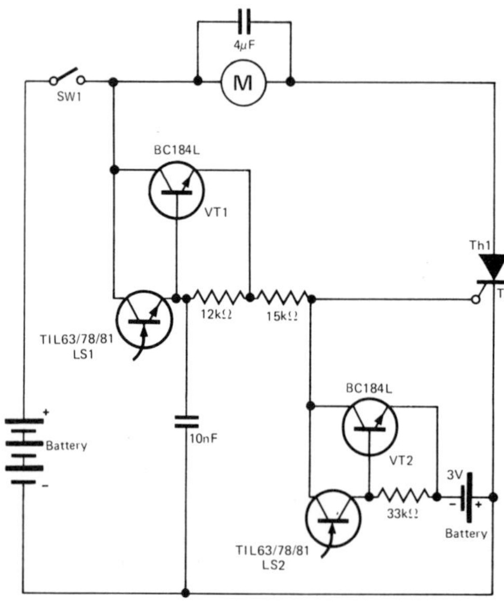

FIGURE 7

On shining the light from a torch on the light sensing device 'LS1' transistor VT1 will turn 'on' and the gate current will flow through thyristor, switching 'on' the supply to the motor. By diverting the light beam from 'LS1' device to 'LS2', transistor VT2, will turn 'on' and apply reverse voltage to the gate of thyristor Th 1. The latter will turn 'off' and hence the supply to the toy motor will also be switched 'off'.

The above circuit can have various applications, especially in the toy manufacturing industry. An example might be walking dolls. The eyes of a doll could be replaced by light sensing devices. By shining the light in one eye of a doll, the doll will move, and by diverting the light beam from a torch to the other eye of the doll will stop moving.

SWITCHING d.c. SOLENOIDS BY THYRISTORS

Additionally, a thyristor can be switched off in d.c. circuits by applying a reverse voltage to the gate. Fig. 8 shows a circuit with an array of solenoids L_1 being controlled individually by applying signals to the thyristors Th_1 gates.

All the solenoids are switched off by a signal applied to th base of transistor VT_1. This circuit required two d.c. supplies, 24V and 8V. The capacitor C_1 is charged to 24V and its energy is used to give an extra boost to the solenoid at the time of switching on. As soon as the solenoid is energised its holding current is taken, then from a 8V d.c. supply. Diode D_1 ensures that the capacitor C_1 is charged to 24V at the time when thyristor Th_1 is off. Diode D_3 is a blocking diode, preventing interaction between the gates of thyristors. Diode D2 discharges energy stored in the solenoid at the time when thyristor Th_1 is switched off.

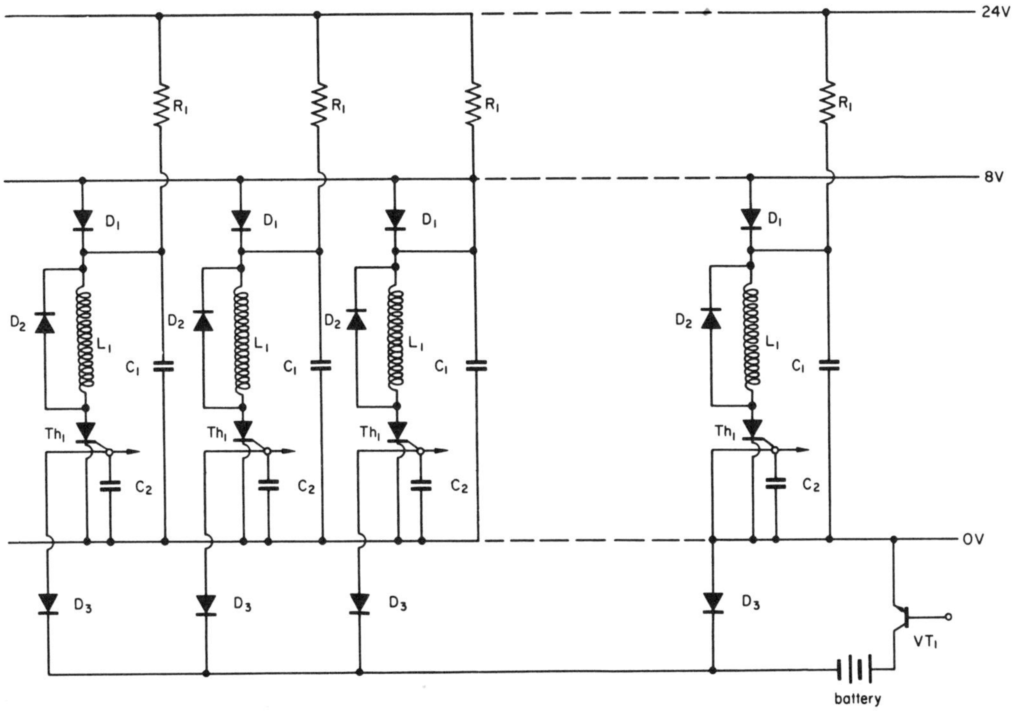

FIGURE 8

There are two points which it will be useful to remember, when designing a solid state contactor:—

(a) The neutral of the supply voltage is not at the EARTH potential.

(b) The triac gate pulses should have the same polarity as the main triac voltage or alternately the gate should be always negative with respect to the MT1 terminal for either main voltage polarity.

IV BURST FIRING TECHNIQUES USING TRIACS

by Jurek Budek

Thyristors and Triacs used in a phase control mode produce considerable radio frequency interference (r.f.i.) due to the step change in current. A solution to the problem of suppression becomes more difficult and expensive as the load increases. In domestic appliances, particularly, the r.f.i. is acute and must be suppressed according to B.S. specification 800. For domestic use phase controllable power up to 500 Watts may be acceptable. However, above this value the ratio between the phase controllable power to the total consumed power becomes significant and in addition to the expensive suppression circuit, the power factor may cause some concern to the Electricity Board.

Electric heaters and other loads with a long time constant may be controlled by passing through them a selected number of full or half cycles. The current, then will be in phase with the voltage and there will be no problem of r.f.i., the switching taking place at the point when the voltage crosses zero. Such control is called Burst Firing or Zero Voltage Switching. Burst firing is not suitable for lamp dimming due to the flicker, or for Motor Speed Control and transformer input control.

This chapter discusses two circuits. Power control can be achieved by blocking individual half cycles in an established pattern, or by passing a selected number of cycles in a mark-space mode.

VOLTAGE AND POWER

A simple substitution in the following equation gives the value of output r.m.s. voltage, V.

$$V = (V_o^2 \cdot x/t)^{1/2} \quad \text{where}$$

V_o – maximum r.m.s. output voltage (uncontrolled)

t – period expressed in number of half cycles

x – number of conducting half cycles

As an example:–

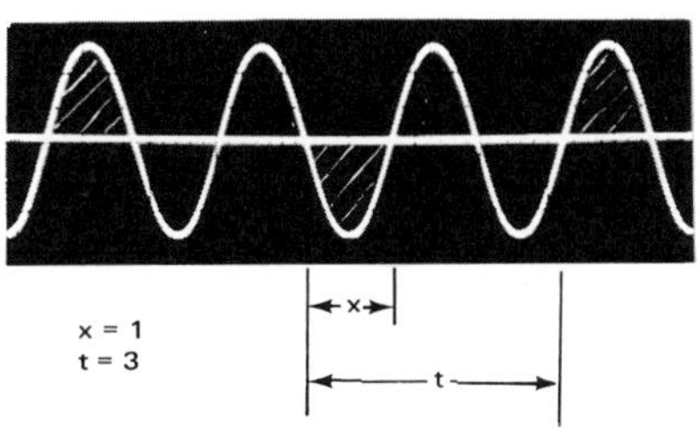

FIGURE 1

$$V = (V_o^2/3)^{1/2} = V_o/3^{1/2}$$

and the output power (Resistive load), P:-

$$P = P_o \cdot x/t \quad \text{where}$$

P_o – maximum output power (uncontrolled)

$$P = P_o/3$$

If the control is in full cycles, t will represent the period consisting of the number of full cycles and x will give the number of full conducting cycles, as shown in Figure 2.

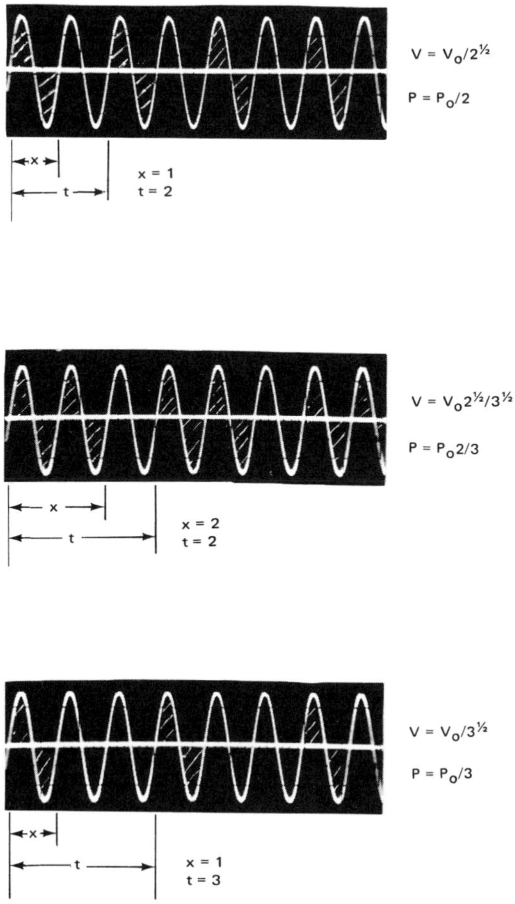

FIGURE 2

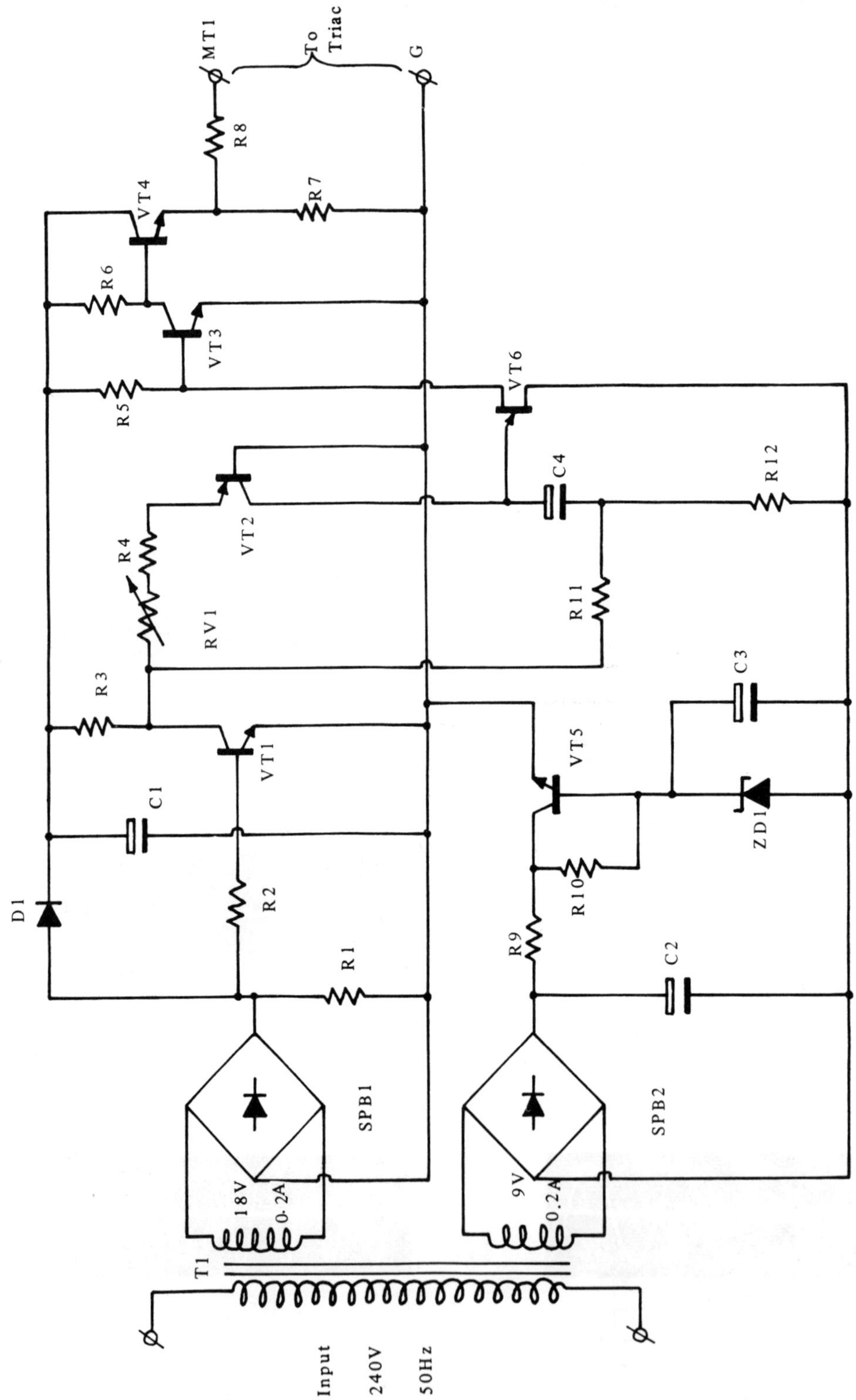

MT1
To
Triac
G
R8
R7
VT4
R6
VT3
R5
VT6
R4
VT2
RV1
C4
R12
R11
R3
C3
VT5
VT1
C1
D1
R2
R10
ZD1
R9
R1
C2
SPB1
SPB2
18V
0.2A
9V
0.2A
T1
Input
240V
50Hz

FIGURE 3

BURST FIRING USING A TRANSISTOR PUMP

The circuit shown in Figure 3 provides Burst Firing controlled in steps of half cycles down to one firing in fifteen half cycles. Table 1 shows the power reduction for this type of control with 1 half cycle conduction.

The positive DC supply is derived from an 18V winding of transformer T1 (Figure 3) via a single phase bridge rectifier, SPB1, diode D1 and the 100μF smoothing capacitor C1. Diode D1 is included in order to obtain unsmoothed full wave rectified voltage at the positive terminal of the bridge for the porpose of pulse generation. The negative supply is similarly generated from a 9V winding of transformer T1. A series stabilizing transistor, VT5, and zener diode, ZD1, provide a smoothed constant voltage reference necessary for accurate firing of the unijunction transistor, VT6.

The full wave unsmoothed voltage is applied to the base of transistor VT1 which acts as a pulse generator. At the collector of VT1, the positive pulses appear each time the base voltage drops to zero, and hence synchronize with the mains at the time when the supply voltage is virtually zero. Then the pulses from the collector of transistor VT1 are fed via the potentiometer RV1 and fixed resistor R4 to the emitter of transistor VT2 which works as a transistor pump. Capacitor C4 is charged with constant current pulses giving voltage steps of equal increments, as shown in Figure 4.

When the increments of step voltage between emitter and base 1 reach approximately 4.5V the unijunction will fire, pulling down the base voltage of transistor VT3, thus turning it off. Capacitor C4 will discharge through the emitter of the unijunction and resistor R12 until insufficient current can be supplied into the emitter to keep it conducting. The unijunction then switches off the voltage at the base of transistor VT3, rises, and the latter turns on.

Transistor VT4 acts as an emitter follower and provides sufficient current to drive a 25 A r.m.s. Triac. Therefore positive pulses will appear between the Triac's main terminal MT1 and gate, G, each time the unijunction transistor fires (See Figure 5).

By adjusting the value of resistor, R12, the unijunction's 'on' time is made sufficiently long enough to ensure that the width of the output pulses are at least 500μs for any point of control.

Figure 4 shows that the voltage across the capacitor C4 rises in approximate linear steps (with a slope of about 0.2V per

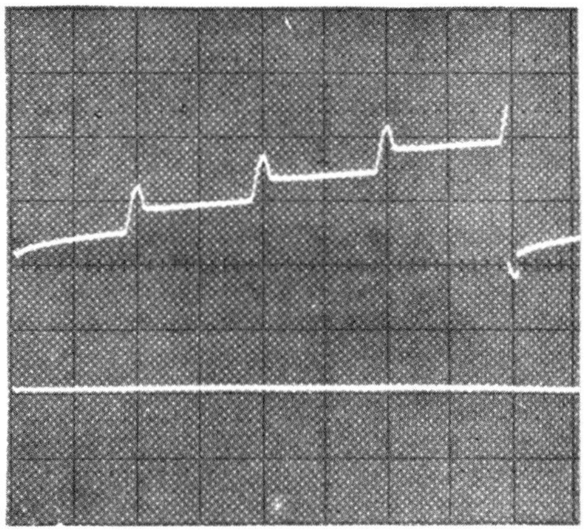

TRANSISTOR PUMP
V_{BE} OF VT6

1V/DIV ↑
5ms/DIV →

FIGURE 4

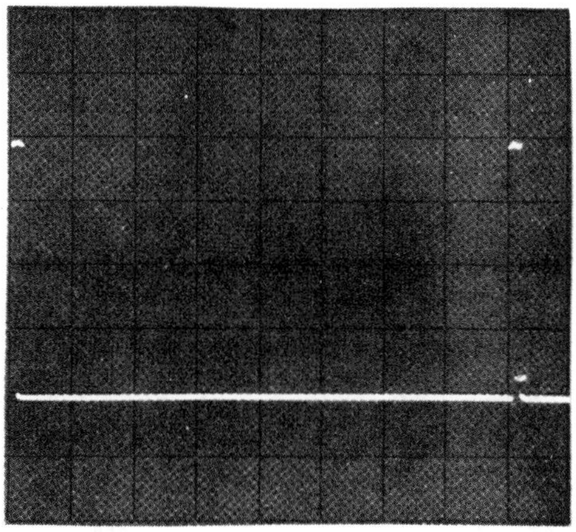

OUTPUT AND PULSES (MT1-G)

5V/DIV ↑
5ms/DIV →

FIGURE 5

Table 1

Total No. of half cycles	1	2	3	4	5	6	7	8	9	10	11	12	13	14	15
% Power Output	100	50	33	25	20	16.7	14.3	12.5	11.1	10	9.1	8.4	7.7	7.1	6.7

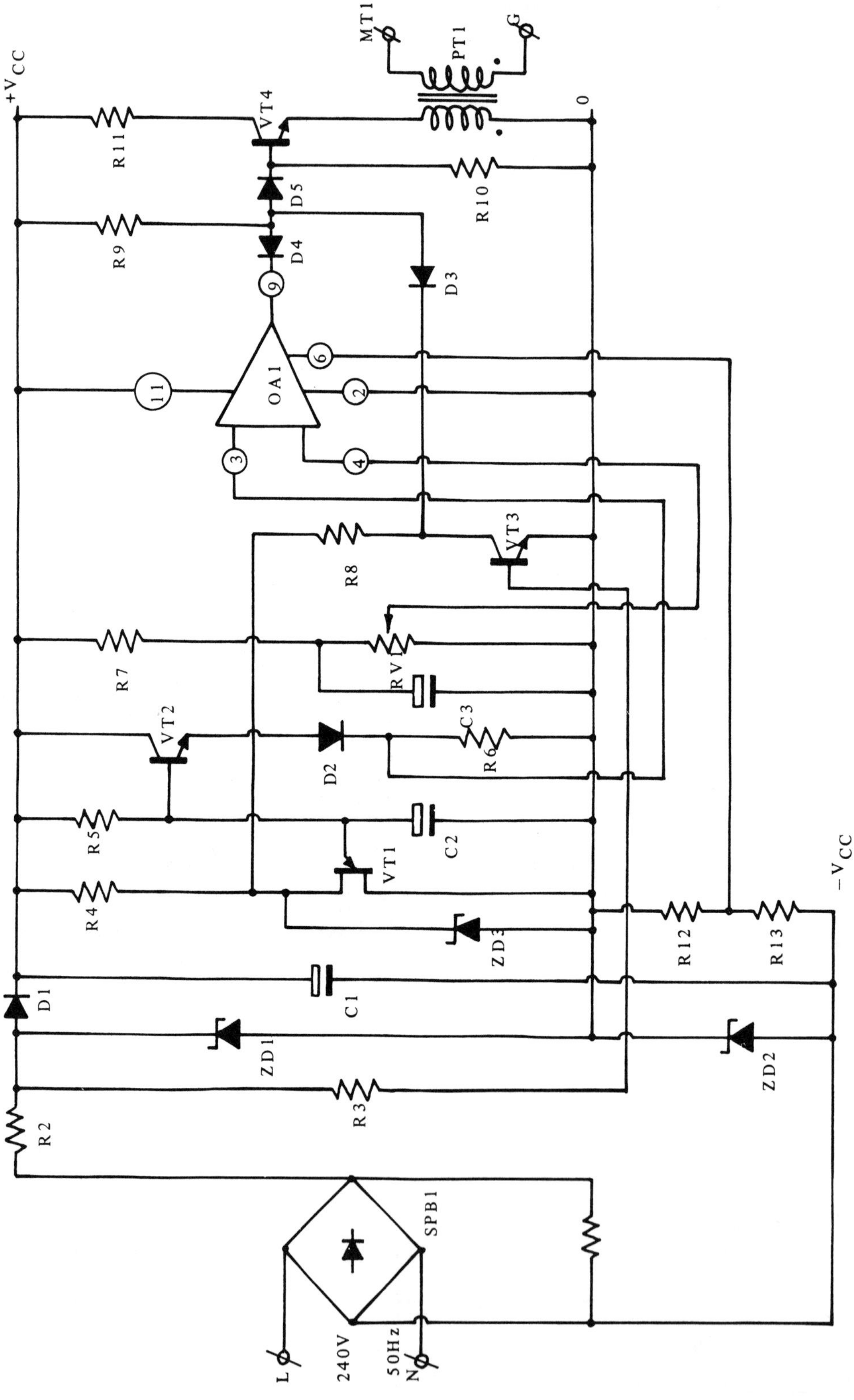

MT1
PT1
G
$+V_{CC}$
R11
VT4
D5
R10
0
R9
D4
D3
9
OA1
6
11
2
3
4
R8
VT3
R7
RV1
C3
VT2
D2
R6
R5
C2
VT1
$-V_{CC}$
R4
ZD3
R12
R13
D1
C1
ZD1
ZD2
R2
R3
SPB1
L
240V
50Hz
N

FIGURE 6

step). If the unijunction were fired at some point along the linear step instead of at the beginning of the step (quite possible with unijunction transistors), indeterminate firing would occur resulting in unwanted phase control. The potential divider chain R11 and R12 is introduced to provide, at the beginning of each step. a narrow pulse, whose magnitude is small but greater than the slope, to ensure that the unijunction can fire only at this time.

If the output pulses need to be slightly delayed to ensure correct Triac firing, a O.1μF capacitor can be added between the base and emitter of transistor VT3.

In this circuit the power varies in steps of 100%, 50%, 33% and smaller decrements down to zero. The control step pattern can be rearranged so that the smaller steps are around 100% of full power and the larger jumps at the lower end. This can be achieved by the addition of an npn transistor, three resistors and a capacitor to the circuit shown in Figure 3.

BURST FIRING BY MARK SPACE CONTROL.

The circuit is shown in Figure 6. Bridge SPB1 provides full wave rectification and an input to the two zener diodes ZD1 and ZD2, for a constant voltage supply of 12V ($+V_{CC}$) and 7.5V ($-V_{CC}$). Using the unijunction transistor, VT1 a sawtooth waveform is generated with a frequency of about 1 second period determined by resistor R5 and capacitor C2. Transistor VT2 brings the sawtooth voltage down so that its optimum value is below one volt, thus offering better control when working in conjunction with the differential amplifier 'OA1'. The pulses at zero voltage crossing points are generated by transistor VT3, whose base is connected to the unsmoothed supply across zener diode ZD1. The base voltage drops to virtually zero each time the mains voltage crosses zero turning the transistor 'off' and providing positive pulses at its collector.

The sawtooth is then injected to the differential comparator OA1 – non inverting pin (3), and a DC voltage level control from the potentiometer RV1 is connected to the inverting pin (4). The output (pin 9) will give a mark-space control by varying the DC level as shown in Figure 7.

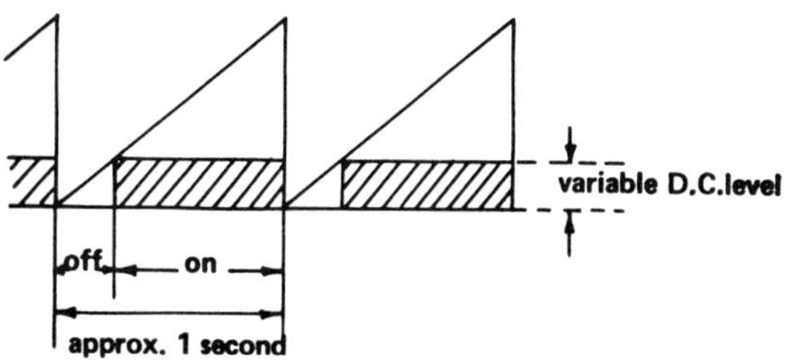

FIGURE 7

As the differential comparator is very sensitive to the input voltage, a 100μF capacitor, C3, has been added across potentiometer RV1 in order to obtain a satisfactory mark-space waveform, without intermittent switching when in the 'ON' position. The differential comparator needs two supply voltages, a +12V and a –6V; the latter being taken from a resistor chain R12 and R13.

The output from the differential comparator and the train of pulses at zero voltage crossing points from the collector of transistor VT3 are applied to an 'AND' circuit consisting of diodes D3 and D4 and transistor VT4. When both the cathodes of D3 and D4 are at a positive level, pulses will appear in the output of transformer PT1. (This transformer is included to give gate isolation protection for the Triac). There will be no pulses if either or both cathodes of diodes D3 and D4 are at zero level, as shown in Figure 8. Therefore pulses at zero voltage crossing are controlled by the mark-space-mode covering one half cycle in one second (i.e. 1%) to 100%.

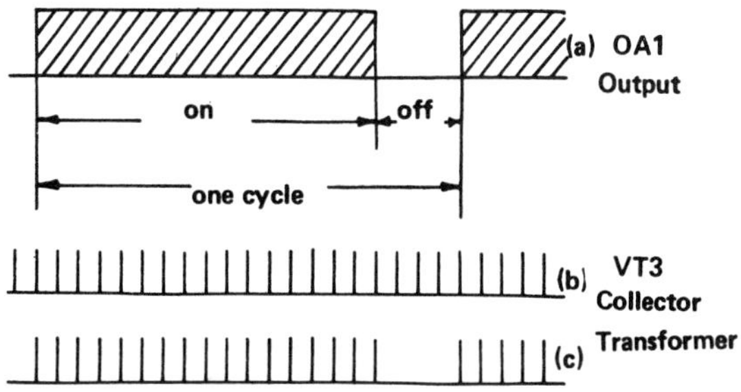

FIGURE 8

List of Components (Figure 3)

C1, C2	100μ F, 25V
C3, C4	4 μF, 25V
D1	1N4001
SPB1, SPB2	1B05J05
ZD1	1S3075A
VT1, VT3, VT5	BC183L
VT2	2N3702
VT4	2N3704
VT6	2N2160
R1	1.5k Ω, ½W
R2	12k Ω ½W
R3	470Ω, ½W
R4	680 Ω, ½W
R5	8.2k Ω, ½W
R6	1.2k Ω, ½W
R7	1k Ω, ½W
R8	100 Ω ½W
R9	330 Ω ½W
R10	3.3k Ω, ½W
R11	6.8k Ω ½W
R12	120 Ω ½W
RV1	100k Ω, ½W Potentiometer
T1	240V/18V, 0.2A; 9V, 0.2A transformer

List of Components (Figure 6)

C1, C3	100μF, 25V
C2	5μF, 25V
SPB1	TI04
ZD1	12V, zener diode, 1W 1S3012A
ZD2	7.5V, zener diode, 1W 1S3007A
ZD3	5.1V, zener diode, 400mW, 1S2051A
VT1	Unijunction transistor, TIS43
VT2, VT3	Transistor, BC184L
VT4	Transistor, 2N3704
R1	100kΩ ½W
R2	5k Ω 10W
R3, R6, R8	12kΩ ½W
R4	270Ω, ½W
R5	470k Ω, ½W
R7	820Ω, ½W
R9	8.2kΩ, ½W
R10	5.6kΩ, ½W
R11	33Ω, ½W
R12	4.7kΩ, ½W
R13	680Ω, ½W
RV1	500Ω, ½W
OA1	Differential comparator, type SN72710N
PT1	Pulse transformer, ITT type PTATA
D1-D4	1S921

V THYRISTOR REVERSIBLE D.C. SUPPLY

by Jurek Budek

A need exists for a circuit which will provide a reversible D.C. supply each time the power is switched on. This means that when the equipment is switched on a certain D.C. polarity will appear across the load. After the equipment has been switched off and switched on again, the D.C. polarity across the load will be reversed.

This circuit has wide applications in reversible D.C. drives and fluorescent lamp supplies. It is believed, for instance, that the life of a fluorescent lamp will not decrease if the lamp is used with D.C. supplies rather than a.c., and the quality of light will improve as the stroboscopic flicker present with a.c. supplies is greatly reduced. However, one condition must be met, namely, that the next switching on of the lamp must provide the opposite polarity to the previous one. This is to prevent a mercury condensation at one end of the lamp.

CIRCUIT OPERATION

The circuit shown in Figure 1 gives the details of reversible D.C. supply to two fluorescent lamps connected in series.

The main supply to the high leakage reactance auto transformer (T1) is via a double pole switch. The series capacitor provides power factor correction and the fuse protection against accidental D.C. short circuits. The a.c. is then rectified by a single phase bridge, SPB1, and fed into the thyristor single phase bridge SPB2. When thyristors '1' and '3' are energized a positive supply will appear on the bottom section of the load and when '2' and '4' are energized, the positive supply will be on the top section of the load. Care must be taken to ensure that all four thyristors are not energized simultaneously, the correct pair of thyristors are energized at all times, otherwise a short circuit will be imposed across the bridge SPB1.

In order to achieve an 'electronic memory' to remember the D.C. polarity after the previous switching, the Integrated Circuit (IC) networks N1 and N2 must have a continuous voltage supply. A step down transformer (T2) provides this supply which also serves the pulse generation circuit for the thyristors. This supply is limited by a zener diode, ZD1, and then smoothed by capacitor C2. A relaxation oscillator consisting of resistor R2, capacitor C3, and unijunction transistor VT1, provides a 15kHz train of pulses which are fed via transistors VT2 and VT3 into two pulse transformers PT1 and PT2. The thyristor bridge gate pulses (G1 K1 – G4 K4) are controlled via two auxiliary thyristors Th1 and Th2. Provided that these thyristors can be energized only one at a time, only one pair of bridge thyristors will turn on. The inputs to the gates of Th1 and Th2 are taken from the outputs of bistable N2. If the $\overline{Q}$ output, pin 6, of the bistable is at 'high' level, the Q output, pin 5, must be at low level and vice versa. The $\overline{Q}$ output is fed back to the D input, pin 2, so that each time a positive pulse appears at the clock input, pin 3, the state of the bistable reverses.

Therefore, while pulses from bridge SPB4 appear at the input of mono-stable, N1, (pins 1 and 2) the $\overline{Q}$ output, at pin 6, will remain at the 'low' level and thus will not change the state of the bistable. When the 100V supply is switched off the clock input of N2 will rise and thus change the states of the bistable outputs. Thus the change in the firing mode of bridge SPB2 takes place when its supply is off. By adding capacitor C10 and resistor R11 to the monostable N1, the output pulse at Q will become longer than the time difference between the two input pulses. Hence this output will stay at a high level as long as the input pulses are present. When the 100V supply is switched off, the output Q will drop to a low level after a time delay given by C10 and R11. This time delay is necessary to provide sufficient time for the load current to drop to zero before the thyristor firing change over takes place. The output $\overline{Q}$ of N1 is connected to the clock, pin 3, of N2, so that the change over from one state to another, takes place at the time when the supply is off.

CYCLOCONVERTER

By replacing the mechanical switch in Figure 1 with an electronic device and rearranging the circuit to all the required switching pattern, it is possible to produce a cyloconverter type of frequency changer. Figure 2 shows some of the waveforms obtained from the unit constructed as in the circuit diagram shown in Figure 3. The reduction in frequency is given in Table 1.

Table 1

Frequency division	Output Frequency Hz	Shown in Fig. 2
1	50	a
2	25	b
3	16.6666	c
4	12.5	d
5	10	e
6	8.3333	f
7	7.1466	
8	6.5	
9	5.4444	
10	5	

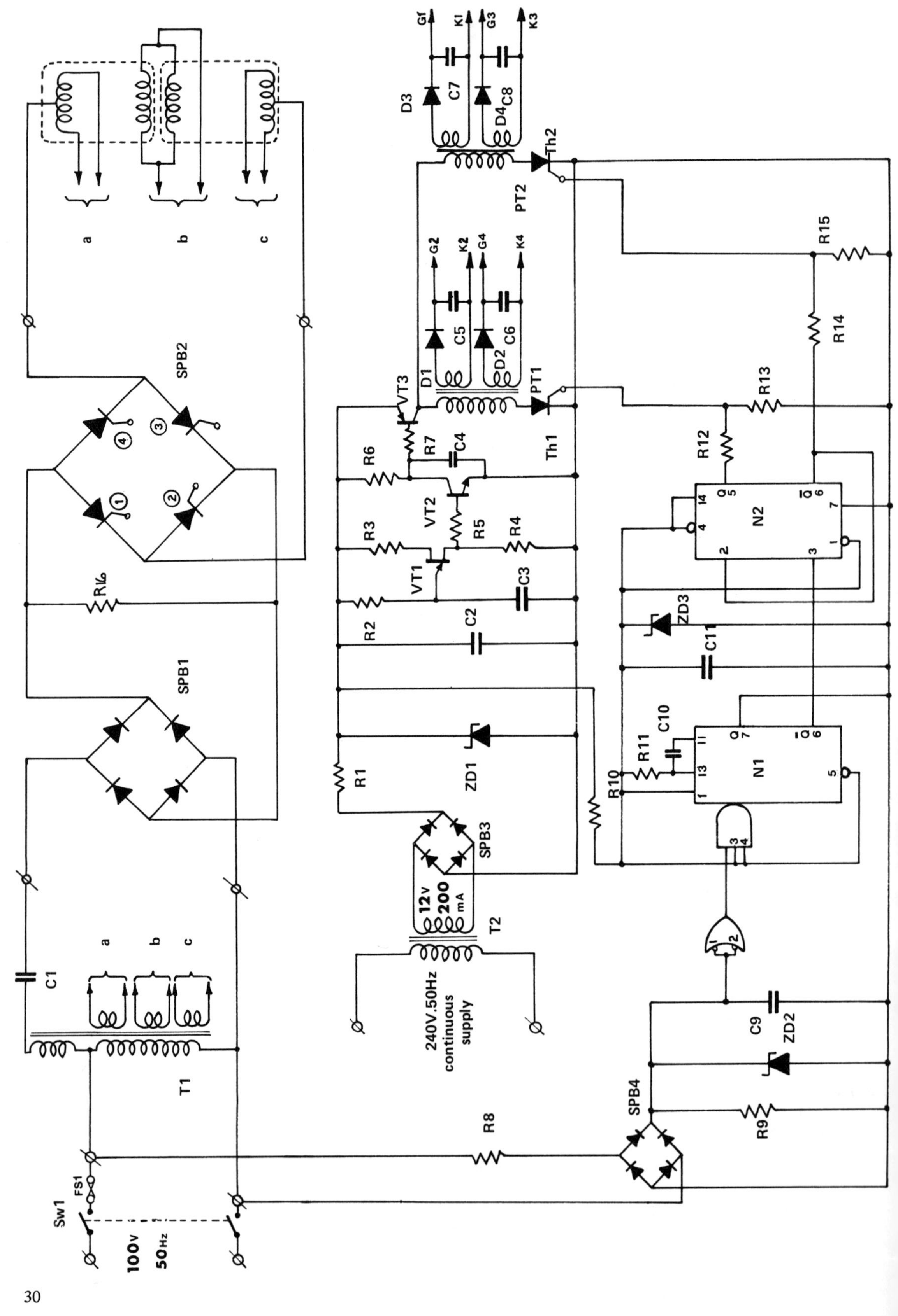
SPB1
SPB2
SPB3
SPB4
T1
T2
240V.50Hz continuous supply
12V 200 mA
100V 50Hz
Sw1
FS1
N1
N2
PT1
PT2
Th1
Th2
VT1
VT2
VT3
ZD1
ZD2
ZD3

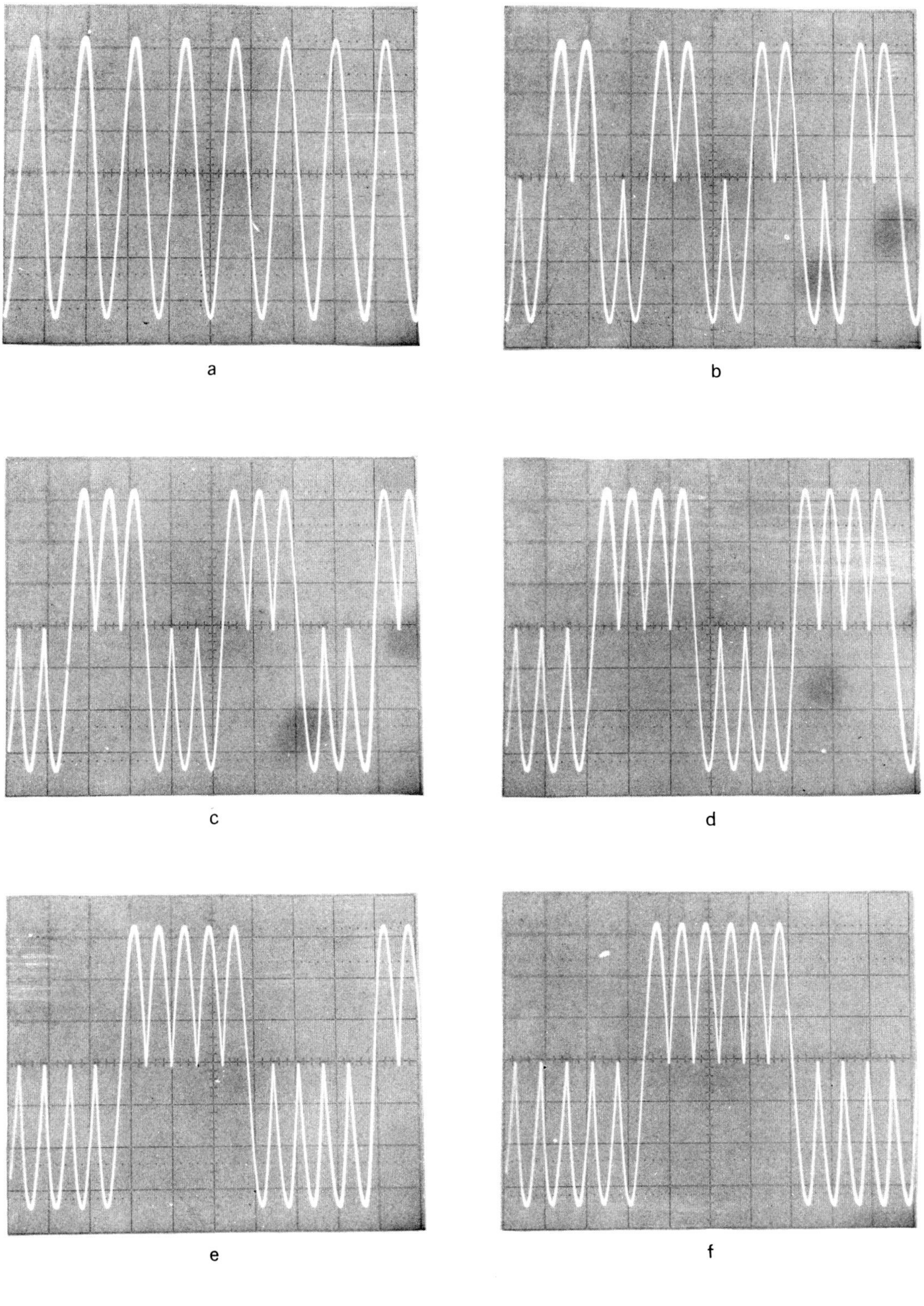

a b c d e f

FIGURE 2

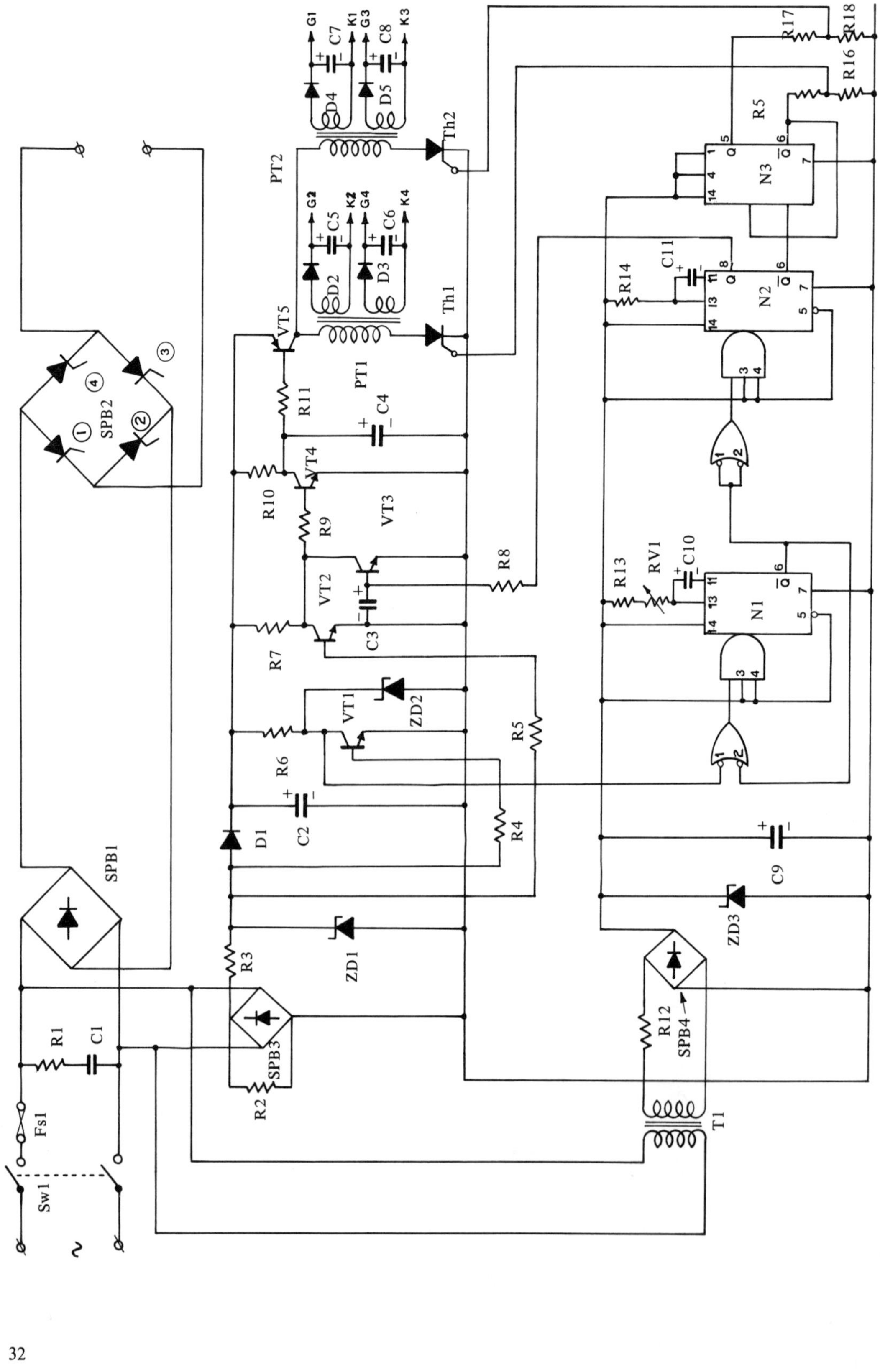

SPB1
SPB2
SPB3
SPB4
Sw1
Fs1
~
R1
C1
R2
R3
D1
C2
ZD1
ZD2
ZD3
R4
R5
R6
R7
R8
R9
R10
R11
R12
R13
R14
R16
R17
R18
RV1
VT1
VT2
VT3
VT4
VT5
C3
C4
C5
C6
C7
C8
C9
C10
C11
D2
D3
D4
D5
PT1
PT2
Th1
Th2
T1
N1
N2
N3
G1
K1
G2
K2
G3
K3
G4
K4

FIGURE 3

Zero voltage switching has been chosen in order to avoid random switching and to maintain pulse symmetry. Bridge SPB3 provides full wave rectification and an input to zener diode ZD1. Transistor VT2, whose base is connected to the unsmoothed supply across the zener diode ZD1, generates the pulses at zero voltage crossing. The base voltage drops to virtually zero each time the mains voltage crosses zero, turning the transistor 'off' and giving pulses on its collector. These pulses are then fed via transistors VT4 and VT5 into two pulse transformers PT1 and PT2. Auxiliary thyristors Th1 and Th2 control these pulses to the thyristor bridge SPB2 in the same manner as described in the previous section. The switching logic has been implemented by using integrated circuit monostables N1 and N2 and bistable N3. Figure 4 illustrates the sequence of operation. The mains synchronising pulses, generated at zero voltage crossing by transistor VT1 are supplied to one of the gate inputs, pin 1, of monostable N1. The negative edge of these pulses triggers N1 resulting in Q changing to a logical '1' and $\overline{Q}$ to a logical '0' for a period determined by the time constant C10 and R13 + RV1). By altering the value of potentiometer RV1 the length of the output pulses from N1 can be varied, hence a variable division ratio between the input pulses and its output is determined. In order to prevent further input pulses from triggering N1 during its timing period, output $\overline{Q}$ is taken to the second gate input, pin 2, to inhibit the input pulses during this time.

Output $\overline{Q}$ of N1 is also connected to the gate inputs, pins 1 and 2, of N2 so that each time $\overline{Q}$ goes to a logical '0', N2 is triggered. The output pulse length of N2 is proportional to the fixed time constant of C11 and R14, approx. 0.7ms. Output $\overline{Q}$ of N2 is connected to the clock input, pin 3, of the bistable N3, so that each time it goes positive, the outputs from N3, i.e. Q and $\overline{Q}$ change state and thus let the supply through to the gate of thyristor Th1 or Th2.

In order to avoid turning 'on' the thyristors in SPB2, accidentally possible for certain set values of potentiometer RV1, output Q from N2 is connected to the base of transistor VT3. Thus each time a positive pulse appears at the base, transistor VT3 will turn 'on' shorting out the trigger pulses.

In conclusion, by simply varying potentiometer RV1 in the circuit, as shown in Figure 3, the output waveforms across a load will progressively change in the manner illustrated in Figure 2. The cycloconverter shown and explained works satisfactorily on resistive loads, the change from one frequency to another being smooth and simple. Inductive loads, however, will require further circuitry to change over from one polarity to another while the load current is zero.

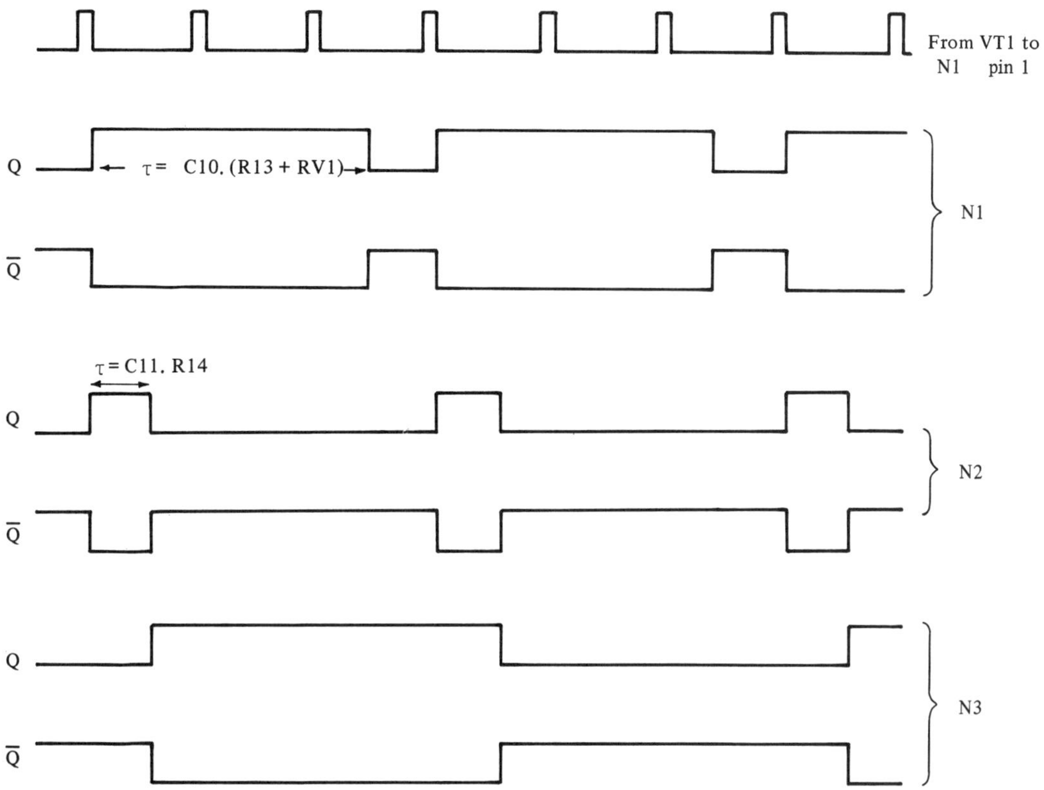

FIGURE 4

List of Components for Figure 1

C1 – Power factor correction capacitor, 18μF
C2 – 1000μF, 25V
C3 – 4700pF, 25V
C4 – 0.05μF, 25V
C5, C6, C7 0.1μF, 12V
C8, C9, C11
C10 – 8μF, 6V
D1, D2, D3, D4 Diode 1N4001
SPB1 1B40K80
SPB2 Thyristor Bridge (Thyristor rating 1000V, 5A mean)
SPB3, SPB4 TI06
ZD1 1S3007A
ZD2, ZD3 1S3051A
VT1 TIS43
VT2 BC184L
VT3 2N2905
R1 33, ½W
R2 10k, ½W
R3 100, ½W
R4 56, ½W
R5, R6 1k, ½W
R7, R9 330, ½W
R8 3.3k, 5W
R11 47k, ½W
R12, R13, R14, R15 3.9k, ½W
T1 High reactance auto transformer
T2 240V/12V, 0.2A transformer
PT1, PT2 Pulse transformers, ITT, PTATB
N1 Monostable IC SN74122N
N2 Bistable IC ½ of SN7474N

List of Components for Figure 3

C1 – 0.1μF, 400V
C2 – 100μF, 25V
C3, C4, C5, C6, C7, C8, C11 – 0.1μF, 12V
C9 – 1000μF, 25V
C10 – 4μF, 12V
D1, D2, D3, D4, D5 – 1N4001
SPB1 – 1B40K60
SPB2 – Thyristor bridge
SPB3, – 1B08T40, SPB4 – TI06
ZD1 – 1S3007A
ZD2, ZD3 – 1S3051A
VT1, VT2, VT3, VT4 – BC183L
VT5 – 2N2904A
R1 – 68Ω, ½W
R2 – 22k, 5W
R3 – 10k, 10W
R4, R8 – 10k, ½W
R5 – 2.7k, ½W
R6 – 3.3k, ½W
R7, R9, R10 – 1k, ½W
R11 – 330Ω ½W
R12 – 6Ω, ½W
R13 – 4.7k, ½W
R14 – 33k, ½W
R15, R16, R17, R18 – 3.9k, ½W
RV1 – 50k, ½W
T1 – 240V/6V, 0.2A transformer
PT1, PT2 – Pulse transformer, ITT, PTATB
N1, N2 – Monostable, SN74122N
N3 – Bistable, SN 7474

VI UNIJUNCTION TRANSISTORS THEORY, OPERATION AND CIRCUITS

A unijunction transistor (UJT) is a three-terminal device exhibiting a stable incremental negative resistance region under certain conditions. This negative resistance makes possible the design of unique switching circuits comprising fewer components than comparable transistor circuits. The input impedance of the UJT in the OFF state is high – in the order of 5 megohms – making the UJT suitable for high-input-impedance voltage-sensing circuits.

From the standpoint of its theory, the UJT is not a new device. In 1948 Heinrich Welker applied for a patent in France on a unijunction-type device. The patent was granted in the U.S.A. in 1954.* As early as 1949 Shockley and Haynes had written a paper describing the basic operation of the UJT. In the early days, this device was known as the double-base diode and older literature refers to the UJT by this name.

The UJT can be operated in a number of different circuit configurations such that any of the three terminals can serve as a signal input or load output. This makes the UJT ideal for timing circuits, triggering circuits, oscillators, voltage sensors and power-controlling devices. Circuit variations yield excellent counter and memory systems.

This chapter describes some of the unique features and characteristics of the UJT. The first portion deals with the general features and the developments of equivalent circuits. Next, theory of operation is discussed, followed by a definition of the basic parameters and terms used in UJT work. In addition, two sections – device theory and method of construction – have been added for completeness. Readers interested primarily in circuit work may wish to omit these two sections and proceed to the discussion of circuit stability which includes a general explanation of astable, bistable, and monostable operation. This chapter concludes with a collection of practical UJT circuits that illustrate some of the possible uses of this device.

DEVICE SYMBOL

The symbol for a UJT is given in Fig. 1. Base 1 and base 2 leads are shown at right angles to the base because they are non-rectifying, ohmic contacts. However, the emitter connection is represented by an arrow because it is a rectifying or PN junction, and the arrow is slanted to indicate the emitting properties of the junction. The arrow head is pointing toward the base, signifying a P-type emitter and an N-type base.

* U.S. Patent No. 2,683,840.

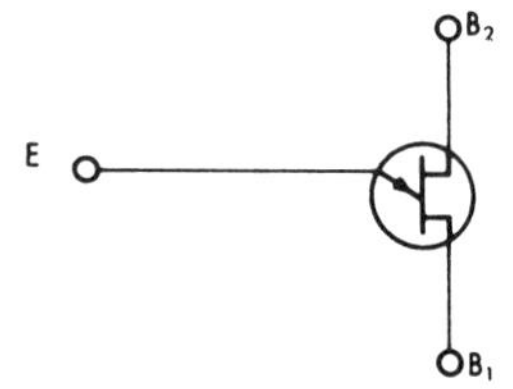

FIGURE 1 Unijunction Transistor Symbol

DEVICE CHARACTERISTICS

Bias polarities for a PN-type UJT are shown in Fig. 2. If V_{BB} is held constant as I_E is increased from zero, an emitter characteristic similar to Fig. 3 will be obtained. The negative-resistance region is where the slope of the V_E, I_E curve is negative. End points of this region are marked by the peak point (I_P, V_P). and the valley point (I_V, V_V).

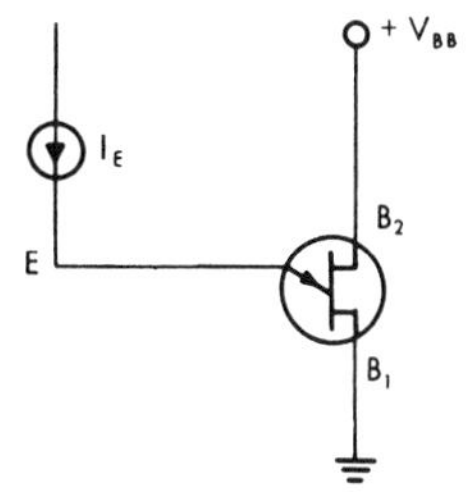

FIGURE 2 Unijunction Bias Polarities

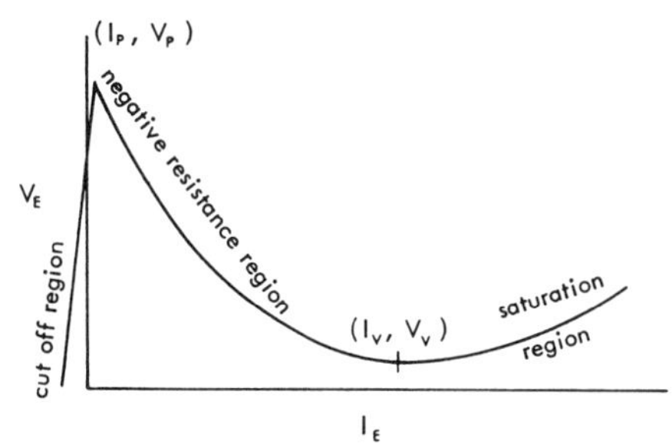

FIGURE 3 Unijunction Emitter Characteristic

The UJT is a current-controlled device. For each specific emitter current there is a unique emitter voltage; however, as many as three separate values of emitter current may exist for a specific emitter voltage.

Cut-Off Region

The input impedance in the OFF region is that of the reverse biased emitter diode. This reverse bias causes a diode leakage current to flow. This leakage explains the negative emitter current near the origin of the emitter characteristic. Currents in the OFF region are typically in the order of 1 to 10μA at 25°C for an alloyed construction and a maximum of 10nA for planar construction.

Negative Resistance Region

The incremental resistance in this region exhibits a negative characteristic; that is, as voltage increases, current decreases. At low current levels, the incremental resistance is in the order of −100k and approaches zero at the minimum voltage point. Currents in the negative resistance region run typically from 10μA at the low end to 30mA at the high end or from 5μA to a mA with planar construction.

Saturation Region

The high-current, positive-incremental-resistance region is called the saturation region. Saturation voltage is measured from emitter to base 1 at some specified emitter current, usually 50mA. Typical saturation voltages are in the range of 3 to 5 volts.

EQUIVALENT CIRCUITS

Several equivalent circuits are shown to aid in understanding UJT operation. Since the UJT is usually used in the switching or large-signal mode, different values for the equivalent circuit components must be used for each area of operation. Figure 4 shows a general equivalent circuit.

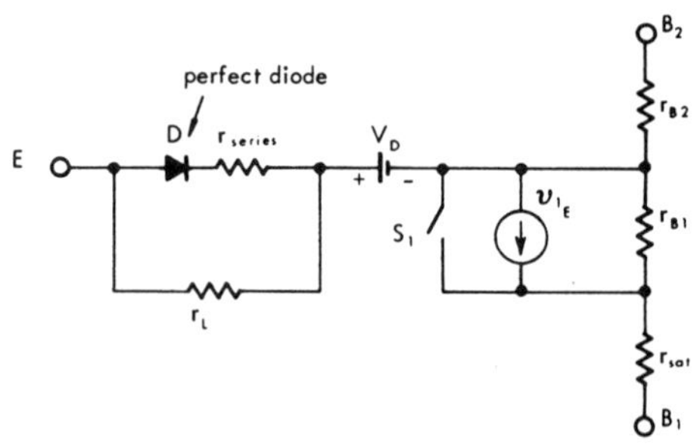

FIGURE 4 General Equivalent Circuit

The simplified equivalent circuit of Fig. 5a represents UJT behaviour in the OFF region. Because it is small compared to r_{b1}, r_{sat} is omitted; r_{series} is omitted since D is reverse biased. The solid line of Fig. 5b shows the portion of the characteristic curve represented by this equivalent circuit. In Fig. 5b (Ip, Vp) is the point at which the emitter diode becomes forward biased.

$$V_P = \frac{r_{B1}}{r_{B1} + r_{B2}} V_{BB} + V_D = \eta V_{BB} + V_D \qquad (1)$$

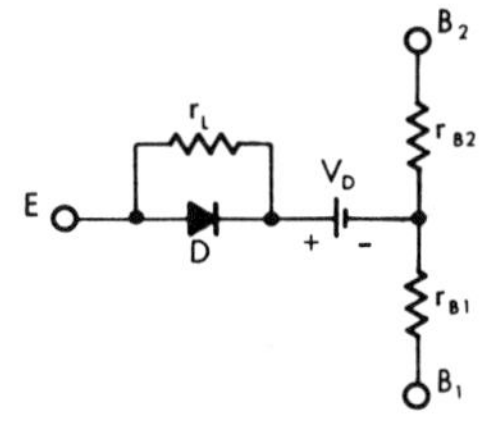

FIGURE 5a

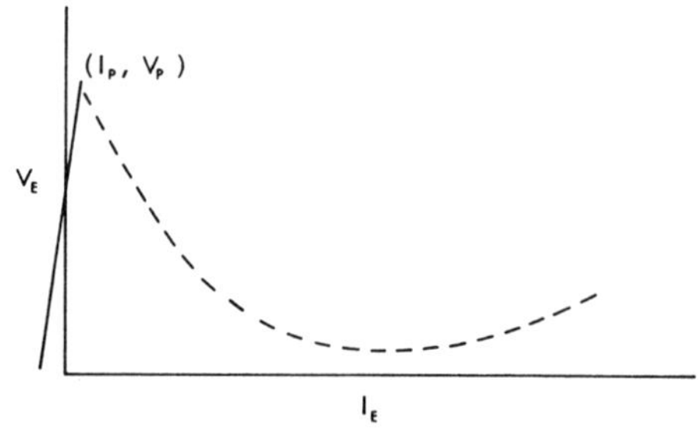

FIGURE 5b

If the UJT is operated in the negative-resistance region as a small-signal device (such as a negative-resistance amplifier), the general equivalent circuit must be used. Small-signal measurements must be made at the operating point for parameters such as v in the current generator. This equivalent circuit is useful in representing the negative-resistance region as indicated by the solid curve in Fig. 6.

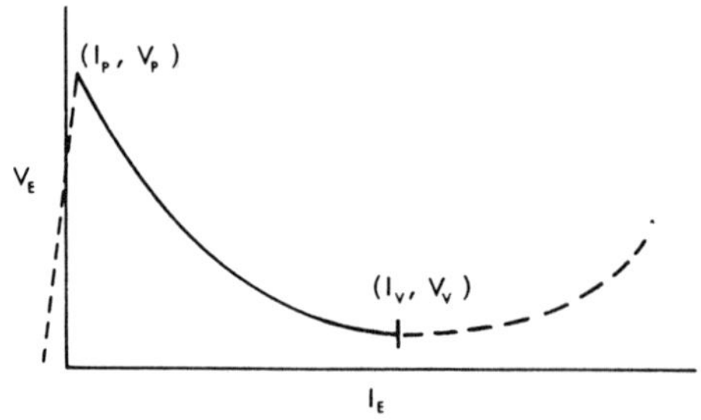

FIGURE 6

The saturation region equivalent circuit is illustrated in Fig. 7a. Because of the heavy hole injection from the emitter, the base-1 resistance has been reduced to r_{sat}, which is typically 15-30 ohms (incremental at I_E = 50mA. R_L has been omitted since the diode is forward biased. The bottom curve of Fig. 7b shows the emitter – base-1 diode characteristic curve with base 2 open. Saturation voltage of the UJT can only approach the voltage of the lower curve. The difference in voltage between the two curves is the voltage drop across r_{sat} due to the additional current from base 2.

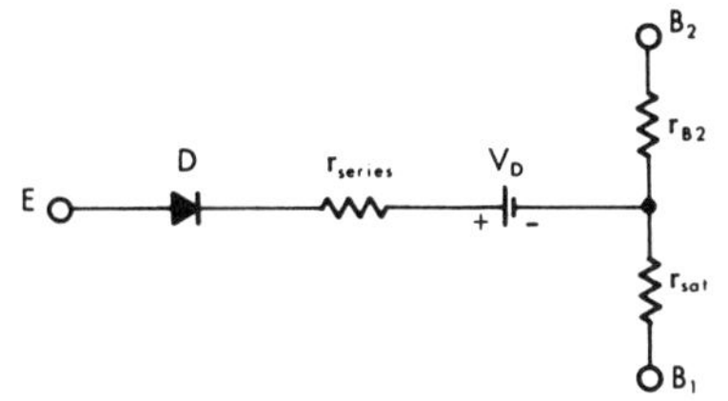

FIGURE 7a

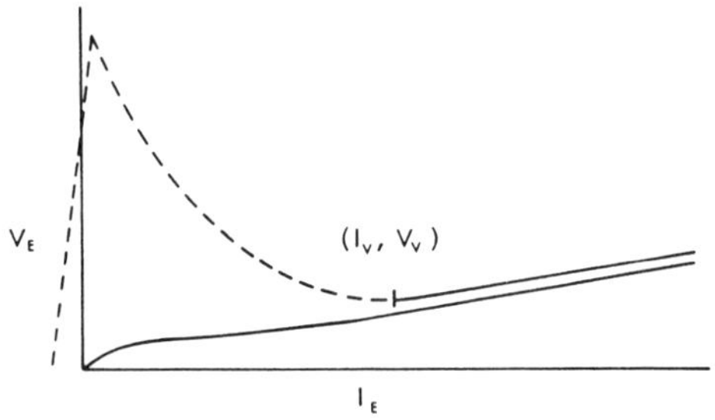

FIGURE 7b

DEFINITION OF PARAMETERS AND SYMBOLS

r_{BB}, the static interbase resistance, is measured between the two base terminals with the emitter open. This is simply the ohmic resistance of the silicon bar (r_{bb} is the incremental interbase resistance); r_{BB} is slightly dependent upon the applied voltage and is specified at a low voltage and current level to elimate the effects of self heating. It is dependent upon temperature and varies between +0.8 and +0.9%/C°. The value r_{BBO} is the value of r_{bb} at 25°C. A curve of r_{BB} vs temperature is given in TI UJT data sheets.

r_{B1} is the static resistance of the silicon bar from the point of emitter contact to the base-1 terminal. This resistance is a function of emitter current decreasing as I_E increases. r_{b1} as a function of I_E is shown in Fig. 8.

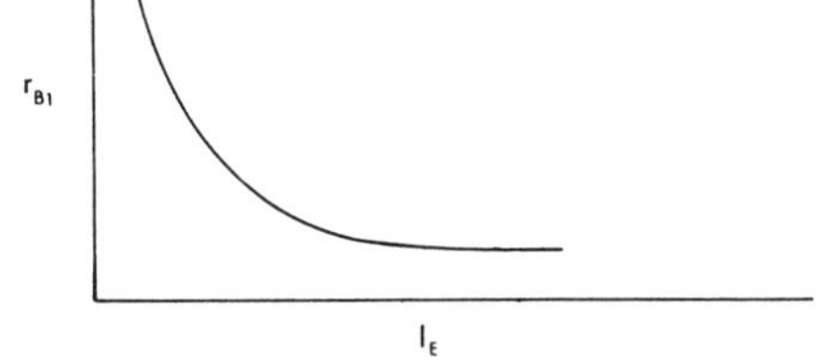

FIGURE 8

r_{B2} is the d-c resistance of the silicon bar from the point of emitter contact to the base-2 terminal. The a-c value is given as r_{b2}; it is usually assumed to be constant with emitter current. When the UJT is in saturation, a significant portion of the device dissipation occurs in this resistance, which occupies a small volume. This situation can cause a hot-spot – especially at high base supply voltages – unless an external base-2 resistor is added. The sum of r_{B1} and r_{B2} is equal to the interbase resistance r_{BB}.

I_V and V_V designate the emitter current and voltage at the valley point, as shown in Fig. 3. $\Delta V_E/\Delta I_E = 0$ at the valley point. From this point to the peak point the incremental resistance is negative. The valley point is a function of temperature and V_{BB}.

I_P and V_P designate the peak-point current and voltage. The peak point is approached as the emitter diode becomes forward biased causing the device to go into its negative resistance region. V_P is a function of the interbase voltage and η.

η, the intrinsic stand-off ratio, is defined in the following equation: $V_P = \eta\ V_{BB} + V_D$, where V_D is the forward voltage drop of a silicon diode. The ratio η determines the firing voltage of the device for a constant interbase voltage; it is relatively constant with changes in interbase voltage and temperature, and may be thought of as the attenuation ratio of the base resistance as shown in the equivalent circuit.

I_{EB20} is the leakage current of the emitter – base-2 diode when biased in the reverse direction with base 1 open. Since I_{EB20} is defined as emitter current and flows out of the device, it is given a negative sign in the data sheet. I_{EB20} is similar in behaviour to the leakage of any silicon diode. This leakage current will affect the charging rate of a capacitor tied to the emitter, causing timing error if not taken into account.

I_B (mod) gives an indication of the current gain from the emitter to base 2. I_{B2} (mod) is specified at a constant V_{B2B1} and a specific emitter current. With a resistor in base 2 and with I_{B2} (mod) known, the pulse amplitude at base 2 resulting from the emitter firing may be calculated.

V_{BB} is the total base supply voltage.

V_{B2B1} is the voltage that appears at the base-2 terminal with respect to the base-1 terminal.

V_{EE} is the emitter supply voltage.

R_{B1} and R_{B2} are external resistors placed in series with base 1 and base 2, respectively.

THEORY OF OPERATION

A physical equivalent circuit of the UJT is shown in Fig. 9. The bar is made from high-resistivity (lightly doped) N-type silicon. This means there are relatively few free carriers (electrons) in the silicon. Resistance between B_1 and B_2 with the emitter open or reverse-biased is in the order of 5k to 10k. As the voltage between the emitter and base 1 is increased, a portion of the (P-type) emitter becomes forward biased and holes are injected into the base. As holes are injected, electrons are pulled into base 1 to maintain charge neutrality. The injection of holes and the corresponding electron flow cause a decrease in the emitter-base 1 resistance. Accompanying the decrease in resistance is a decrease in voltage on the base side of the emitter diode. This reduced voltage tends to forward bias

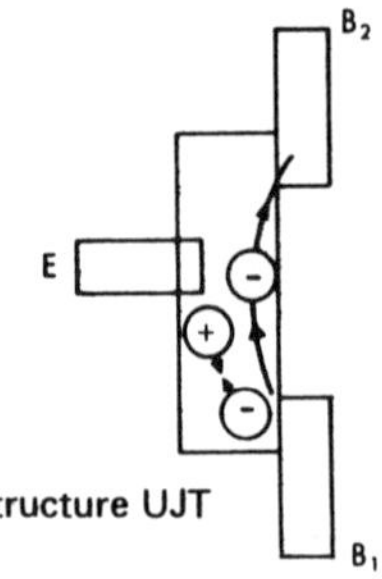

FIGURE 9

more of the emitter which, in turn, injects more holes, further lowering the resistance. This regenerative action gives rise to the negative-resistance portion of the curve. The regeneration continues until saturation or limiting occurs and the incremental resistance again becomes positive. Base 2 is usually biased at a higher potential than the emitter, thus any injected holes are repelled by base 2.

Mathematically, the change in conductivity of r_{b1} may be expressed by the following relation:

$$\sigma = q(p\mu_p + n\mu_n) \quad (2)$$

where σ = conductivity
q = electronic charge
p = hole concentration
μ_p = hole mobility
n = electron concentration
μ_n = electron mobility

Since p and n must change by equal amounts, Eq. (2) may be rewritten as:

$$\Delta\sigma = \Delta pq(\mu_p + \mu_n) \quad \text{where } \Delta p = \Delta n \quad (3)$$

Thus, as Δp increases $\Delta\sigma$ also increases, causing a decrease in the emitter –base– 1 resistance. In this short explanation, both μ_p and μ_n are assumed constant for small variations in p and n. However, the mobilities *are* affected by the total carrier concentration and become important in the limiting process near the valley point.

CONSTRUCTION

At present, there are three different process types for manufacturing unijunction transistors. The 2N489 series bar unijunction consists of an "N" type silicon bar the ends of which are the base 1 and base 2 contacts. The Silicon bar is alloyed onto a slotted metallized ceramic which controls the base 1 base 2 distance. The emitter consists of a "P" type aluminium wire which is alloyed into the bar above the base 2 contact.

The cube silicon alloyed structure used in the 2N2646 is similar to the 2N489 device except for a 33% reduction in size. In this approach, the emitter is an aluminium wire alloyed into the side of the cube. Base 1 contact consists of a gold wire making ohmic contact with the top of the cube. It has a reduced emitter base 1 distance and small active area size.

After being somewhat neglected, diffused techniques have now been applied to unijunction transistors. The base 2 area is the bottom of the planar unijunction chip as was the case in the cube structure. The base 1 contact area consists of an "N" plus diffused area into the "N" type silicon chip. The emitter area is a "P" type diffusion. All junctions in the planar device are created internally within the chip and are protected by oxide passivation. The size of this transistor chip is only 20 mils square. In the planar process, the emitter-to-base dimensions are established by photo-lithographic techniques which permit a high degree of dimensional accuracy. Because of this precision, planar UJTs feature shorter emitter-to-base 1 distances and much smaller base 1 area than the bar or cube alloy type. The latest expanded-contact techniques are used to make this possible. The shorter emitter-to-base-1 dimensions reduce the carrier transit time and provide faster and more uniform turn-on times. The smaller base 1 area combined with the reduced interjuction distances decrease the effective volume in which conductivity modulation takes place. This reduction has the effect of lowering peak-point current, valley-point current, and emitter saturation voltage. The oxide-passivation expanded-contact technique has allowed planar UJTs to offer a substantially smaller emitter reverse current over three orders of magnitude less than the alloy or cube structures.

The low peak-point current provides a high degree of sensitivity at low trigger-current levels. This characteristic is particularly valuable in designing sensitivity time-delay circuits and precise voltage and current-level sensing circuits. In some oscillator circuits, however, the lower valley-point current can be a disadvantage as it limits the average emitter current or load power that can be handled. The higher value of the peakpoint voltage of the planar UJT – which is typically 9 volts – permits it to be used with more loosely specified and more economical SCRs as a triggering device. This allows positive triggering of higher-current SCRs even at very low temperatures.

MULTIVIBRATORS.

The practical circuits shown in the last portion of this chapter operate the UJT in an astable, bistable, or mono-stable mode, depending upon the relationship of the d-c load line to the UJT characteristic curve. General features of these three modes will be discussed here before proceeding to the practical examples in the circuits section.

Stability

A basic UJT relaxation oscillator (astable multivibrator) is shown in Fig. 10.R_E is the emitter load whose load line crosses the UJT characteristic.

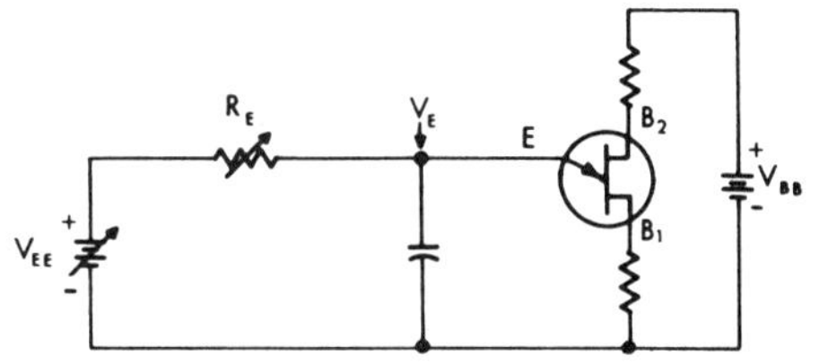

FIGURE 10

A typical UJT characteristic curve and four possible load lines are illustrated in Fig. 11. Point A represents an intersection in the saturation region. Point B intersects in the negative-resistance region, while point C crosses in the cutoff region. Operation at points D and E cannot be obtained with this circuit.

In the absence of reactive elements, points A, B, and C are all stable. However, if a capacitor is added from emitter to base 1, point B becomes unstable. The stability of this point can be examined by considering the effect of a small disturbance at that point. For example, consider a positive voltage disturbance ΔV in Δt time at point B on load line III. This incremental change of voltage decreases the current coming from V_{EE} by $\Delta V/R$, and causes a current $(\Delta V/\Delta t)C$ to flow into C, resulting in a decrease in emitter current of $\Delta V/R + (\Delta V/\Delta t)C$. The increase in voltage and decrease in current are compatible with the characteristic curve which tends to make the action regenerative. In a practical case, the high-frequency characteristic of the UJT determines a minimum or critical capacitance which will cause the point to be unstable. Thus, when a UJT is biased in its negative-resistance region, a capacitor added across its input will tend to make it unstable.

Unconditionally stable operating points are located at points A and C since they lie on a slope of positive resistance. A positive ΔV at point A or C creates a demand for more emitter current, bur less current is available (due to the decrease of the current $\Delta V/R_E$), making points A and C non-regenerative.

Stability of a point may also be tested mathematically. The incremental emitter current is given by

$$i_e = -\frac{V_e}{R_E} - \frac{dv_e}{dt} C = \frac{V_e}{r_e} \qquad (4)$$

where r_e is the incremental impedance of the emitter characteristic. Solving Eq. (4) for V_e yields an equation of the form

$$V_e = Ke^{-t/\tau}$$

where K = a constant

$$\tau = \frac{R_E r_e}{R_E + r_e} C$$

$$\frac{R_E r_e}{R_E + r_e} = R_E \parallel r_e$$

If $R_E \| r_e$ is negative, then the exponent is positive and increases without limit. This point is unstable. If $R_E \| r_e$ is positive, the exponential becomes negative and decreases toward zero. This point is stable. Thus, it can be seen that the stability of a circuit depends upon the sign of r_e and its magnitude with respect to R_E.

Astable Multivibrator

Stability criteria already presented may be used to determine the load line for an astable multivibrator. Load line III of Fig. 11 with its intersection in the negative-resistance region of the UJT curve, together with a capacitive reactance, produces an astable circuit. Such a load line is obtained with the circuit of Fig. 12, whose wave forms are shown in Fig. 13. The load line for this circuit, with the ideal and actual paths of operation, is shown in Fig. 14.

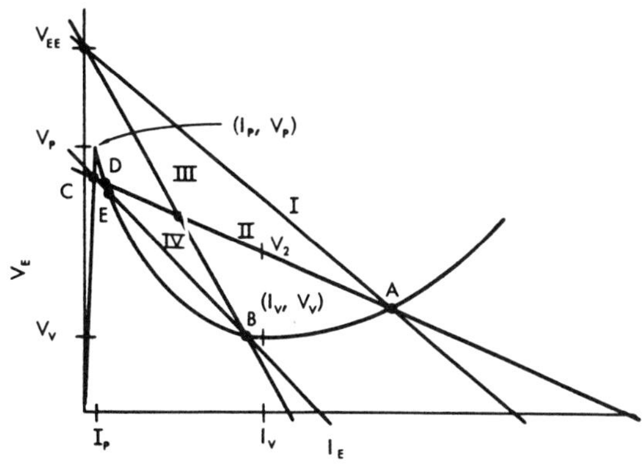

FIGURE 11

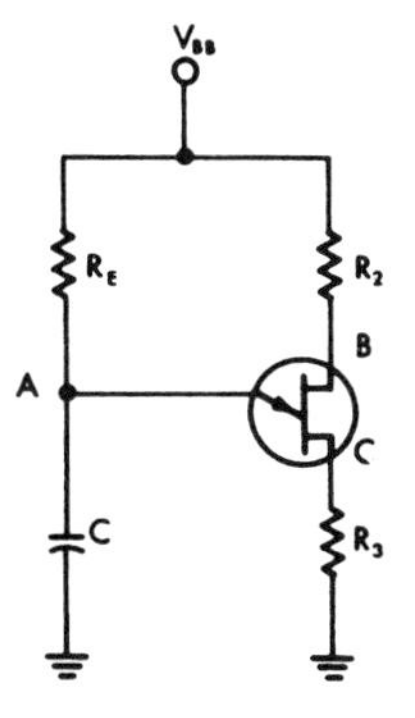

R_2 and R_3 are usually made small and in most cases neglected.

FIGURE 12

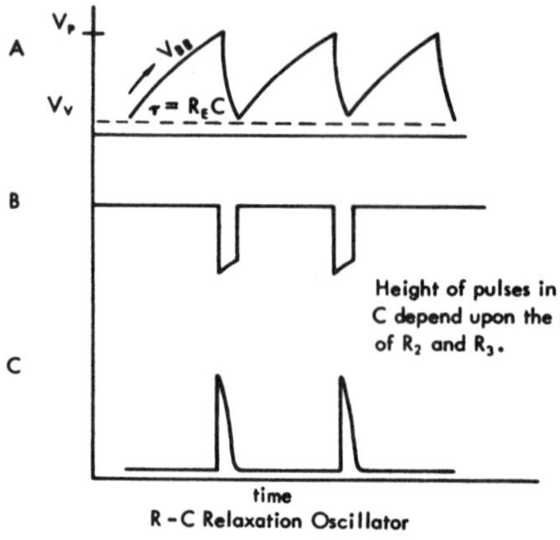

FIGURE 13

The ideal path of operation described by the dotted line may be explained as follows: Assume Q_1 is in the cutoff region (P_1). The capacitor C is being charged toward V_{BB} through R_E. When V_P is reached, operation which can no longer be sustained on this portion of the curve changes instantaneously to point P_2. V_E does not change during the transition period since the voltage across C cannot change

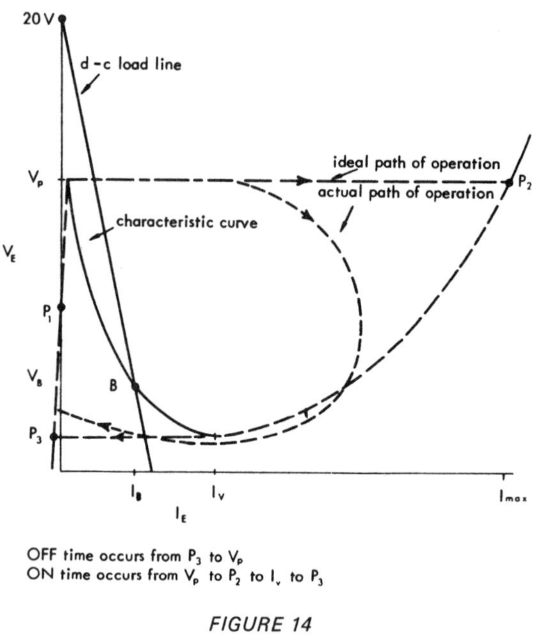

FIGURE 14

instantaneously. C discharges from V_P to V_V along the characteristic curve with a time constant approximately equal to ($r_{sat} + r_{series} + R_3$) C. (It should be remembered that r_{sat} is a function of I_E). When the operating point enters the negative resistance portion of the UJT curve, regeneration occurs which tends to turn the device OFF. Operation is moved to P_3. Ideally, this transition to the OFF stage occurs instantaneously and the capacitor voltage remains at V_V.C begins to charge toward V_{BB} again. When P_1 is reached, one cycle has been completed.

In an actual circuit the current does not change instantaneously; a time interval exists which allows C to discharge partially before the maximum current is reached, causing the rounding effect at high currents. The path of operation intersects the UJT curve to the right of point b, but not necessarily to the right of the valley point. Again, sufficient current is not available to sustain operation in this region, and conduction ceases. The load line crosses the characteristic curve at point B, which is located well into the negative resistance region. This assures oscillation. The condition for oscillation is given in the following expressions:

$$\frac{V_{BB} - V_V}{I_V} < R_E < \frac{V_{BB} - V_P}{I_P} \qquad (5)$$

The total time for a single period is

$$t_{total} = t_{OFF} + t_{ON} \qquad (6)$$

Where:

$$t_{OFF} = R_E C \ln \frac{V_{BB} - V_V}{V_{BB} - V_P};$$

$$\text{if } V_V \ll V_{BB},\ t_{OFF} \cong R_E C \ln \frac{1}{1 - \eta} \qquad (7)$$

t_{ON} = emitter voltage fall time.

The ON time is given in TI data sheets as Emitter Fall Time vs Capacitance.

For the case where $t_{OFF} \gg t_{ON}$, the frequency of oscillation is given by:

$$f = \frac{1}{R_E C \ \ln \frac{1}{1 - \eta}} \qquad (8)$$

Frequency of oscillation as given by Eq. (8) is stable because it depends upon stable circuit components (R_E,C) and a stable UJT parameter (η).

An improvement may be made in the relaxation oscillator by making t_{ON} a function of known parameters so it can be controlled in the same manner as t_{OFF} is controlled. This can be accomplished by adding a diode and a resistor to the basic relaxation oscillator circuit. This modified circuit and its associated wave forms are given in Fig. 15. Addition of the diode and resistor (D, R_1) allows control of both the ON and OFF time by providing an external resistive charge and discharge path for C. The path of operation superimposed on the characteristic curve is shown in Fig. 16.

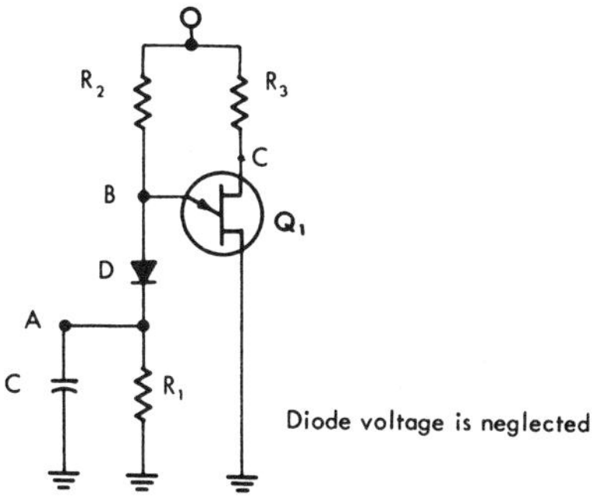

FIGURE 15a

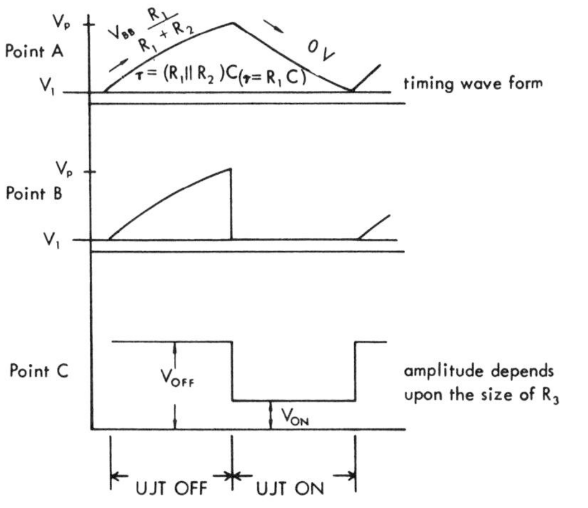

Astable Multivibrator

FIGURE 15b

Circuit operation is as follows: Assuming Q_1 to be OFF and at P_1, C begins to charge toward $\dfrac{V_{BB}\,R_1}{R_1 + R_2}$ through D. When Vp is reached, Q_1 fires, lowering the emitter potential and reverse biasing D. Q_1 now locks up at point P_2, the intersection of R_2 and the characteristic curve. Point A, now isolated from the rest of the circuit by the reverse-biased diode D, is at a potential of Vp and discharges toward zero volts through R_1. When point A reaches the same potential as point B, D conducts. This diverts some of the emitter current into the capacitor, causing Q_1 to turn OFF and the operating point to return to P_1. The conditions for oscillation are that load line II intersects the voltage axis at a point >Vp and that line I intersects in the negative resistance region. This is stated mathematically by the following two equations:

$$\frac{R_1}{R_1 + R_2} V_{BB} > V_P \qquad (9)$$

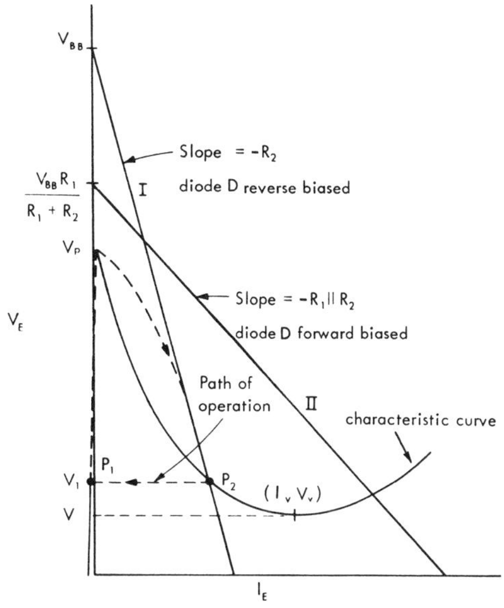

UJT OFF time	-	From P_1 to V_P
UJT turn ON time	-	From V_P to P_2
UJT ON time	-	At P_2
UJT turn OFF time	-	From P_2 to P_1

FIGURE 16

$$R_2 > \frac{V_{BB}}{I_V} \qquad (10)$$

The equations for the ON and OFF times are given below:

$$t_{ON} = R_1 . C . \ln\left\{\frac{V_P}{V_1}\right\} \qquad (11)$$

$$t_{OFF} = (R_1 \| R_2).C.\ln\left[\frac{\dfrac{V_{BB}.R_1}{R_1 + R_2} - V_1}{V_{BB}.\dfrac{R_1}{R_1 + R_2} - V_P}\right] \qquad (12)$$

where V_1 is the saturation voltage at points P_1 and P_2. V_1 can be approximated by using V_{sat} or V_V from data sheet information.

if $V_1 \ll V_{BB} \dfrac{R_1}{R_1 + R_2}$

then

$$t_{OFF} \cong (R_1 \| R_2).C.\ln\left\{\frac{1}{1 - \eta\left(\dfrac{R_1 + R_2}{R_1}\right)}\right\} \qquad (12a)$$

Values were given to the components in the astable multivibrator circuit of Fig. 15a, and the ON and OFF times were measured and then calculated using Eqs. (11) and (12).

$R_1 = 10k$	$V_P = 12V$
$R_2 = 4k$	$I_V = 7.5mA$
$R_3 = 100\Omega$	$V_V = 2.5V$
$C = 0.1\mu F$	$V_{sat(50mA)} = 3V$
$V_{BB} = 20V$	$V_1 = V_{sat}$
$D = 1S120$	
UJT = TIS43	
Calculated	Measured
$t_{ON} = 1.39ms$	1.5ms
$t_{OFF} = 0.454ms$	0.4ms

At point C, $V_{OFF} = 20V$ and $V_{ON} = 19.5$

Bistable Multivibrator

As the name implies, a bistable circuit must have two stable operating points. The bistable circuit of Fig 17 may be in either the ON or OFF state and may be triggered into either of these states. Figure 18 shows three load lines superimposed upon a UJT characteristic curve. Note that the $V_{EE}<V_P$ so that the OFF point for all three load lines is P_1. Depending upon the slope of the line, the ON operating point may be P_3, P_3', or P_3''. P_3 is unconditionally stable; i.e., any capacitance on the emitter (stray or triggering) will not cause the circuit to oscillate. P_3' is conditionally stable; however, there is an advantage in biasing at this point. When turning the UJT OFF, less power is required for triggering at P_3' that at P_3 because the trigger power tends to be augmented by the UJT. P_3'' is the point at which III becomes tangent to the characteristic curve. The UJT will lock up here, but any shift of the characteristic curve to the right will leave only one stable point, P_1. At points P_2 and P_2' the negative incremental resistance is greater than the external emitter resistor. Thus, operation is never at these points but is always in a path around them.

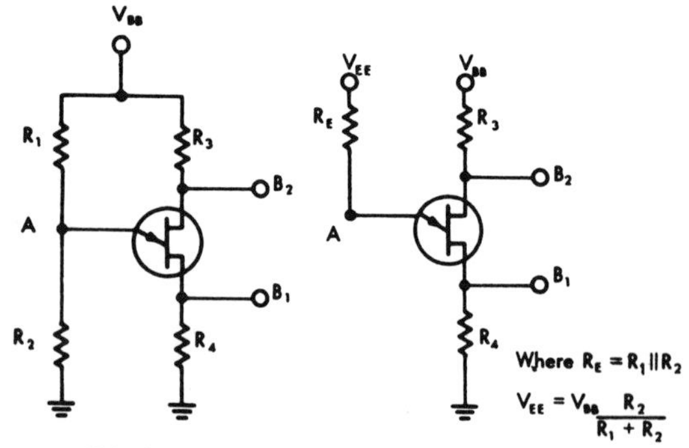

FIG. 17a FIG. 17b

FIGURE 17 Bistable Circuits

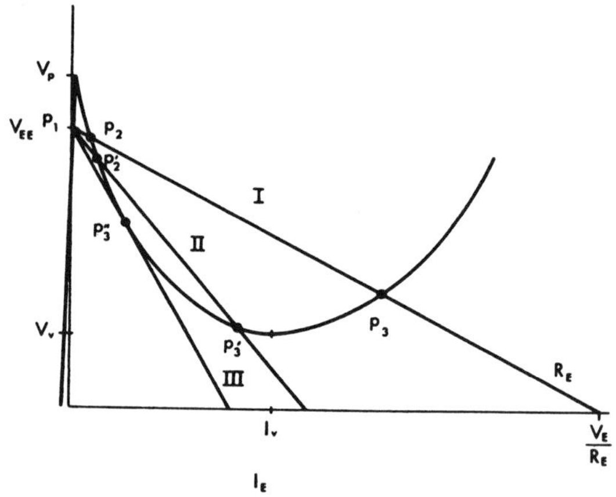

FIGURE 18 Characteristic and load line curves of bistable circuit

Triggering may be accomplished at several places in the bistable circuit shown in Fig. 17. Point A or V_{EE} can be triggered with a positive pulse of a magnitude greater than V_P-V_{EE}. This effectively raises point P_1 above V_P, and shifts operation to P_3. To turn the device OFF, point A or V_{EE} may be lowered to where the load line crosses the characteristic curve at only one point, in the OFF region. This will shift operation from ON to OFF. A negative pulse at B_2 that exceeds

$$\frac{V_P-V_{EE}}{\eta}$$

effectively lowers V_P to a point below V_{EE} so the device turns ON. A positive pulse at B_2 lifts the UJT characteristic so the load line intersects only at P_1, turning the device OFF. Triggering at B_1 requires the same polarity as at B_2.

As in astable circuits, a load line may be shaped by the use of a diode to obtain more useful characteristics. Such a circuit is given in Fig. 19a. When Q_1 is OFF, operation is at P_1 (Fig 19b) and the emitter is clamped to V_{EE} by D_1. Operation can be shifted to the ON condition by the application of a negative pulse at point A. This lowers V_P to below V_{EE} and causes the circuit to switch to P_2 in the negative resistance region. When B is returned to the negative supply, D_2 is reverse biased and C_2 is isolated from the circuit. If Point B is raised to some positive voltage, the incremental resistance of D_2 decreases. When this is reduced to a certain limit, C_2 will cause the circuit to be unstable and flip back to P_1.

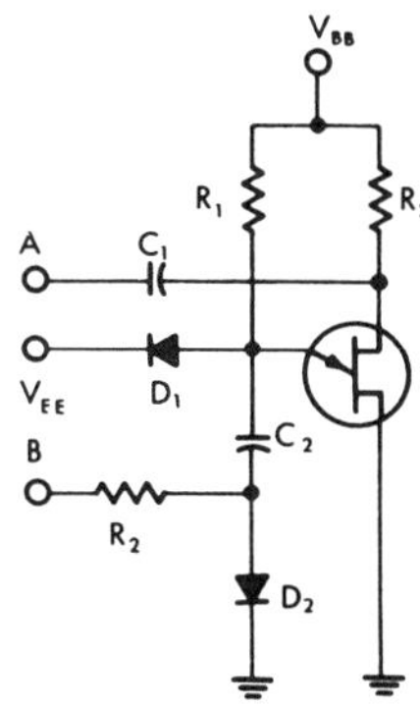

FIGURE 19a Modified Bistable Circuit

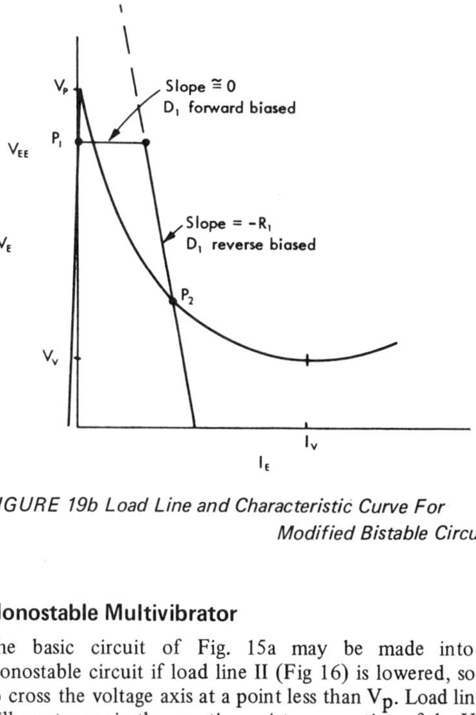

FIGURE 19b Load Line and Characteristic Curve For Modified Bistable Circuit

Monostable Multivibrator

The basic circuit of Fig. 15a may be made into a monostable circuit if load line II (Fig 16) is lowered, so as to cross the voltage axis at a point less than V_P. Load line I still must cross in the negative resistance portion of the UJT curve. These two conditions for monostable operation are given in the next two inequalities:

$$\frac{V_{BB}R_1}{R_1 + R_2} < V_P \tag{13}$$

$$\frac{V_{BB} - V_V}{I_V} < R_1 < \frac{V_{BB} - V_P}{I_P} \tag{14}$$

The first expression states that the operating point in the cutoff region is stable while the second states that the operating point in the active region is unstable. Thus, there is only one stable operating point. Circuit operation is similar to astable operation. The UJT will lock up somewhere in the OFF condition just below V_P. When a positive trigger pulse is applied to the emitter, the UJT turns ON, D becomes reverse biased, and C begins to discharge through R_1. The UJT stays on until the voltage across C is equal to the emitter voltage, and when D becomes forward biased. This causes the UJT to turn OFF. The ON time or the width of the output pulse is given by the equation:

$$t_{ON} = R_1 C \ln \frac{V_P}{V_1} \tag{15}$$

After the output pulse falls, the circuit must be allowed to recover for a given time before the next input pulse is applied. This recovery time allows the operating point to move from the ON condition to the OFF condition just below V_P:

$$t_{recovery} \cong 3(R_1 \| R_2)\, C$$

PRACTICAL CIRCUITS

Method For Decreasing The Emitter Fall Time of an R-C UJT Relaxation Oscillator

In an R-C UJT oscillator, as shown in Figure 20, a limitation exists in that the fall time of the emitter voltage during discharge is relatively slow, as shown in Figure 20b. In other words, at high frequencies the fall time may be a considerable portion of the total time. The slow fall time limits the frequency of oscillation. The modification, shown in Fig. 21, decreases the emitter fall time, thereby increasing the maximum frequency of operation. Circuit operation is as follows: Capacitor C charges toward V_{EE} through R_1 and R_2. When the firing potential of Q_1 is reached, the capacitor starts to discharge through Q_1. The resulting voltage across R_3 turns Q_2 ON. This provides another discharge path for C through Q_2, thus discharging C faster than usual. With the values given in Figure 21, the fall time was observed to decrease by a factor of four.

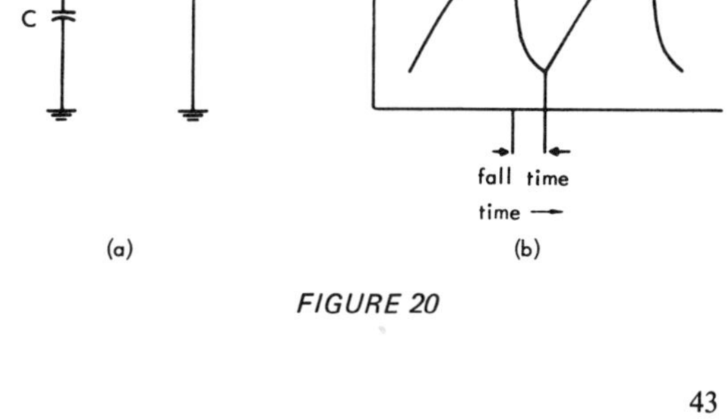

FIGURE 20

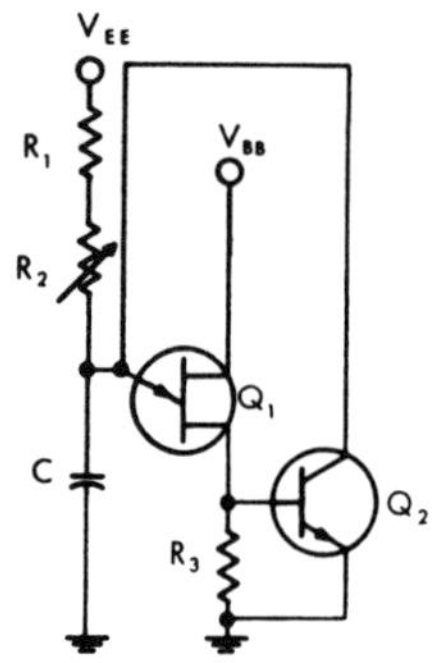

FIGURE 21

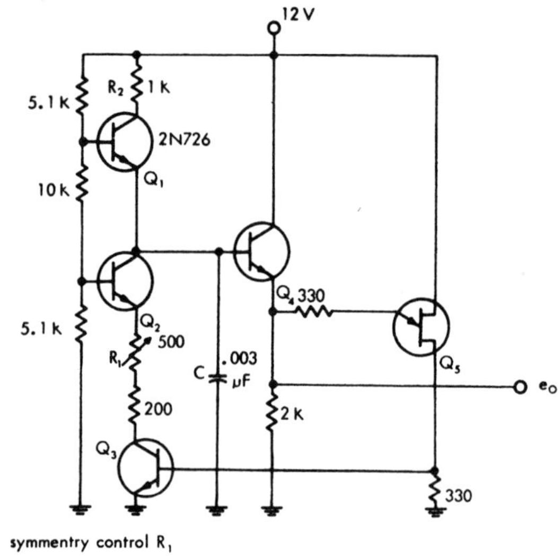

FIGURE 22 (a)

Triangular Wave Generator

The triangular wave form generator shown in Figure 22a with its wave form works in the following way: Q_1 is a current generator that supplies a current I to capacitor C, producing a linear ramp voltage. An emitter follower Q4 is used to couple the capacitor to the UJT Q5. When V_P is reached, Q_5 fires, turning Q_3 and Q_2 ON. Q_2 is a current generator of value 21 adjusted by R_1. This discharges C at the same rate as that at which it was charged; therefore, a symmetrical triangular wave is formed. Q_2 must carry a current of 21 because half of its current is derived from C and half from Q_1 which is always ON. The frequency of this circuit may be varied by changing C, or by changing I with R_2. An increase in I increases the frequency, while an increase in C decreases the frequency. A variable frequency ramp generator using alternative devices is shown in Figure 22c. Capacitor C_2 is charged by the constant current source to provide a ramp signal for V_{OUT}. Each time the UJT fires, C_2 is discharged to nearly ground potential. The ramp output begins anew when Q_2 turns off.

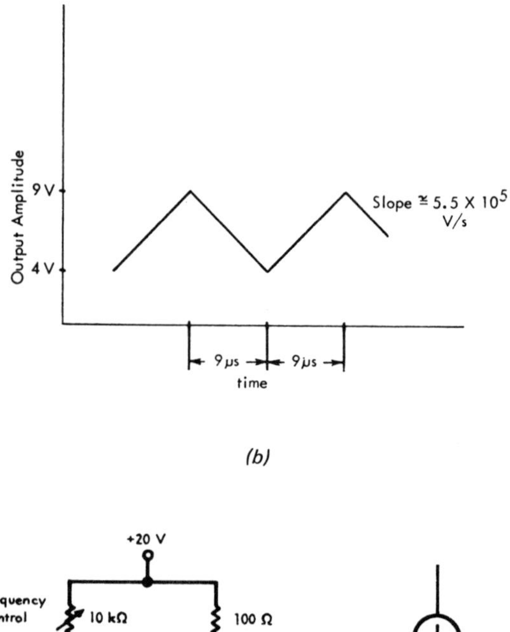

(b)

Variable-Frequency Variable-Slope Ramp Generator

The circuit shown in Figure 23a is capable of generating a ramp voltage with a variable slope that is independent of frequency. The frequency of oscillation may be varied independently of the slope. Circuit operation is as follows: Q_1 in conjunction with R_1 and C_1 forms a relaxation oscillator that controls the frequency of operation. When Q_1 fires, Q_2 is driven into saturation, causing the voltage across C_2 to return to zero. When Q_1 ceases to conduct, Q_2 turns OFF. The current generator consisting of Q_3, D_1, and R_2 begins charging C_2 at a constant rate, producing a linear voltage increase. This wave form is shown in Figure 23b. As soon as Q_1 fires, the cycle

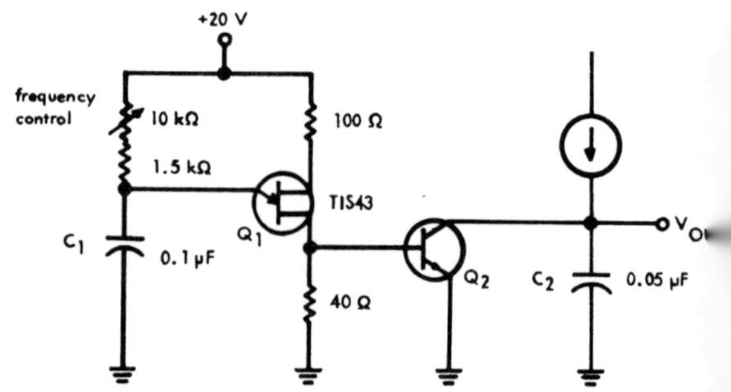

(c)

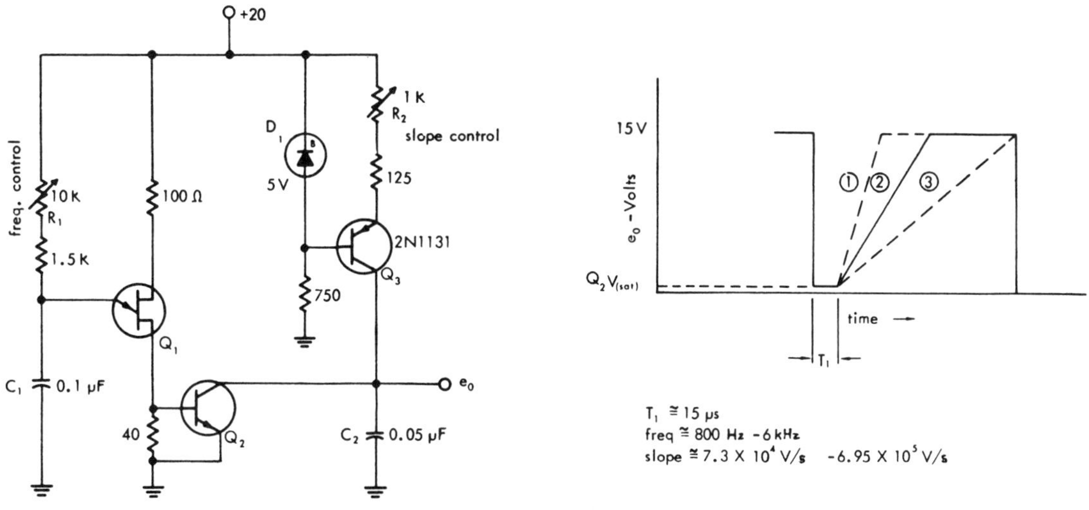

(a) *(b)*

FIGURE 23

starts over again. Curves 1, 2 and 3 from Figure 23b shows three possible slopes. Curves 1 and 2 level off at 15V because Q_3 has reached saturation before Q_1 has ended the cycle. The ramp voltage can be extended to more than 15 volts by increasing the 20-volt supply of the current generator. This will allow a greater swing across C_2 before Q_3 saturates. T_1 corresponds to the time Q_2 is held ON by C_1 discharging. T_1 can be decreased by decreasing C_1.

Staircase Wave Form Generator and Counter

This circuit (Figure 24) is capable of generating a staircase wave form at the emitter of Q_4. It can also divide the input frequency by a desired ratio. The circuit successfully operated in temperature environments from 0° to 100°C. Operation is possible over a wide variation in supply voltages; however, for accurate counting, the supply must be regulated.

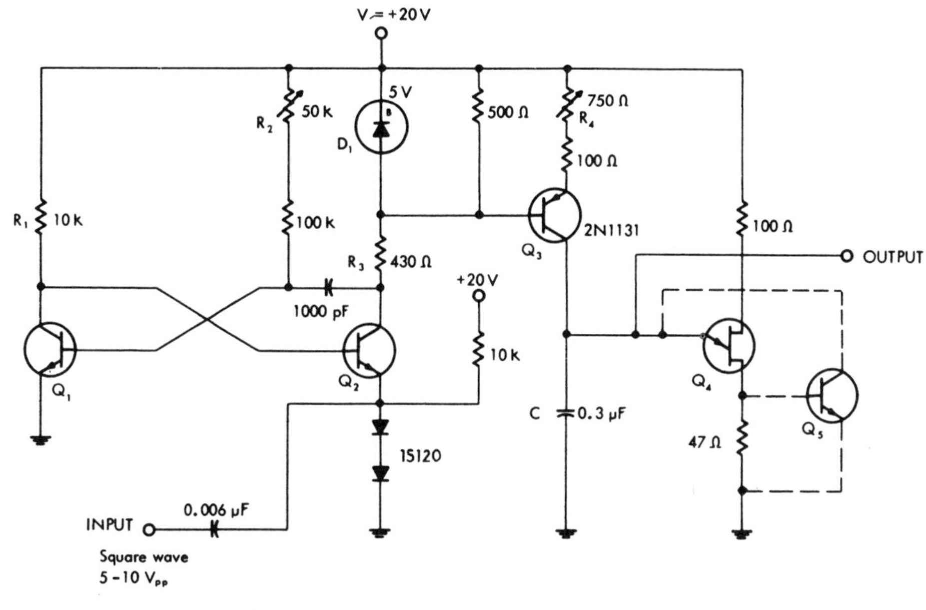

FIGURE 24

Circuit operation is as follows: A negative input pulse triggers Q_2 into conduction for a period of 15μs as set by R_2. This furnishes drive for D_1 and Q_3 which forms a current source that is adjusted by R_4. C is charged to some voltage by the current source according to the relation E = IT/C. The height of each voltage step on the capacitor is determined by R_4, which sets the amount of current for the 15μs period. When the voltage across C reaches the firing potential of the UJT, it fires and discharges the capacitor, producing a positive output pulse across the 47-ohm output resistor. The output pulse is greater than 5 volts with a rise time equal to approximately 5μs.

This counter will operate satisfactorily with input frequencies up to 10kHz. The addition of Q_5 discharges C faster, more than doubling the maximum frequency of operation. Low-frequency operation is limited by the amount C is discharged by the leakage of Q_3, Q_4 and Q_5. At 25°C the frequency of operation can be extended below 100Hz.

Wave forms containing approximately 5-24 steps may be obtained by adjusting R_4. Various setting of both R_2 and R_4 result in a wide variation in the number of steps across capacitor C. Maximum and minimum frequency rates may be altered by adjusting R_2.

The voltage wave form across capacitor C looks like the wave form in Fig. 25. Some of its features have been magnified to illustrate certain points. Point 1 shows the constant charging rate of the capacitor during the 15μs ON period. Point 2 illustrates how leakage paths can decrease the stored voltage across the capacitor. Point 3 is the peak, or firing, voltage of the UJT. This point starts the reset of the counter by discharging C. The discharge time of C through the emitter of the UJT is indicated by Point 4.

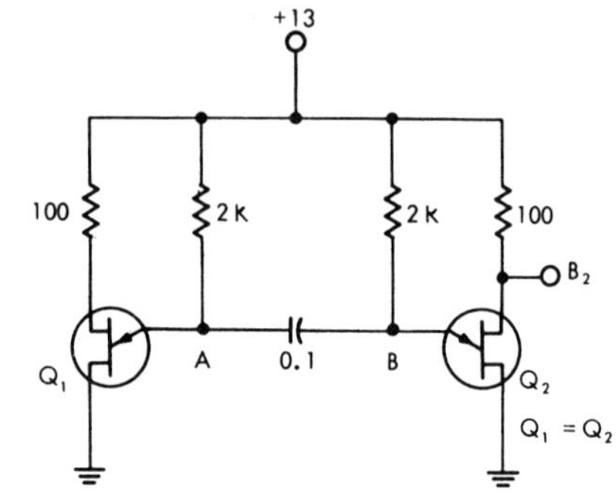

FIGURE 26 (a) Unijunction Flip-Flop

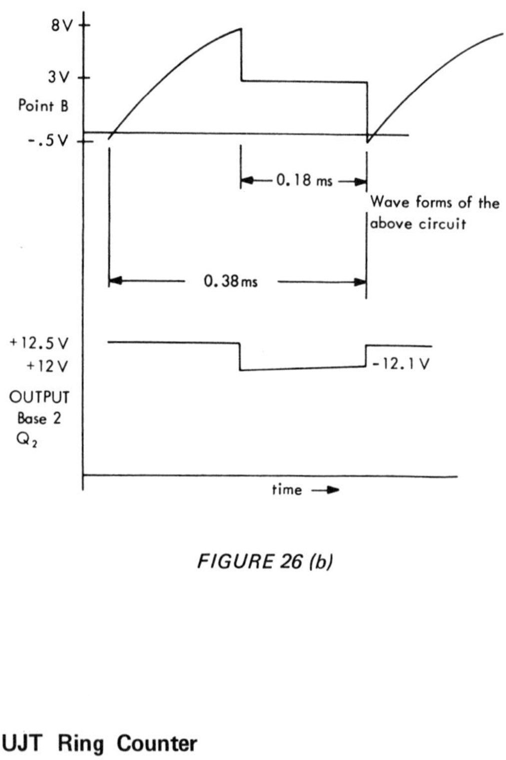

FIGURE 26 (b)

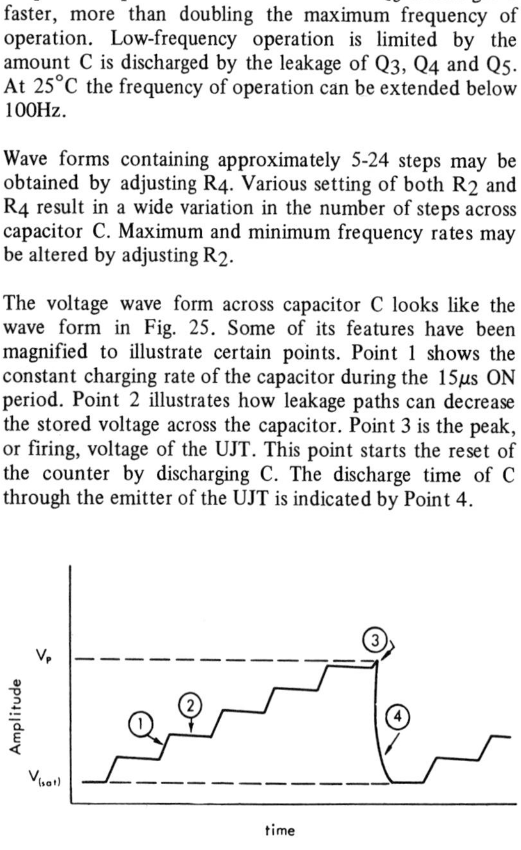

FIGURE 25

Astable Multivibrator

Circuit operation of the flip-flop circuit may be explained in the following way: Assume that Q_1 is ON with point A held at V_{sat}. Point B will be charging toward +12V. When V_P is reached, Q_2 will fire, bringing point B to V_{sat}. Because of the voltage on the capacitor, point A is reserve biased and Q_1 turns OFF. Now, point A begins to charge toward +12V and the cycle repeats itself. Wave forms of the circuit in Figure 26a are given in Figure 26b.

UJT Ring Counter

A ring counter can be designed to count and to read out directly in the decimal system, so the information can be easily interpreted. Ring counters using UJT s can be designed that give excellent performance and component reduction when compared to a conventional transistor counter. Table I compares the number of parts of a UJT ring counter with that of a typical transistor flip-flop counter having diode logic for an output. Both counters have a decimal output and drive NIXIE tubes for their readout. If can be seen from Table 1 that the sizable reduction in parts will result in a simpler and less expensive counter.

Table 1

	UNIJUNCTION TRANSISTORS		CONVENTIONAL TRANSISTORS	
	UJT COUNTER	READOUT	TRANSISTOR COUNTER	READOUT
Transistors	11	10	8	10
Diodes	10	0	17	32
Resistors	22	21	35	31
Capacitors	11	0	9	0
	Total Parts = 85		Total Parts = 142	

The basic operation of any ring counter may be described in the following manner: Referring to Figure 27 and assuming that count 1 is ON, an input pulse is fed into the centre of the circle. The pulse is directed to count 1, turning if OFF. As this count turns OFF, it generates a trigger pulse which turns ON count 2. Thus, the ON count has progressed from count 1 to count 2. The sequence of operation continues to move the ON stage progressively around the circle in a clockwise direction. Since the circle is closed, it is called a ring counter.

The operation of the circuit in Figure 28 depends upon the storage of a charge on the 510-pF coupling capacitors so that the resulting voltage V_{cap} plus the voltage V_{EE} exceeds the firing voltage of the UJT.

One UJT in the ring is held ON by V_{EE} and the 360-ohm emitter resistor. The emitter of the ON UJT will be held at some low level, approximately 4V.

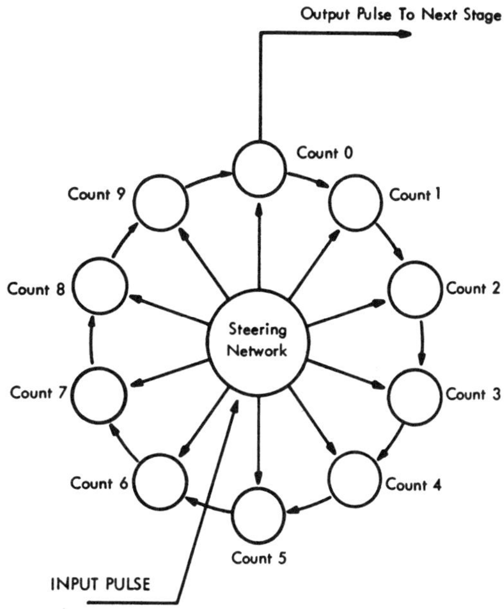

FIGURE 27

Q_O is the trigger transistor. When a pulse is applied through C_O, Q_O turns OFF, removing the voltage V_1 from all of the UJT's. This turns OFF the ON UJT. Assuming that Q_1 is ON, the emitter voltage of Q_1 = 4V. Capacitor C_1 charges to 12–4 = 8 volts with the polarity shown in the schematic.

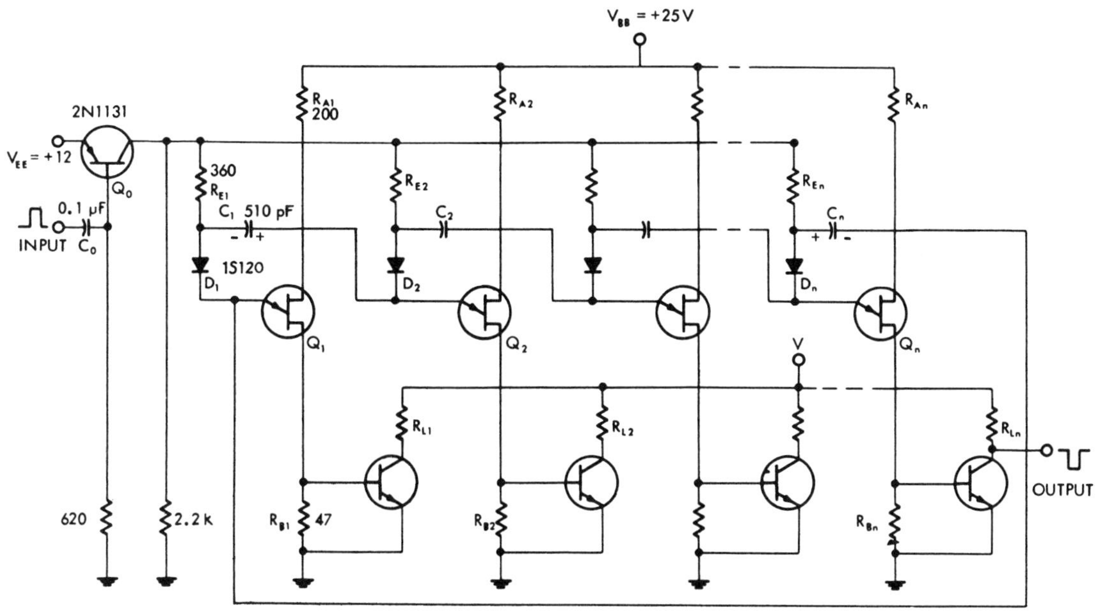

FIGURE 28 Unijunction Ring Counter

The last capacitor in the counter (C_n) charges to the same voltage, but with an opposite polarity. When the input pulse turns Q_o OFF, C_n discharges through D_1, R_{E1}, and R_{En}. C_1 remains charged because of the reverse-biased diode D_2 and the reverse-biased emitter of Q_2. Q_o must remain OFF long enough for C_n to discharge. In this circuit Q_o is OFF for about 10 to 30μs, $(R_{E1} + R_{En}) C_1 = \tau = 510 \times 10^{-12} \times 720 = 0.37 \mu s$. At the end of the 10 to 30μs pulse, 12 volts is applied to the V_{EE} bus (12 volts is below the firing voltage of the UJT, which is approximately 15V). However, the emitter voltage of Q_2 is $V_1 + V_{C1} = 12 + 8 = 20V$. Since this is greater than the firing voltage, Q_2 turns ON. The cycle repeats at the next stage when another input pulse is applied. The working temperature environment of this counter is from 0° to +55°C. Supply voltages are fairly critical and should be obtained from regulated supplies.

The readout from the counter may be taken from transistors placed in base 1 of the UJT's. R_{L1} to R_{Ln} may be any desired type of load such as incandescent lamps or NIXIE tubes. The trigger for another ring counter may be taken by differentiating the positive-going portion of the output pulse from R_{Ln}.

Figure 29 illustrates the emitter waveforms of Q_1 and Q_2. Assume Q_1 ON at t_o, (V_{E1} = 4V). Q_2 is OFF ($V_{E2} = V_{EE}$ = 12V) at t_o. An input pulse at t_1 turns OFF Q_o causing the V_{EE} bus voltage to fall to zero. C_o discharges through D_1 causing V_{E1} to fall rapidly to zero. Since C_1 has no discharge path, it holds V_{E2} at + 8 volts. At the end of the 13 μs input pulse, 12V is again applied to the V_{EE} bus. V_{E1}, V_{E3}, V_{E4}, . . . V_{En} all rise to 12 volts, less than the minimum value required to fire the UJT's. However, V_{E2} simultaneously rises toward 12 volts + 8 volts (potential on C_1) which is more than is required to turn on Q_2. Q_2 turns ON and V_{E2} now rapidly falls to + 4V. The ON stage has now been shifted from Q_1 to Q_2. Additional pulses will shift the ON state from Q_2 to Q_3 and so on. The nth- input pulse will return the ON state to Q_1.

The power supplies must be arranged so the 25 volt supply always comes on before the 12 volt supply. This requirement ensures that the emitters will be reverse biased when V_{EE} is applied so that a number of stages will not come ON at the same time. To turn a stage ON initially, one need only increase the voltage level on an emitter to V_P by a 1k resistor returned to V_{BB}.

The ring counter may be used to scale down to any desired ratio by adding as many stages within the ring as desired. As many sets of ring counters as desired may be added in series to increase the counting capacity. If it is desired to count up to 999 by the decimal system, for example, it would be necessary to use three rings of ten each, in series.

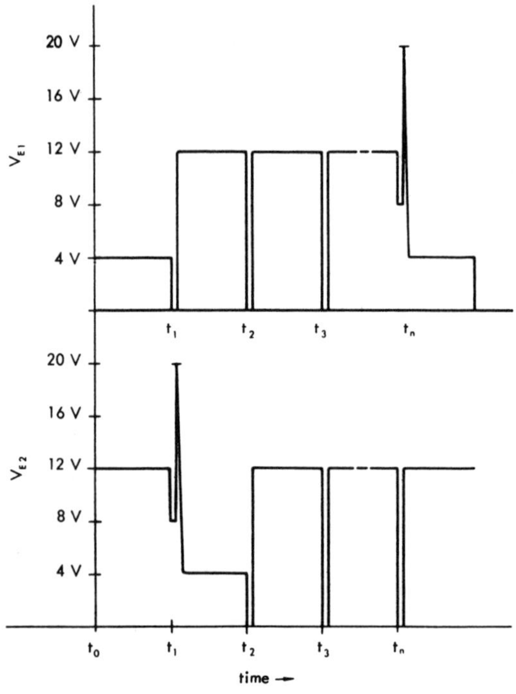

FIGURE 29

Voltage Comparator Circuit

The basic block of the voltage comparator circuit in Figure 30 is two R-C relaxation oscillators consisting of Q_2 and Q_3. Firing voltages for Q_2 and Q_3 are controlled by the differential amplifier consisting of Q_1 and Q_4. D_1 serves as a voltage reference against which the input signal on the base of Q_1 is compared. As V_{in} is increased, the voltage on base 2 of Q_2 is lowered and the voltage on base 2 of Q3 is increased. Thus, V_P of Q_2 is lowered at Q_2 will oscillate before Q_3. Under this condition, the output will be a series of pulses from point A. When V_{in} is lowered, base 2 of Q_3 is lowered with respect to base 2 of Q_2 and the output will be a series of pulses at point B. R_3 is used to balance unequal characteristics of Q_2 and Q_3. R_2 adjusts the bias of Q_1 and Q_4 so various levels of V_{in} may be used for triggering. Changes – as small as 2mV on the base of Q_1 – can switch the output from A to B.

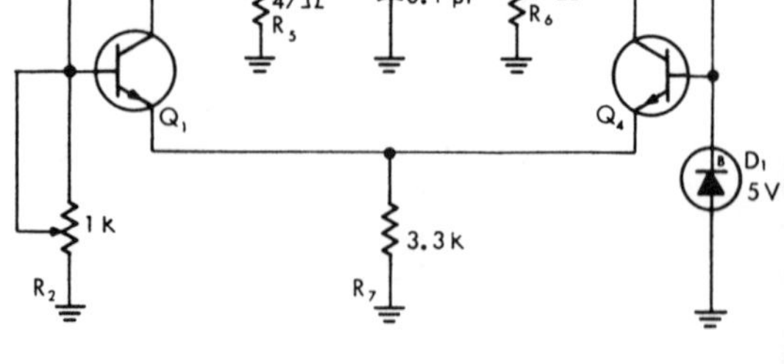

FIGURE 30 Voltage Comparator Chart

Pulse Generator

The simple and inexpensive circuit of Fig. 31 yields a pulse generator whose pulse width and frequency may be varied to a certain degree. The circuit consists of two parts, a free-running R-C relaxation oscillator and a transistor output stage. Normally, Q_2 is held in the OFF state by the 47-ohm resistor from base to emitter. The voltage drop across this resistor resulting from r_{BB} and V_{BB} is small, and decreases as temperature increases due to the increase of r_{BB}. When V_c reaches V_p, Q_1 fires and C discharges. Most of the discharge current passes through the base-emitter diode of Q_2 because of its low impedance. This saturates Q_2 and brings the output from +15V to approximately 0V.

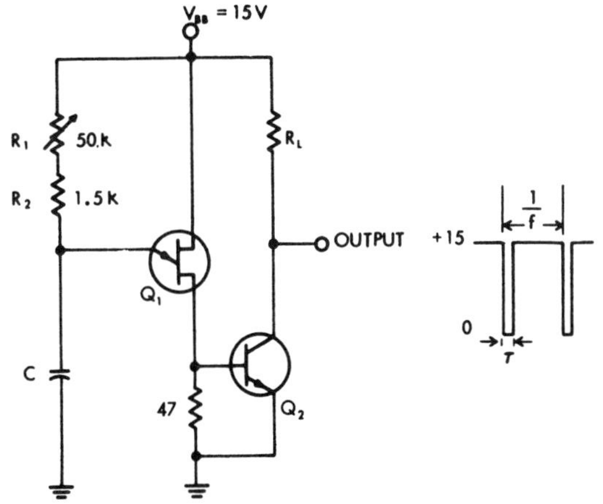

FIGURE 31

When the capacitor has discharged sufficiently, Q_1 and Q_2 turn OFF, thus returning the output to a +15V level. With selected UJT's the output rise and fall times are in the order of 1μs. The associated table shows the frequency and pulse width one can expect with this circuit when various capacitors are used. The circuit, as shown, operates over a temperature range of −25°C to approximately +60°C. The temperature limit can be extended to +125°C by increasing the value of R_2 to a value greater than 3.6k. Variation in R_1 varies the frequency. Changing C varies both frequency and pulse width. A variable resistor added in series with the emitter of Q_1 will vary the pulse width.

Table 2.

C	freq. f.	Pulse width T
0.2 μF	150 Hz -5.0 kHz	18 μs
0.4 μF	70 Hz -2.5 kHz	25 μs
0.6 μF	40 Hz -1.7 kHz	32 μs
0.8 μF	35 Hz -1.2 kHz	40 μs
1.0 μF	30 Hz -1.0 kHz	50 μs

A-C Power Control Circuit

Controlling a-c power in a load may be done effectively by using a UJT to trigger an SCR. Controlling power by switching methods is more efficient that adding resistance in series with a load to dissipate power. The circuit of Figure 32 can be used to change the amount of current flowing through R_L and consequently control its power. R_L may be any sort of load such as a heater in an oven or a high-power spotlight. Circuit operation is as follows: Assuming that a half-cycle wave across the SCR is just beginning, the voltage across the UJT control circuit is at zero. As time increases, the voltage across the control circuit increases and C begins to charge toward V_p through R_1 and R_3. When V_p is reached, the UJT fires, triggering the SCR. The voltage across the SCR is reduced to the saturation voltage. This also shorts out the UJT supply and prevents the capacitor C from charging any more during that half cycle. At the end of the half cycle, when the supply voltage returns to zero, the SCR is extinguished. When the a-c voltage begins to increase, C starts to charge and the cycle of operation begins again. The amount of

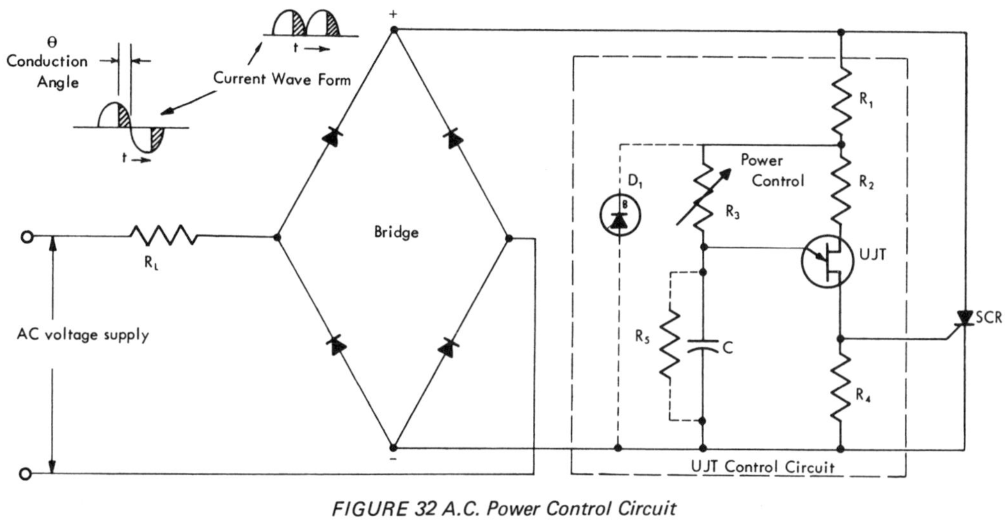

FIGURE 32 A.C. Power Control Circuit

power controlled is limited only by the current and voltage capacities of the SCR and the bridge diodes.

To prevent low frequency flicker when R_3 is large, R_5 can be included. The addition of R_5 will drain off current as fast as it is supplied through R_3 so the UJT will never fire. This will enable R_3 to turn the power OFF without using excessive values of resistance. If the a-c supply voltage becomes large enough to cause possible destruction to the UJT, a breakdown diode may be added (such as D1 in Figure 32) to clamp the voltage level. Figure 33 shows a typical curve of the conduction angle vs the rms value of current through the load.

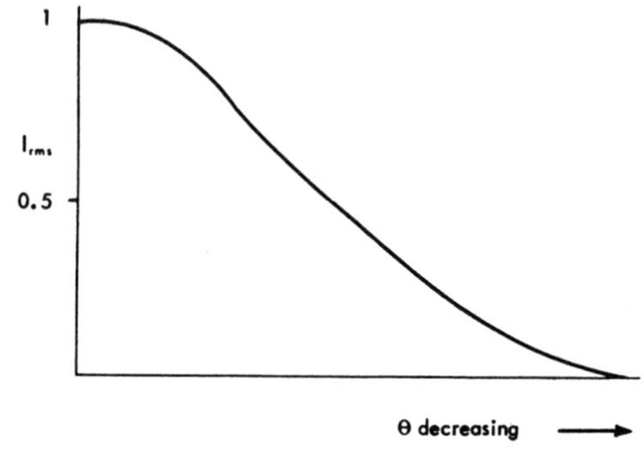

FIGURE 33

Motor Control Circuit

UJT s are well suited for triggering SCR s. The UJT is used in the low-power timing circuitry while the SCR controls the load power. In the circuit in Figure 34, the UJT-SCR combination is used in a speed control drive for a small permanent-magnet motor such as the type found in model train motors. The load in series with the SCR receives a varying amount of power because of the changing of the conduction angle. The circuit operates in the following manner: Assuming that the motor is at a standstill, C charges (through R_1 and R_2) to some voltage as determined by the output of the transformer. Since this voltage is fairly high, causing a rapid frequency of oscillation, the SCR conducts for a large portion of the cycle (depending upon the setting of R_1). This provides high starting power. As the motor begins to rotate, a back emf is generated in such a polarity as to decrease the charging voltage. This tends to reduce the frequency of oscillation of the UJT and cause the SCR to decrease its conduction angle. At a given setting of R_1, some equilibrium will exist between conduction angle and motor speed. As a load is applied to the motor it will tend to slow down, thus reducing the back emf. This reduction in emf will increase the charging voltage of the UJT, increase the frequency of oscillation, increase the conduction angle, and supply more power to the load. Thus feedback is present that tends to maintain the motor at a constant speed for variation in load.

SCR-UJT Switching Mode Voltage Regulator

Low power loss in the control element makes switching-mode regulators ideal for high efficiency. The circuit of Figure 35 uses a UJT as a variable frequency oscillator whose output triggers a one-shot SCR circuit. The output level is changed by increasing or decreasing the number of constant width output pulses from the SCR. These pulses are filtered by L and C and appear as a d-c voltage across R_L.

Output sampling and amplification is done by Q_2 and Q_3. The sampled output is compared to a 6V reference diode in the emitter of Q_2. Depending upon whether the output voltage is too low or too high, Q_2 varies the frequency of Q_1. As the frequency is varied, the output pulses from Q_1 vary the duty cycle of the power control switch, thus adjusting the output level. Output voltage variation is approximately 0.1V for a 50 volt charge at the input, and is approximately 0.2V from zero to full load.

Oscillator and Pulse Height or Width Detector Circuits

Several applications exist for a matched pair of UJTs. The following circuits utilize two UJTs which are matched for η, r_{BB}, I_{EB20}, and frequency of oscillation. In the circuit of Figure 36 UJT Q_1 oscillates at a constant frequency f_1, whereas Q_2 oscillates at a frequency f_2 which is dependent upon the value of V_X. Figure 37 shows a plot of $f_2 - f_1$ as a function of V_X.

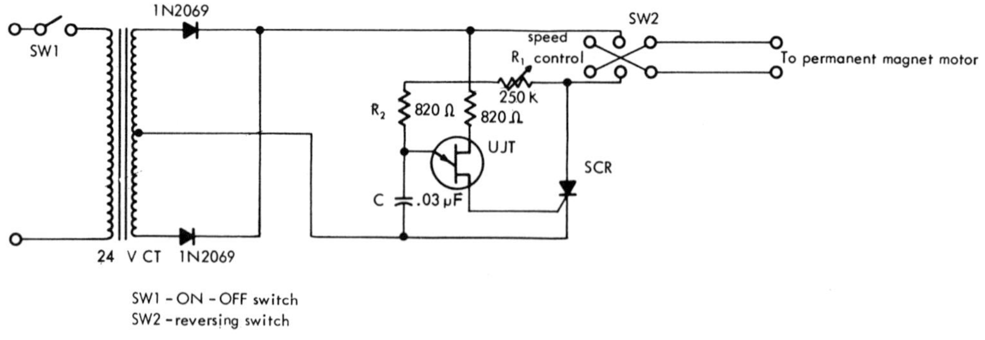

FIGURE 34

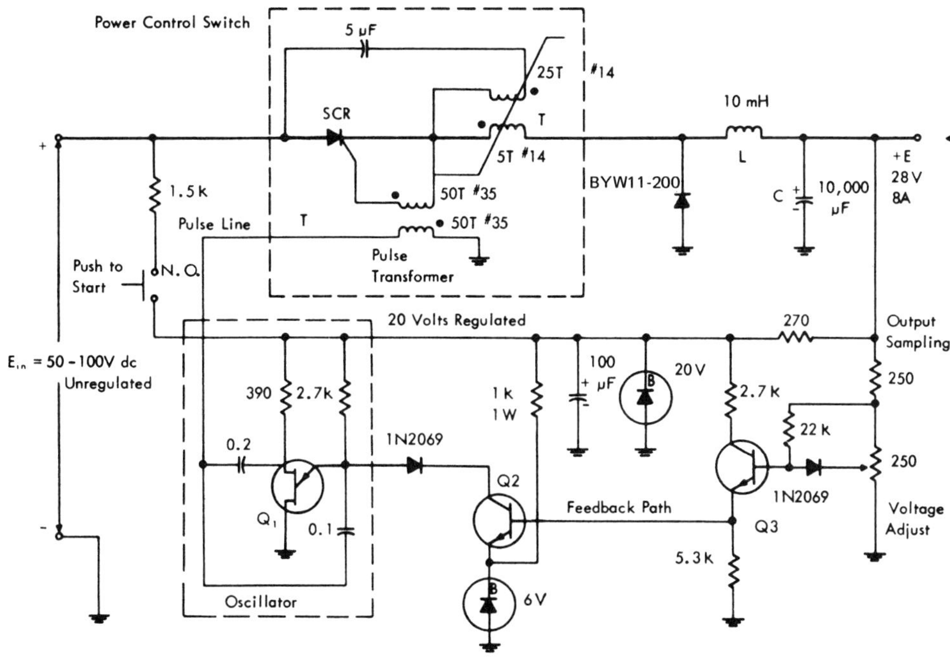

FIGURE 35

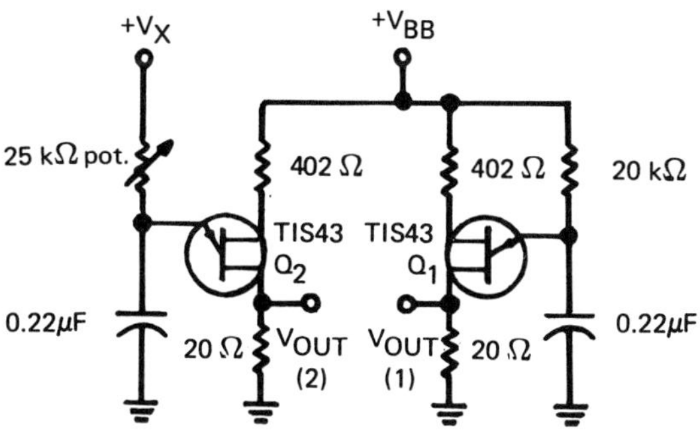

FIGURE 36. Matched Pair of UJTs Connected as Oscillators

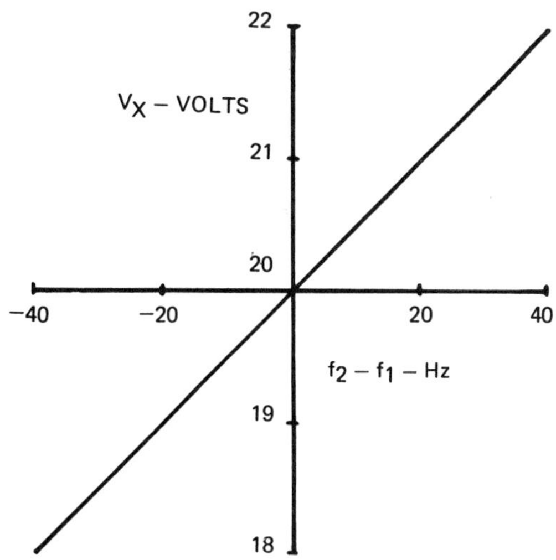

FIGURE 37. Plots of Difference in Output Frequency vs. V_X

The matched UJT pair may be used as a pulse-width detector, or as a pulse-height detector. Figure 38 (a) shows a circuit configuration which performs either of these functions. Each time V_{IN} appears, capacitor C_1 charges toward the peak level of V_{IN}. If the input pulse falls to ground potential before transistor Q_1 turns on, transistor Q_2 will turn on (and discharge C_1) during the trailing edge of V_{IN}. The circuit is a pulse-width detector when V_{IN} has a fixed positive level, and is a pulse-height detector when V_{IN} has a fixed width. Plots of pulse width vs. V_{BB} are shown in Figure 38 (b) for various values of pulse amplitude.

In Figure 38 (a), supply voltage V_{BB} can be replaced by a pulse source V' which is synchronized to the V_{IN} pulse train. As long as the width of V' is greater than that of V_{IN}, the circuit functions as if V' were a fixed voltage level. When V' becomes less wide than V_{IN}, transistor Q_2 fires during each cycle.

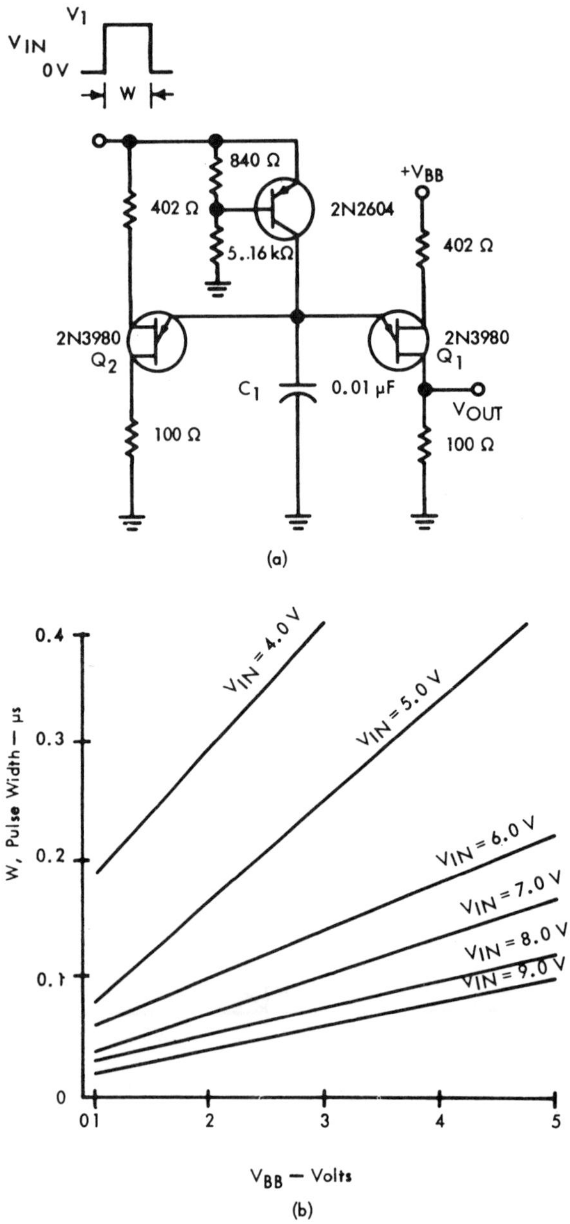

FIGURE 38. Circuit for detecting Pulse Width or Pulse Height (a); Plots of Pulse Width vs. Reference Voltage for Various Levels of Pulse-input Voltage (b).

APPENDIX

Stabilizing V_p Against Temperature Changes

The period of a UJT timing circuit will vary with variations in temperature. One cause of this variation is the change in the emitter firing voltage V_p. Proper selection of an external base 2 resistor can greatly reduce variations in V_p.

The UJT is represented in the OFF condition by the equivalent circuit of Figure 1a. Firing voltage is given as:

$$V_p = V_D + \frac{r_{B1} V_{BB}}{r_{B1} + r_{B2}} = V_D + \eta V_{BB} \qquad (A1)$$

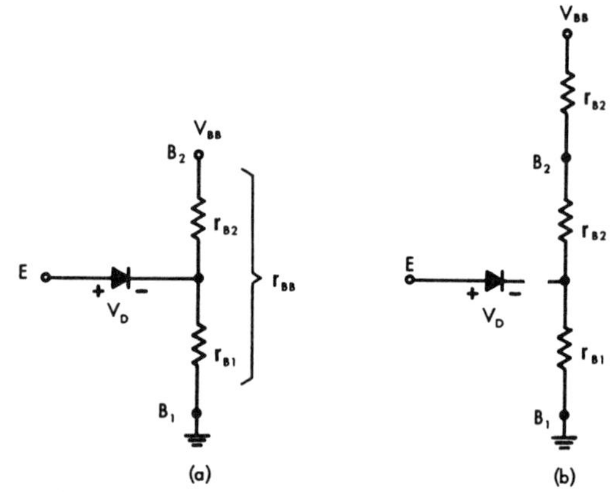

FIGURE 1

The diode voltage V_D changes approximately -2.5×10^{-3} V/C°. The interbase resistance $r_{BB} = r_{B1} + r_{B2}$ changes approximately 0.8%/C°. Since changes in r_{B1} and r_{B2} are proportional, η is constant and the change in V_p with temperature T is due to the diode voltage variation with temperature. Then,

$$\frac{dV_p}{dT} = \frac{dV_D}{dT} = -2.5 \times 10^{-3} \frac{V}{C^o} \qquad (A2)$$

A temperature-stabilizing base-2 resistor (R_{B2}), illustrated in Figure 1b, changes the firing voltage to

$$V_p' = V_D + \frac{r_{BB} V_{BB} \eta}{r_{BB} + R_{B2}} \qquad (A3)$$

It is desirable to find a value of R_{B2} that will cause the variation in Vp' to be zero with respect to temperature.

Differentiating (A3) with respect to temperature,

$$\frac{dV_p'}{dT} = \frac{dV_D}{dT} + \eta \frac{dr_{BB}}{dT} \frac{R_{B2} V_{BB}}{(r_{BB} + R_{B2})^2} \qquad (A4)$$

Over a wide temperature range, V_D will change a fraction of a volt, and r_{BB} may change by a factor of 2:1.

Intuitively, (and from past experience), for optimum compensation

$$R_{B2} << r_{BB} \tag{A5}$$

Assuming (A5) to be true,

$$\frac{dV_p'}{dT} = \frac{dV_D}{dT} + \eta \frac{dr_{BB}}{dT} \frac{R_{B2} V_{BB}}{r_{BB}^2} \tag{A6}$$

for optimum compensation $\frac{dV_p'}{dT} = 0$

Then

$$R_{B2} = - \frac{dV_D}{dT} \frac{r_{BB}^2}{\frac{dr_{BB}}{dT} \eta V_{BB}} \tag{A7}$$

Substituting numerical values into (A7)

$$\frac{dV_D}{dT} = -2.5 \times 10^{-3} \text{ V/C}^o;$$

$$\frac{dr_{BB}}{dT} = -8 \times 10^{-3} r_{BBO}/\text{C}^o$$

where $r_{BBO} = r_{BB}$ at 25°C

$$R_{B2} = \frac{-(-2.5 \times 10^{-3}) r_{BB}^2}{8 \times 10^{-3} r_{BB} \eta V_{BB}}$$

$$R_{B2} = \frac{0.312 (r_{BB})^2}{\eta V_{BB} r_{BBO}} \tag{A8}$$

Since r_{BB} is a function of temperature, $r_{BB} = r_{BBO} [1 + H(T - 25°C)]$

where $H = 8 \times 10^{-3} / C°$ T = temperature in °C

the optimum value of R_{B2} will vary with temperature. A value of R_{B2} obtained with (A8) will be exact only through a limited temperature range. A special case of (A8) occurs near 25°C where $r_{BB} = r_{BBO}$.

Then

$$R_{B2} = \frac{0.312 r_{BBO}}{\eta V_{BB}} \tag{A9}$$

Substituting typical values (V_{BB} = 15V, r_{BBO} = 10k, η = 0.5) into (A9) yields a value of R_{B2} of 416 ohms.

In practice, V_p is not the only UJT variable. It has been found experimentally that good compensation of an R-C UJT oscillator may be obtained over a wide temperature range by using one value of R_{B2}. Figure 2 shows a simple relaxation oscillator and its frequency performance over a temperature range. Data were taken on one UJT that was placed in an oven with the other components external to the oven.

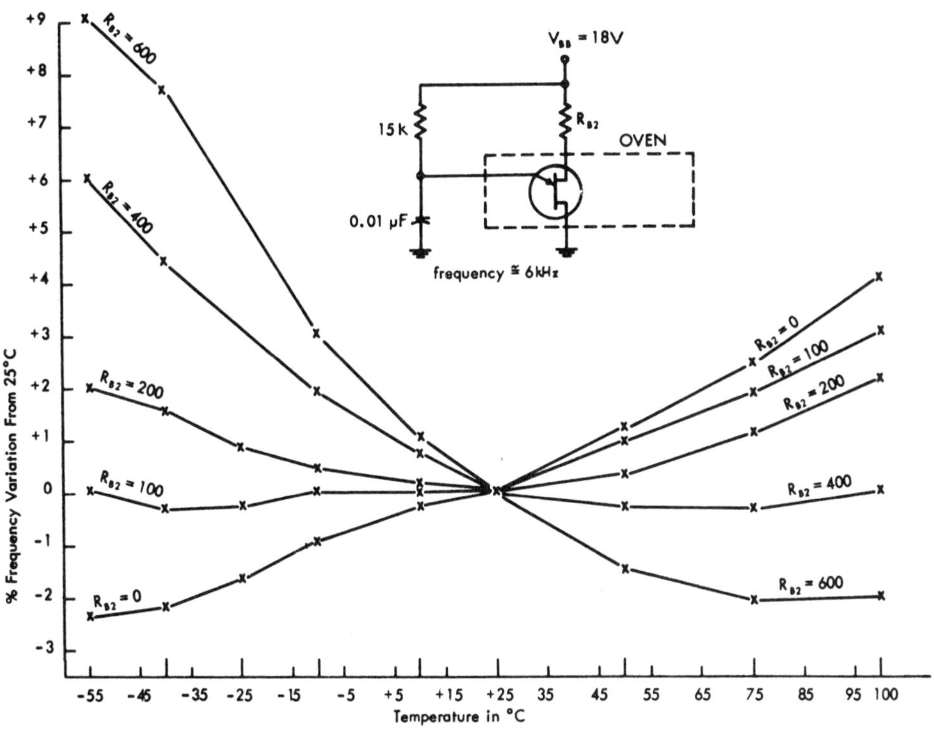

FIGURE 2

SECTION 2

POWER TRANSISTOR APPLICATIONS

VII SWITCHING MODE POWER SUPPLIES

by Kuldip Rupra

The concept of switching mode power supplies is to produce a more efficient, cheaper and less bulky unit than the conventional power supply. The basic principle of the switching mode supply is simple. A switch, operated at a fixed frequency, has its ratio of on time, t_{on}, to off time, t_{off} varied according to the power required at the output. A transistor is suitable as the switch and will feed into an integrating low pass filter to give a reasonable dc output. When compared to the conventional series linear regulator the overall efficiency of the switching mode supply is higher because, when fast switching is used, power dissipated in the device is very much smaller. With ideal switching, the power out P_o is:

$$P_o \triangleq [P_{in}.t_{on}/\tau].xI_o/I_{in} \quad \text{or} \quad V_o = V_{in}.t_{on}/\tau.$$

where τ is the period and equal to $t_{on} + t_{off}$

In the series regulator, the power dissipated in the device, ie, $(V_{in}-V_o).\ I_o$, can be extremely large and unacceptable, eg, for V_{in} = 340V and V_o = 60V at 2A, dissipation in the series element would be 560W.

Due to the comparatively low power dissipation in the switching transistor, its full current and voltage capabilities can be used without having to worry too much about its thermal characteristics, eg, to deliver 60V at 2A from an unregulated 340V supply the switching transistor dissipates in the region of 10W. (See Appendix 2).

CIRCUIT THEORY

Fig. 1 shows, in black diagram form, the principle of switching mode power supply. The series transistor switch is operated at the fixed oscillator frequency (ie, 15kHz). The mark space ratio, ie, on-off ratio, is controlled by a dc level fed into the variable mark-space generator. This level is the amplifier error signal obtained by comparing the output from the power supply with a dc reference voltage.

The Low Pass Filter

The integrating filter network, shown in Fig. 2, is used to give a smooth dc output from square power pulses. It is basically an inductor, L, a capacitor C, and a diode D. The load is connected across the capacitor. When the transistor switch is saturated, the point E is stepped up to 340V positive such that the diode D is reverse biased and current builds up through inductor L. Assuming a constant output voltage V_o across capacitor C:

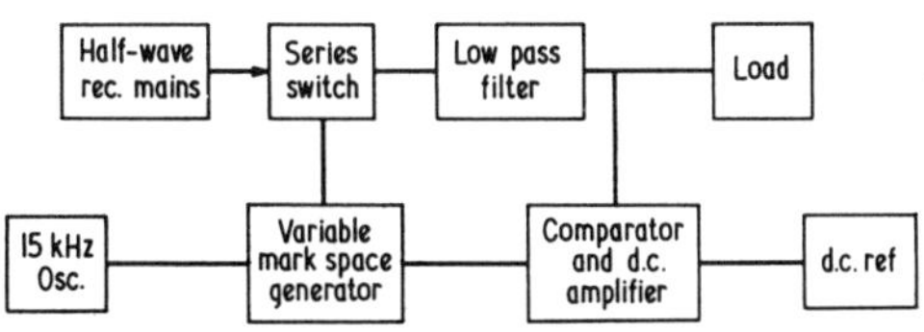

FIGURE 1 Switching mode power supply

$$(V_{in}-V_o) = L.di_L/dt$$

$$V_{in}-V_o \triangleq L.\Delta i_L/t_{on}$$

$$\Delta i_L \triangleq (V_{in}-V_o).t_{on}/L.$$

When the transistor switch is turned off the flow of current is suddenly interrupted and the energy stored in the inductor generates a negative going voltage. This, on exceeding its forward voltage drop (V_F), forward biases diode D and provides a continuous path for the current to flow in the inductor in the normal direction. For a constant load current, the inductor current will decay from $i_{L\ max}$ to $i_{L\ min}$, (Δi_L), in time t_{off}

$$\Delta i_L = V_o.t_{off}L.$$

There are various ways in which an integrating low pass filter may be arranged; this depends on the following factors:

(1) The polarity of V_o required.

(2) The type of transistor to be used (ie, n-p-n or p-n-p).

(3) Whether the filter is employed in the emitter or collector of the transistor.

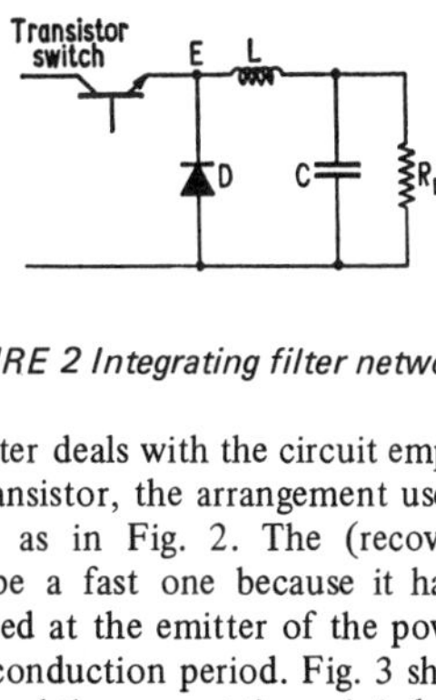

FIGURE 2 Integrating filter network

Since this chapter deals with the circuit employing a silicon n-p-n power transistor, the arrangement used to produce a positive V_o is as in Fig. 2. The (recovery) diode has essentially to be a fast one because it has to clamp the voltage developed at the emitter of the power transistor at the end of its conduction period. Fig. 3 shows the voltage across diode D and the current through inductor L.

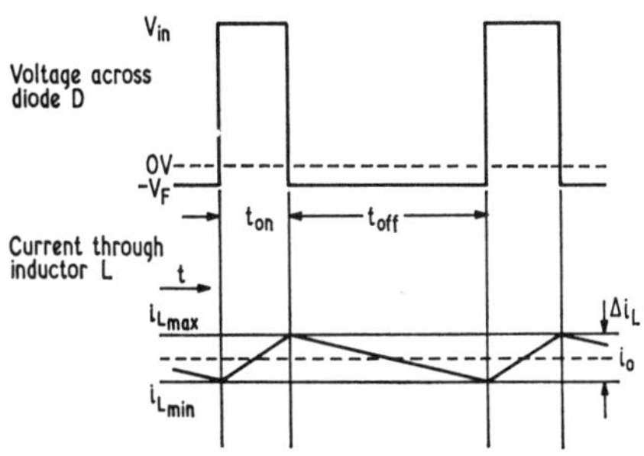

FIGURE 3

Voltage across diode D and current through inductor L

It can be deduced that:

$$i_o = i_{Lmin} + \Delta i_L/2$$
$$V_o = (i_{Lmin} + \Delta i_L/2)R_L$$

and

$$\Delta i_L = V_o.t_{off}/L$$
$$= (V_{in} - V_o).t_{on}/L$$

From the above expressions calculation of the 15kHz ripple at the output can be made (see Appendix 1). The output voltage V_o is a function of pulse width t_{on}, output current (I_o) is limited by the inductance value and its saturation characteristics and also by the current capability of the transistor.

The value of the capacitor may be determined from the switching frequency ripple that can be tolerated.

$$V_{ripple} = (t_{on})^2 . (V_{in} - V_o)/2LC. \text{ (See Appendix 1)}$$

The larger the product of LC the smaller will be the ripple. It is also obvious that for higher oscillator frequencies, smaller values of L and C may be used, making the physical size of the supply smaller. (The circuit in this report uses a 15kHz oscillator which may be replaced by a square wave input derived from a tv line oscillator).

The Driver Stage

In the switching regulators discussed in many publications, emphasis has been placed on application at relatively low input voltages, eg, 60V, where the control circuitry is directly coupled to the switching transistor via a resistor chain. To operate at high voltages of 340V it would be extremely expensive to produce a power supply using this idea because a chain of driver transistors would be required. Efficiency would also be poor.

In this chapter an isolated transformer base drive is employed. This transformer is a current step-up transformer, the primary of which is connected through a 6kΩ resistor to the halfwave rectified mains. The other end of the primary winding is connected to the driver transistor, a BF 259, which is a low current high voltage device. The secondary of this transformer is connected across the base-emitter of the power transistor as shown in Fig. 4. The transformer is so arranged that the driver and output transistor work in an asynchronous mode, ie, during the time when the driver transistor is 'on', the power transistor is "off" and vice versa.

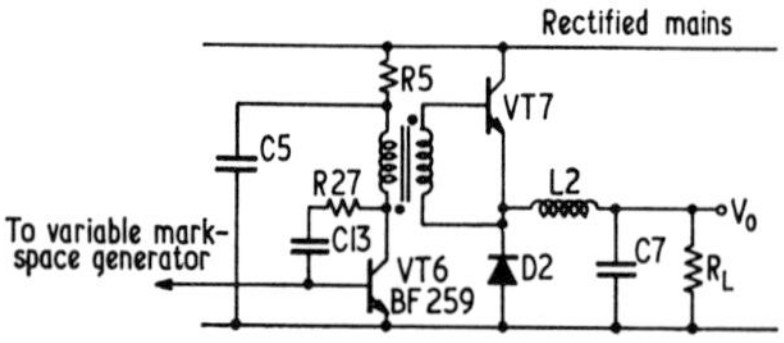

FIGURE 4 Driver

The driver switching is controlled by the variable markspace generator. To prevent an excessive voltage overswing at the collector of VT6 when it is turned "off", some form of damping is essential. Since the device capability is only 300V, two 150V Zener diodes, ZD2 and ZD3, are connected in series across collector and emitter of the driver such that the collector voltage never goes above 300V. The CR network formed by capacitor C13 and resistor R27 connected between the collector and base of the device further damps its collector voltage waveform. Appropriate damping of this circuit is essential.

Fig. 5 shows the driver collector waveform without (A) and with (B) the damping circuit. Also shown is the respective secondary current waveform, ie, the base current of the switching transistor.

The decoupling capacitor C5 helps to provide a high current from the secondary because it smooths the supply to the primary winding by its integrating action and a higher voltage swing across the primary is achieved.

The other feature of this transformer drive is that when the drive pulse width is narrow, eg, 3μs, the base current (secondary current) is relatively large. This is due to the fact that the driver transistor is on for most of the period (ie, 64μs) and the stored energy during this time takes the form of a very short time duration secondary current. It must be noted that if the pulse width becomes too narrow the forward base current will start to drop, because of its initial switching on slope. (Fig. 5 B (c)). The high current during narrow base pulse widths is definitely an advantage, as will be illustrated later.

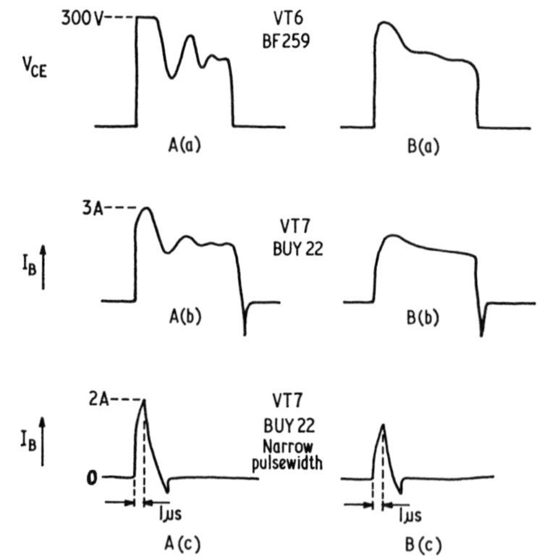

FIGURE 5 Driver collector waveforms

The Variable Mark-Space Generator

This is the part of the control circuit responsible for giving a pulse width proportional to a dc level. An externally triggered monostable may be used to perform this function. Fig. 6 shows a conventional monostable circuit where transistor VT6 is normally on and VT5 is off.

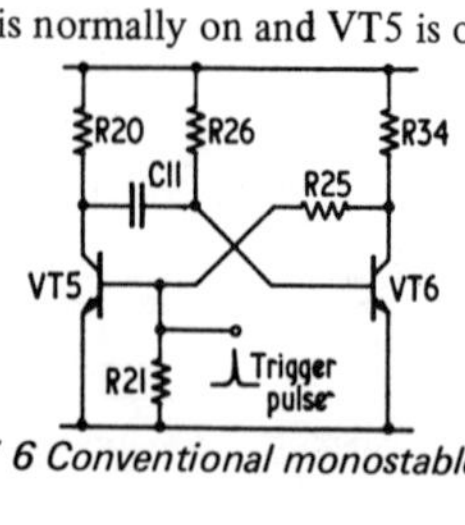

FIGURE 6 Conventional monostable circuit

When a trigger pulse is applied to the base of VT5, it switches on. The time that transistor VT5 can stay on is determined by the time constant of R26 and C11.

$$t_{on(VT1)} \triangleq 0{,}7\ R26\ C11$$

To vary the pulse width one has to change either the value of charging resistor R26 or the capacitor C11. In Fig. 7, the two ways of controlling the pulse width are shown. Fig. 7 (a) shows a monostable circuit which uses a transistor to vary the charging current; the voltage applied to the base of this transistor determines the single shot pulse width. The timing capacitors stay unchanged.

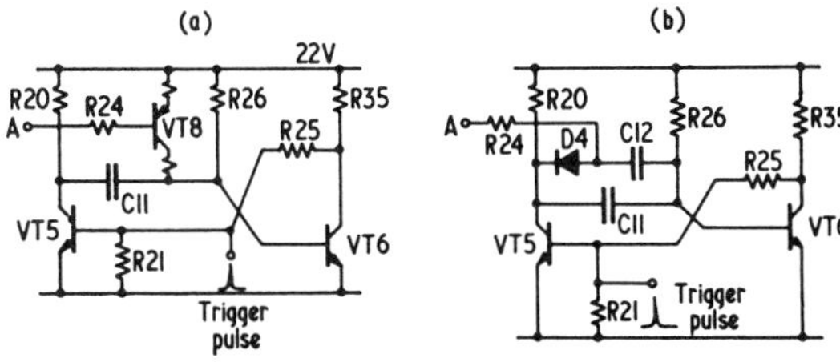

FIGURE 7 Pulse width control

Fig. 7 (b) shows how the time constant may be varied by having the additional capacitor C12 and blocking diode D4 in series with it across the timing capacitor. Capacitor C11 and resistor R26 determine the minimum pulsewidth. To increase the pulsewidth, a dc voltage is applied through a resistor to the junction of the diode D4 and the capacitor C12, to add charge into the capacitor C12 and effectively increase the charge in the timing circuit. As a result, the pulsewidth is increased in proportion to the charge added into the capacitor C12.

Normally the configuration shown in Fig. 7 (a) would be used for the mark-space generator but in this case it has a serious disadvantage. The 22V supply to the control circuit is derived from half wave rectified mains through a 6kΩ resistor and stabilized simply by a Zener. This can, in the worst case, give only 40mA at 22V. For variable charging resistance technique, as shown in Fig. 7 (a), up to 30mA can be used by the monostable alone when a very narrow pulsewidth is required. This would introduce severe regulation problems on the 22V supply. The alternative is to reduce the 6kΩ resistor to a lower value. This is undesirable because power dissipation in the resistor and Zener will be increased. Therefore, the circuit shown in Fig. 7 (b) is employed here.

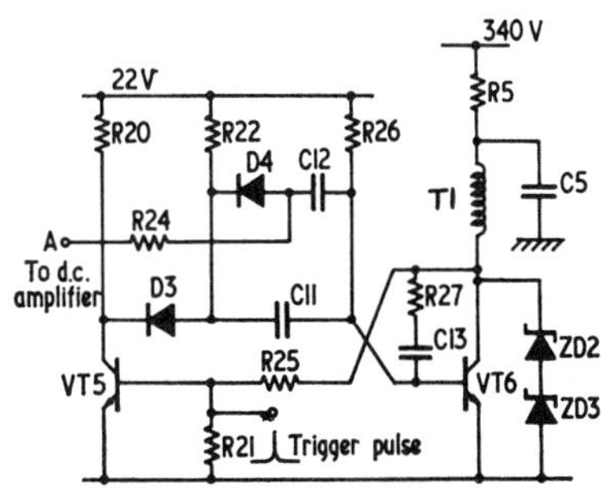

FIGURE 8 Driver and monostable

Fig. 8 shows the driver stage incorporated into the monostable as its stable limb, ie, transistor VT6 is a BF259 with the driver transformer connected as explained in the driver section.

The values of R26 and C11 are so arranged that when there is no voltage at point A, the pulsewidth is minimum, ie, transistor VT6 is off for only about 1μs and the power transistor will be on for little longer than that. To increase pulsewidth at point A, a positive potential is applied.

Diode D3 and resistor R22 are added to make the switching off of transistor VT5 much faster than when capacitor C11 is connected directly to the transistor VT5 collector.

The Comparator and dc Amplifier

This amplifier has a very important function to perform. It compares supply output voltage to a reference voltage, amplifies the error and feeds this information into the variable mark space generator. There are various factors that one has to keep in mind in design of this amplifier. The most important one is the starting up conditions of the regulator and the effect that it has on the power transistor.

Normally, when the power is switched on the output voltage of the power supply which is 0V would demand from the comparator and monostable maximum pulsewidth. This could be fatal to the transistor switch for the following reasons. At switch on the capacitor across the load is free of charge and with maximum pulsewidth, collector current will build up to a very large magnitude, eg, 10A, during the first few cycles. During this period the base current will be very small and because of the wide mark-space ratio, it takes some time to build up to its normal value. As a result, the combination of large collector current and very small base current brings the device out of saturation. For example, a typical device will have developed across it a collector voltage $V_{CE} \triangleq 100V$ at a collector current I_C of 10A and a base current I_B of 0,5A. This means an average dissipation of 500W, assuming a 1:1 mark-space ratio, which would undoubtedly result in catastrophic failure of the device. A solution to this problem is to start the power supply with a mark-space ratio as small as possible, ie, t_{on} is very small, eg, 3μs. This ensures that:

(a) The collector current buildup is much more gradual through the integrating low pass filter.

(b) In case of large collector currents through the device, as the conduction time t_{on} is 1/20th of the whole period, the average power dissipated is 1/10th of that when the mark space ratio is 1:1.

Fig. 9 shows circuits of two dc amplifiers that might be used. Fig. 9 (a) shows a single transistor amplifier in which

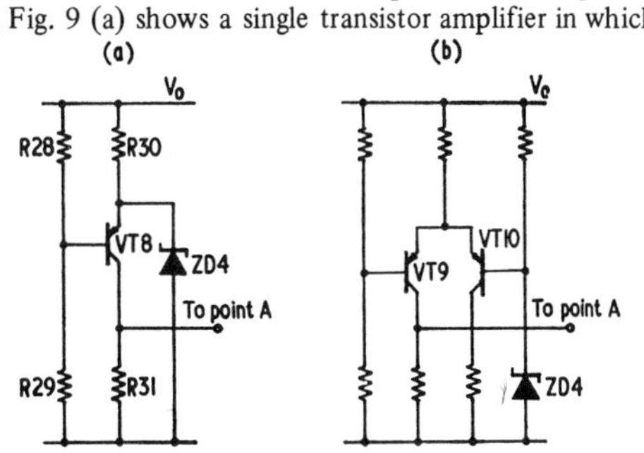

FIGURE 9 D.C. Amplifiers

the dc output from the collector V_C feeding the monostable at point A, is zero when the power supply output voltage, V_O is zero, ie, the pulsewidth is 3μs which is preset by the time constant in the monostable. As the output voltage V_O begins to rise, the voltage V_C also rises. The gain ($\Delta V_C/\Delta V_O$) of this stage in this mode is approximately equal to R31. R28/R30. (R28+R29). The voltage at the collector will continue to rise with increasing V_O until:

$$V_O=(V_Z-V_{BE})\,(R28+R29)/R29$$

where V_Z is the reference voltage and V_{BE} is the forward base-emitter voltage of the transistor.

This is when the emitter voltage is stabilized by the Zener to a selected value. Any further change in voltage V_O will result in the reversal of the mode of operation of the amplifier, ie, with increase in V_O the collector voltage will decrease.

The gain in this mode of operation would be proportional to h_{FE}. R31, ie,

$$\frac{\Delta V_C}{\Delta V_0} \simeq -h_{FE}.R31./R28.$$

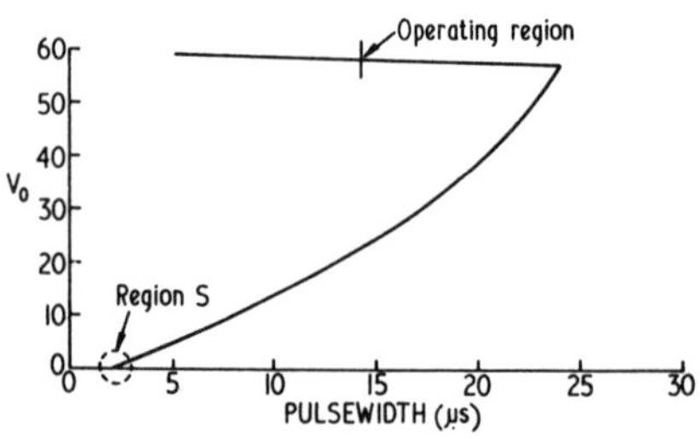

FIGURE 10 Locus of operation of amplifier in Figure 9(a)

Fig. 10 shows the locus of operation of this amplifier when coupled to the monostable. Pulsewidth is directly proportional to the dc output of the amplifier. The reference voltage V_Z= 29V and R28=R29. This amplifier does have a disadvantage. When it goes into its reverse mode (ie, regulating mode), the reference voltage V_Z increases slightly as most of the current flowing through collector resistor R31 is diverted from flowing into the emitter of the transistor to the Zener. This is no problem unless a very high degree of stability is required. However, an alternative arrangement to prevent this is given in Fig. 9(b). This is effectively a log tailed pair circuit and does have a further advantage of better temperature stability. In this report, however, the single transistor amplifier is used to keep down the cost of the power supply.

The Switch

Voltage and current waveforms during normal operation of the 60V 2A power supply are shown in Fig. 11. Also shown is current flowing through diode D2 and choke L2. The device used for the switch is a BUY22 (VT7).

An expanded trace of the switching on and off times of the transistor switch is given in Figs. 12(a) and 12(b). From Fig. 12(a) it can be seen that the transistor's collector current builds up to its peak before the voltage across it starts to drop to its $V_{CE(sat)}$. The current turn on time is typically about 450ns and voltage turn on time is about 150ns. The effective device turn on time for assessing device dissipation is 450+150ns.

Similarly, during turn off, ie, when the transistor base-emitter is reverse biased, the voltage across the device switches from $V_{CE(sat)}$ towards the supply voltage. During this time the device is forced by the inductor L2 to conduct until the emitter voltage is minus the forward drop ($-V_F$) of diode D2 (with respect to earth) when the diode takes over and maintains the current flow through the inductor into the load. The effective device turn off time is the collector emitter voltage turn off time plus the collector current turn off time which are typically 150 and 350ns respectively.

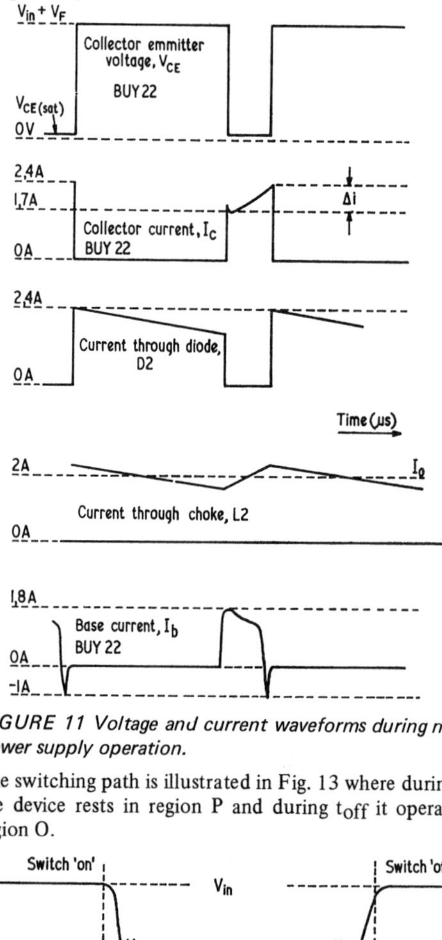

FIGURE 11 Voltage and current waveforms during normal power supply operation.

The switching path is illustrated in Fig. 13 where during t_{on} the device rests in region P and during t_{off} it operates in region O.

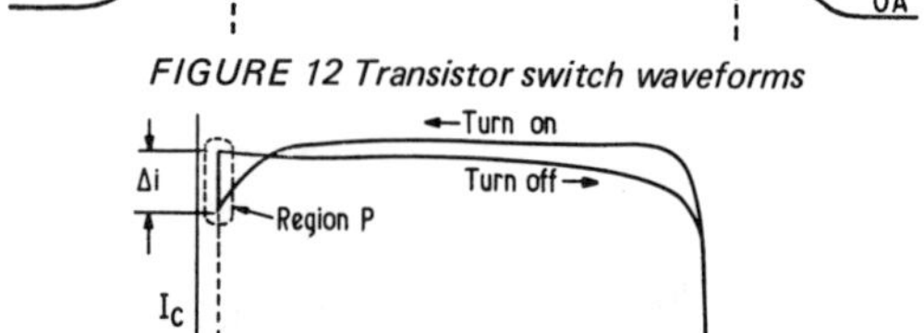

FIGURE 12 Transistor switch waveforms

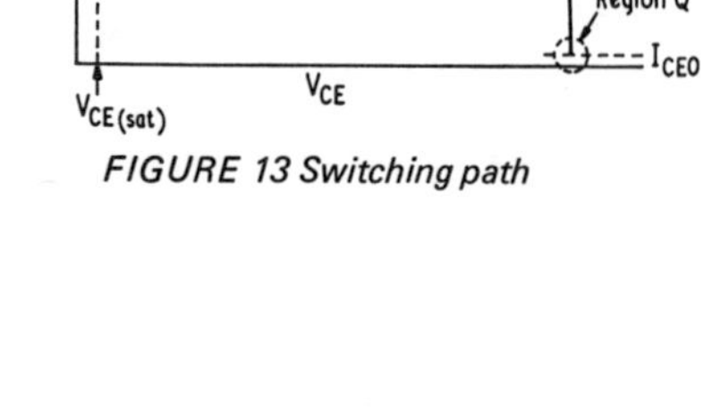

FIGURE 13 Switching path

Breakdown Characteristics

Working at high voltage, extreme care has to be taken so that the breakdown voltage of the transistor is not exceeded. Since BV_{CEO} is much lower than BV_{CER} or BV_{CEX}, as shown in Fig. 14, it is important that the device should never operate in this condition. This can be achieved by connecting an acceptable value of resistor between the base and emitter of the device so that the device operates under BV_{CER} conditions. The compromise value of resistor chosen in this case is 6,8Ω. This will reduce the forward base current, but the driver transformer allows for this.

Parameter	*Collector Emitter Breakdown Voltage*
BV_{CEO}	When base emitter is open circuited
BV_{CER}	When base emitter is shunted by resistance R
BV_{CES}	When base emitter is shorted together
BV_{CEX}	When base emitter is reverse biased by X volts

Protection Circuit

The current limiting resistor 0,5Ω (R2) in the emitter of the transistor, VT7, increases the negative base current and improves device turn off time. This helps to minimize power dissipation in the device during turn off. The more important advantage of this resistor in the emitter is that at higher collector currents, eg, during an output short circuit, it limits collector current. This occurs as a potential is developed across the emitter resistor which opposes the base drive, limiting the base current and in turn, limiting the collector current.

When the load current exceeds the required value preset by potentiometer RV3, the cathode of the thyristor Th1 goes far enough negative to provide gate current for it to latch on. This in turn takes the base of transistor VT4 negative, and as a result partially switches it off. A much smaller voltage excursion at its collector occurs and, as a result, the monostable reverts to a narrow mark space ratio. The dc amplifier also forces the monostable to operate during output short-circuit conditions in region S, shown in Fig. 10. Under this condition, the collector current may be 3A for 3μs only. This current is larger when the power supply load is increasing from normal towards a short circuit condition. A prolonged short will blow the fuse Fs2 which is chosen in accordance with the trip adjustment (say 2A). Initially the wiper of RV3 is set to be at earth potential.

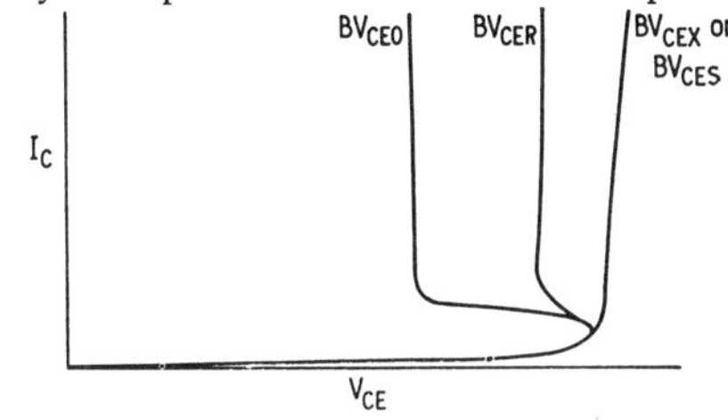

FIGURE 14 Breakdown characteristics

Protection may be needed against the output voltage V rising suddenly as a result of an accidental short between collector and emitter of the switching transistor. This can be achieved with the following trip arrangement.

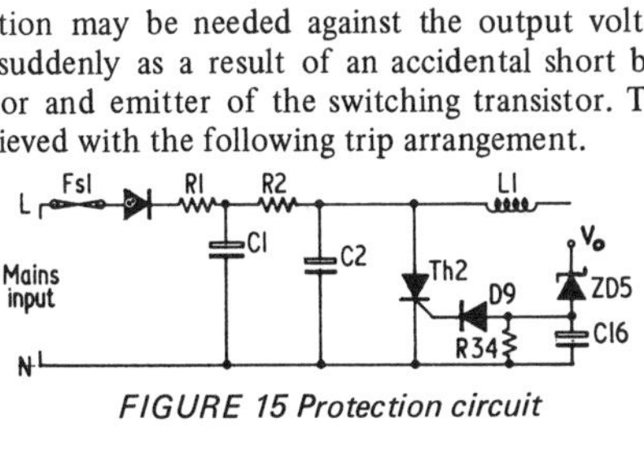

FIGURE 15 Protection circuit

Zener diode ZD5 begins to conduct when V_O reaches the Zener voltage. This provides gate current via diode D to the thyristor Th2 and turns it on, blowing fuse Fs1, which saves the load from being damaged.

THE CIRCUIT PERFORMANCE

The complete circuit, Fig. 16, can provide 60, 30 and 90V. A table showing the value of reference voltage V_Z and bleed resistor R9 to obtain these output voltages is shown.

Output voltage	V_Z:V	*Zener*	*R9*:Ω
60	30	(1S3030A)	330
30	15	(1S3015A)	680
90	47	(1S3047A)	1000

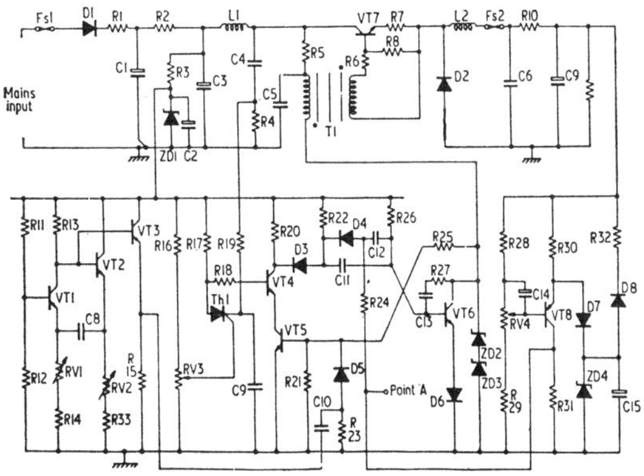

FIGURE 16 Completed circuit

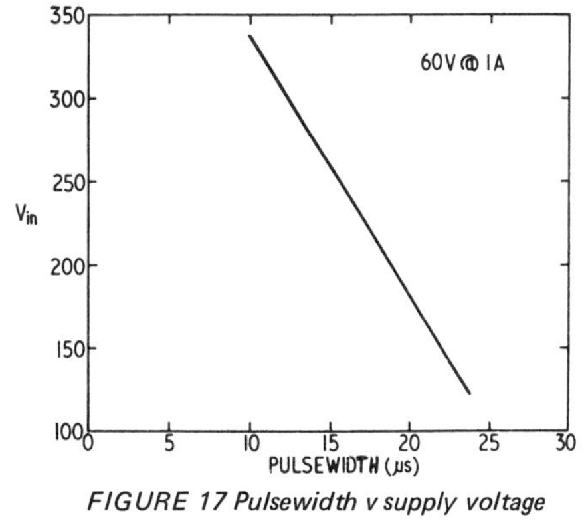

FIGURE 17 Pulsewidth v supply voltage

To provide 90V both the monostable and the driver must work close to their design limits and when large load currents are needed poor regulation will occur if the supply is operated at low mains input voltage.

The relationship of the pulsewidth to the supply voltage for a constant 60V output at 1A load current is given in Fig. 17.

The degree of regulation achieved for 60 and 30V for varying load currents with low and high mains voltages, ie, ±10% nominal mains can be seen from Fig. 18. The bleed resistor R9 is essential because of the fall-off in regulation at small load currents.

Under full-load conditions, ie, 60V at 2A: ripple at 15kHz is 150mV. Ripple at 50Hz is 50mV.

Under no load conditions, using an adequate bleed resistor, ripple at 15kHz+ripple at 50Hz is 10mV.

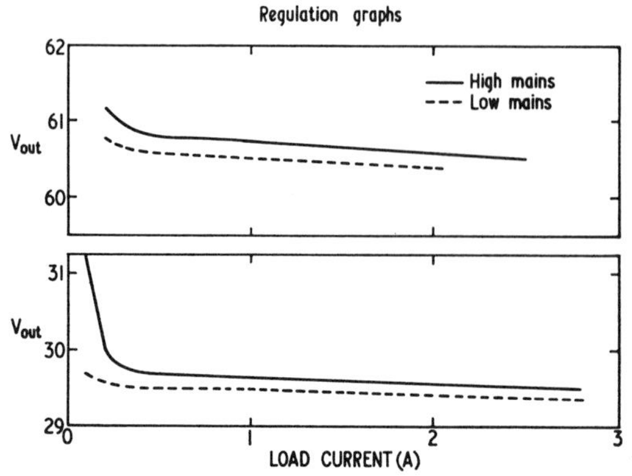

FIGURE 18 Regulation graphs

CONCLUSIONS

The switching mode power supply described will give 60V at 2A, 30V at 3A and 90V at 1,5A. Regulation better than 0,5% has been achieved. Total ripple content at the input is less than 200mV at full load. The supply has built-in safeguards against overload and accidental output short circuit.

The type of arrangement discussed is capable of delivering higher load currents with only small changes in the circuit. For example, to obtain 60V at 5A the inductor L2 in Fig. 17 would have to be such that it does not saturate at 5A and the driver transformer would have to provide higher base current for the transistor VT7.

To deliver output voltages of more than 90V the monostable and driver circuits have to be redesigned.

APPENDIX

1. Switching frequency ripple calculation

The ripple content ΔV_o of the output voltage for particular values of inductance and capacitance in the low pass integrating filter circuit can be calculated as follows:

The voltage across the inductor L2 is:

$$(V_{in}-V_o)=L.\,di/dt.$$

Therefore, the change in current through the inductor in time, t(seconds), is:

$$\Delta i=(V_{in}-V_o).t/L.$$

Assuming that most of the current goes through the filter capacitor:

$$C.dV_o/dt=(V_{in}-V_o).t/L,$$

and the corresponding change in output voltage by integrating this is

$$C\int_0^t dV_o/dt= (V_{in}-V_o).t/L\ .dt$$

$$C\Delta V_o=(V_{in}-V_o).t^2/2L$$

$$\therefore \Delta V_o=(V_{in}-V_o).t^2/2LC$$

neglecting the dc resistance of the inductor.

The time t in this expression is the on time, t_{on}, of the series transistor switch VT7 in Fig. 16.

2. Power dissipated in series switch BUY22

The power dissipated during the conduction period t_{on}:

$$P1 \triangleq I_{C(mean)}.V_{CE(sat)}.t_{on}/\tau$$

The power dissipated during the off period, t_{off}

$$P2 \triangleq I_{C(peak)}.V_{in}(t_1+t_2)2.\tau$$

The power dissipated during the time the transistor is switching from hard off to hard on:

$$P3 \triangleq I_{C(peak)}.V_{in}\ (t_1+t_2)2.\tau$$

where t_1 is the transistor voltage switching on time, t_2 is the transistor current switching on time.

The power dissipated during the time the transistor is switching from hard on to hard off:

$$P4 \triangleq I_{C(peak)}.V_{in}.(t_3+t_4)/2.\,\tau$$

where t_3 is the transistor voltage switching off time, t_4 is the transistor current switching off time.

$\therefore$ total power dissipated in the transistor=P1+P2+P3+P4.

An example:

Typical figures are

τ=1/frequency=64 μs	t_3=150ns
$V_{CE(sat)}$=1,5V	t_4=350ns
t_{on}=14μs	V_{in}=340V
t_{off}=50μs	V_o =60V
t_1=150ns	I_o=$I_{C(mean)}$=2A
t_2=450ns	$I_{C(peak)}$=2,3A

which gives

$$P1+P2+P3+P4=0,6+2,56+3,6+3W$$
$$=9,8W$$

Components List

Capacitors
C1 200μF
C2 25μF
C3 100μF
C4 2,2μF
C5 0,1μF
C6 2,2μF
C7 250μF
C8 0,01μF
C9 0,1μF
C10 470pF
C11 1000pF
C12 0,05μF
C13 1000pF
C14 4μF
C15 12μF
C16 50μF

Resistors
R1 6Ω, 10W
R2 10Ω, 10W
R3 6kΩ, 10W
R4 0,5Ω, 5W
R5 6kΩ, 15W
R6 1Ω, 5W
R7 0,5Ω, 5W
R8 6,8Ω, 5W
R9 330Ω, 680Ω, 1kΩ, 10W
R10 2Ω 5W
R11 2k7
R12 2k7
R13 1k8
R14 1kΩ
R15 8k2
R16 12k
R17 8k2
R18 1k2
R19 27Ω
R20 8k2
R21 4,7k
R22 3k3
R23 150k
R24 4k7
R25 390k
R26 3k9
R27 12k
R28 2k2
R29 2k2
R30 3k3
R31 12k
R32 27Ω
R33 1k
R34 47Ω

Transistors
VT1 BC182L
VT2 BC182L
VT3 BC182L
VT4 BC184L
VT5 BC182L
VT6 BF259
VT7 BUY69B
VT8 2N3702

Semiconductor Diodes
D1 1N4006
D2 1N3883
D3 1S44
D4 1S44
D5 1S44
D6 1S921
D7 1S44
D8 1S111
D9 1S44

Thyristors
Th1 TIC44
Th2 TIC106D

Inductors
L1 1mH
L2 6mH non saturable up to 3A.

Variable resistor
RV1 5k
RV2 5k
RV3 1k
RV4 1k

Fuses
Fs1 2A
Fs2 as required

Zener Diodes

ZD1 1S3027A
ZD2 3 1S3150A
ZD4 1S3015A
or 1S3030A
or 1S3047A
ZD5 as required

Transformer T1

Core FX 2241
Primary turns 300 34 S.W.G.
Secondary turns 10 24 S.W.G.
Gap 0.006"

VIII SWITCHING REGULATOR AND LINE DRIVER

by Kuldip Rupra

The switching regulator offers considerable advantages over other available regulators, especially where low power dissipation is a prime consideration as in, for instance, the 135 – 150V regulated supply required in a monochrome TV for the BU105 line output stage. This chapter therefore describes, in particular, such a TV supply, although a similar system is also applicable to a number of other uses.

The regulated voltage is achieved by using the series switching technique where the mark space ratio of the switching transistor is varied in sympathy with the output voltage change caused by either mains voltage fluctuation (±10%) or load variations. The principle of such a regulator is explained in greater detail in the previous chapter. For a regulator output of 145V, say, and a period (τ) of 64μs (i.e. frequency of operation 15kHz), the conduction period (t_{on}) of the series switch under normal mains condition is 32μs. The absolute limits of t_{on} are 28μs with low mains (–10%) and 38μs with maximum mains (+ 10%). Both these figures are within the permissible drive pulse width tolerance for a BU105 line output stage, and therefore this design uses the low pass filter coil also as the voltage step-down transformer (e.g. 30 : 1) which provides the base current for the BU105. The fact that there is a 50Hz width modulation (1 – 3μs) on the BU105 drive pulses should not worry the line synchronisation section. This will be explained later.

A further advantage of this concept is that any number of auxiliaries low tension supplies may be obtained by additional windings on the filter choke.

The circuit is protected against the 145V rail being shorted to earth and also the series switch failing, i.e. collector-emitter short circuit.

CIRCUIT DESCRIPTION

General

Figure 1 shows, in block diagram form, the principle of a switching mode power supply. The series transistor switch is operated at the fixed oscillator frequency. Its mark space ratio is determined by the variable mark space generator which is in turn controlled by a D.C. level fed to it. This level is the amplifier error signal obtained by comparing the output from the supply with a D.C. reference voltage.

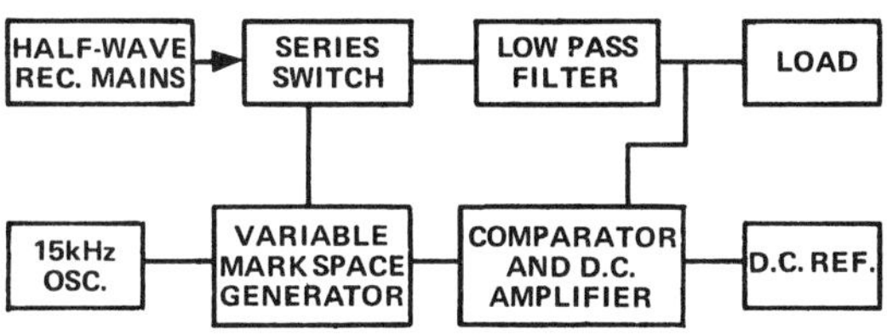

FIGURE 1 Switching Mode Power Supply

The complete switching regulator and line driver circuit is shown in Figure 2.

Series Switch operation

Half wave rectified mains is fed to the collector of the switching transistor VT4. An important point about this regulator design is the use of a BD410 in this position. As the device is in a SOT32 package it provides a considerable price saving over any device in a TO3 package. Its main parameter requirements are:–

1. $BV_{CER} > 400V$ 2. Reasonably fast switching
3. Good saturation characteristics at I_C = 600mA.

As can be seen from the device data sheet, the breakdown voltages required for this application are well within the device's capabilities.

The transistor must switch reasonably fast since the majority of the power dissipation occurs during the switching excursion. Photographs 1 and 2 show the switching speeds of a typical device. Both switch 'on' and switch 'off' are less than 1μs. (Typically t_{on} = 400ns t_{off} = 500ns). This means that dissipation in the device during these transient conditions is less than 3W.

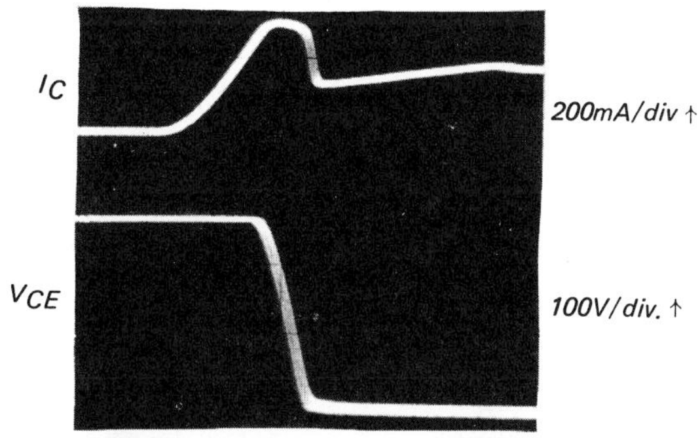

PHOTO. 1 SWITCH ON 500ns/div. →

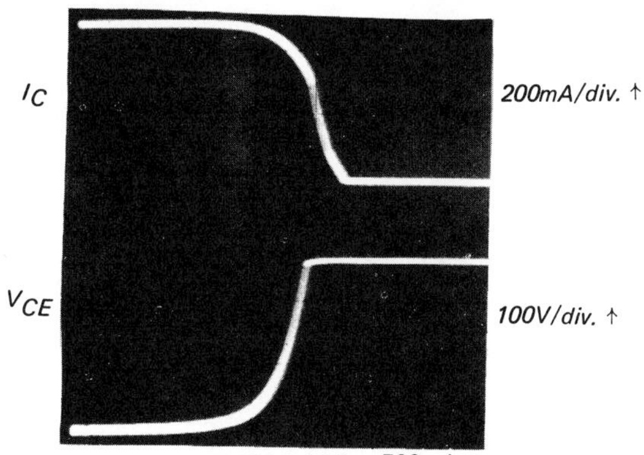

PHOTO. 2 SWITCH OFF 500ns/div. →

Monochrome TV Switching Regulator and Line Driver

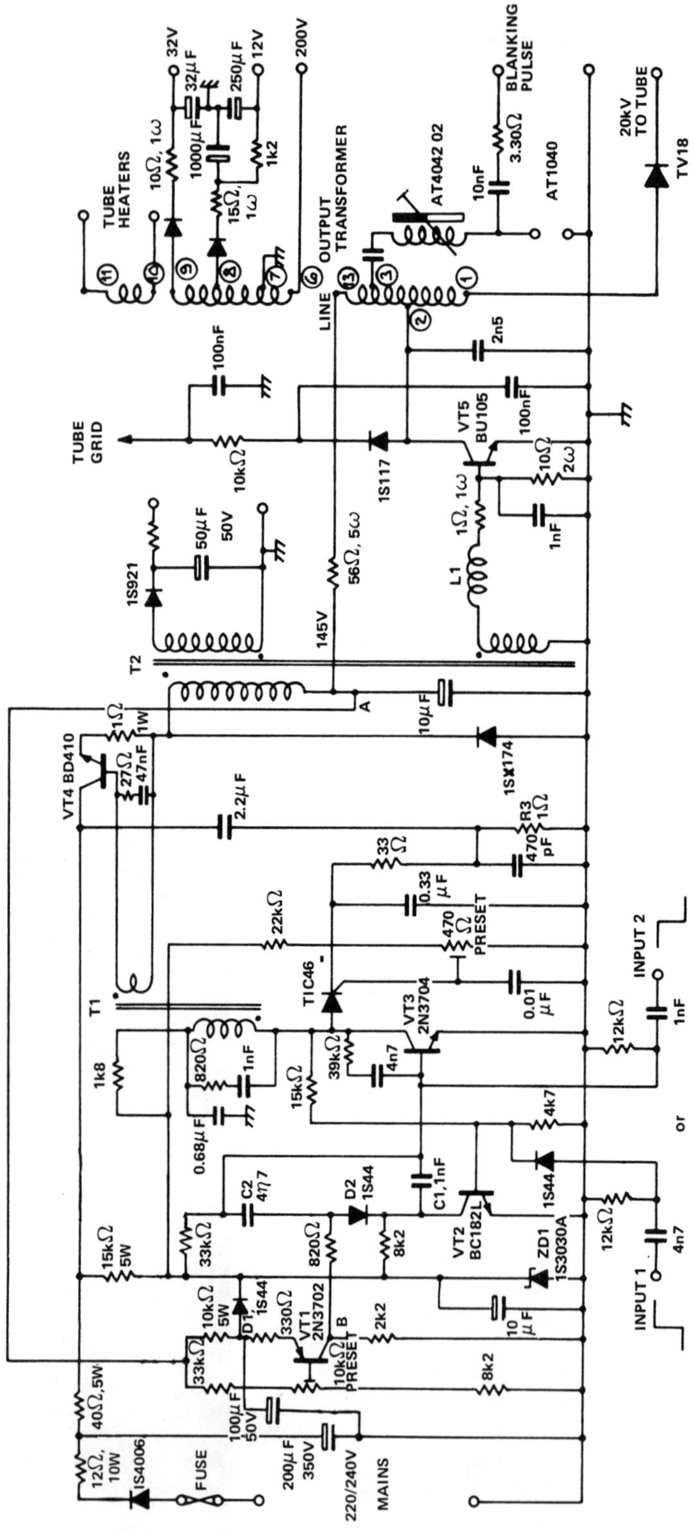

FIGURE 2

The saturation characteristic of the device needs to be sufficiently good to avoid excessive dissipation, i.e. $V_{CE(sat)}$ must not be more than 10V with collector current of 600mA. From Photograph 3 it can be deduced that the base drive required to maintain this condition is 125mA. Extensive measurements show, however, that for such applications this value of base current required may be reduced by 30% to 75mA to maintain $V_{CE(sat)}$ of 10V at 600mA under the dynamic condition.

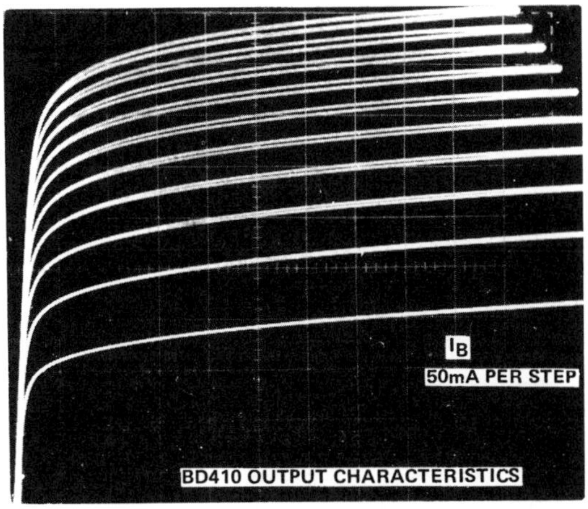

PHOTO. 3 V_{CE} *1V//div.* →

Line Oscillator

The oscillator is provided externally to the regulator circuit shown in Figure 2, and provides a pulse to be fed to the input point. The trigger pulse required by the variable mark space generator from the line oscillator must have a fast triggering edge. A positive trigger pulse, which would be fed to Input 1, must have a width greater than 4μs. If a negative going trigger, fed to Input 2, is used, its pulse width should not be less than 4μs or greater than 10μs. This helps to limit the maximum pulse width out of the monostable to 54μs which, under low mains and overload condition, avoids monostable operation at half the required frequency. The trigger pulse amplitude can be between 4V and 10V, and may be of fairly high source impedance.

In the worst case, low mains and maximum loading, the base drive pulse width to the BU105 would be changing by about 4μs. In a normal, flywheel synchronising system, the flyback voltage pulse is referred to the line oscillator. If the response time of the system is faster than 50Hz, which it normally is, the slight 50Hz modulation occurring on the leading edge of the BU105 base drive is allowed for, so that picture displacement does not occur.

Since the control circuitry for this switching regulator needs line oscillator pulses to drive it, it is essential that the oscillator starts up rapidly and that it is frequency stable from low voltage to the ultimate regulated rail voltage. Also too much current must not be consumed since its supply has to be dropped from the 300V line.

There are various ways in which the line oscillator function can be performed. One of them is to use a SN76533 'Jungle' Integrated Circuit to act as the line oscillator and phase comparator. The circuit arrangement for this is given on page 75.

The 'Jungle' has an internal zener diode supply for the line oscillator, which is frequency stable to within 10% during switch on and switch off.

With this system, the synchronisation pulses are at always a predetermined separation from the leading edge of the output pulse of the SN76533, determined largely by the storage time of the BU105 and, to some extent, the delay caused by the driver switching 'on'.

The waveforms shown in Photographs 4, to 8 are taken with a fifth harmonic tuned line output stage and the oscilloscope triggered by the leading edge of output pulse from the SN76533. The relationship of the sync. pulse to the collector current turn 'off' (or initiation of flyback voltage) is shown by comparing Photographs 4 (top trace) and 8 (bottom trace).

Monostable and Driver

A cheap, effective, simple monostable circuit is employed to perform the pulse width modulation for regulation purposes. The stable element (transistor VT3) of the monostable is also used as the driver stage. This driver has a current step-up transformer (T1) as its load, providing sufficient base drive for the BD410 series switching device. Transistor VT2 is normally 'off' except when a positive going signal is applied to its base. Alternatively, applying a negative going signal to the base of transistor VT3 to switch it 'off' will in turn switch transistor VT2 'on' through the resistor network from transistor VT3's collector. Monostable action starts from this point.

The transformer in the collector of transistor VT3 is so arranged that the BD410 is 'on' during the period the driver VT3 is 'off', i.e. the 'on' time of VT2. In practice, however,

the BD410$-t_{on}$ = VT2$-t_{on}$ + a delay due to transformer leakage inductance.

It is important that, during supply switch on, the t_{on} of the BD410 is very small so that large currents through this device (even though occurring for only a few cycles) are avoided. This will also ensure a gradual build up of the output voltage which gives added stability to the system.

This narrow pulse width during 'start up' is achieved by ensuring that the timing network consists of only resistor R2 and capacitor C1. This in fact is so, since as the output voltage (point A) is zero, the collector of the comparator is also zero and thus the diode D2 is reverse biassed during the timing period, effectively isolating capacitor C2 from the timing network. Waveforms applicable to this section of the circuit are shown in Photographs 4 and 5.

PHOTO. 4

Video input — 2V/div.↑

Line Oscillator Voltage (Input to Monostable) — 2V/div.↑

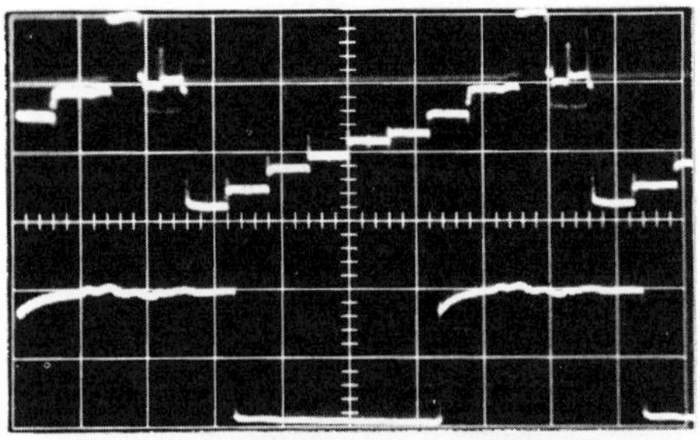

PHOTO.5

V_{CE} VT3 — 5V/div.↑

I_B BD410 VT4 — 100mA/div.↑

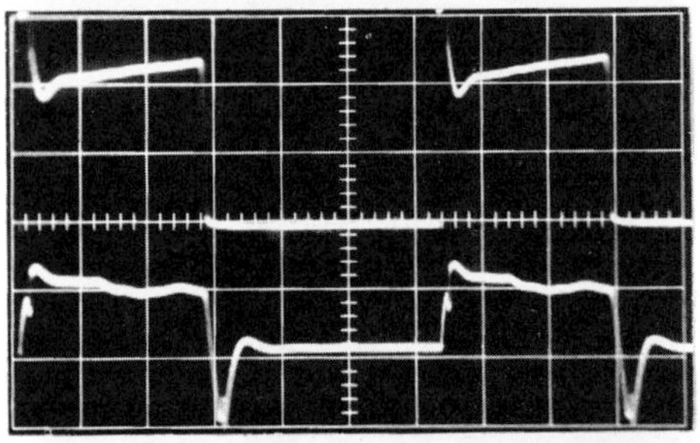

PHOTO. 6

I_C BD410 VT4 — 200mA/div.↑

V_{R3} — 200mV/div.↑.

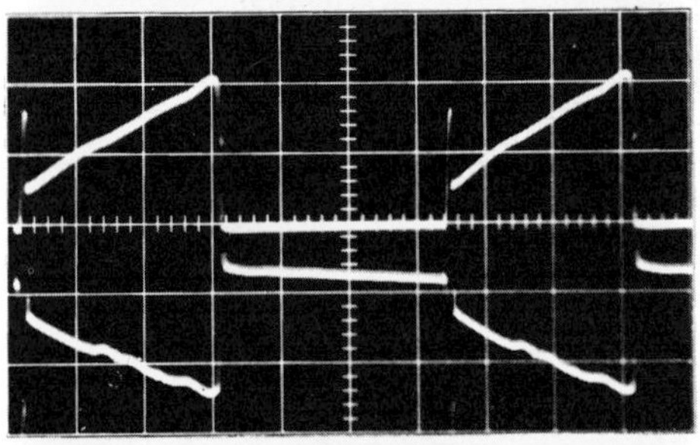

PHOTO. 7

V_{CE} BD410 VT4 — 200V/div.↑

I_B VT5 (BU105) — 1A/div.↑

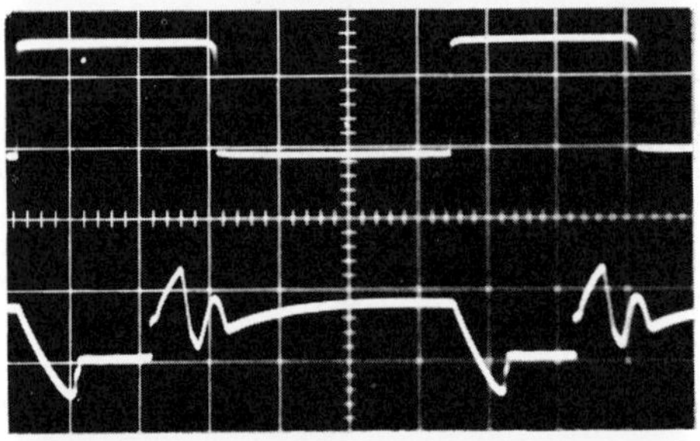

PHOTO. 8

I_C BU105 VT5 — 2A/div.↑

V_{CE} BU105 VT5 — 400V/div.↑

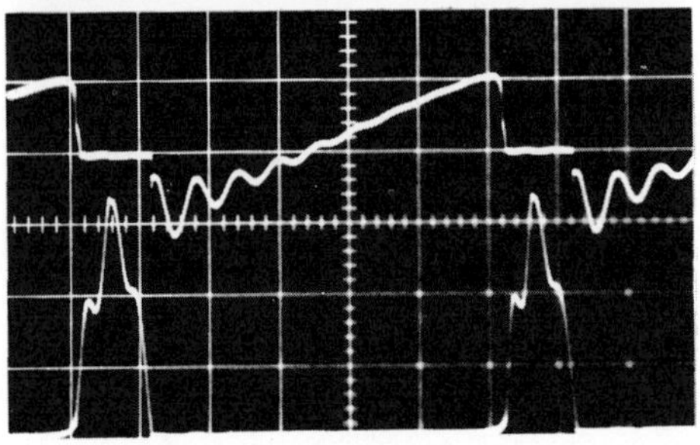

10µs/div.→

PHOTOGRAPHS OF WAVEFORMS DURING NORMAL OPERATION

Comparator

The minimum pulse width from the monostable results in the voltage developed at point A

$V_{out(A)}$	=	$V_S \cdot t_{on} / \tau$
	=	320. 4/64
	=	20V
where V_S	=	voltage at the collector of BD410
t_{on}	=	$4\mu s$
τ	=	$64\ \mu s$

Increasing the voltage at point A increases the voltage at the collector of the comparator transistor VT1, which in turn adds to the t_{on}. The resulting output voltage rise causing a further increase in voltage at point B, and so on. This progressive increase of pulse width, t_{on}, and V_{out}, continues until the required V_{out} is reached. Here the comparator transistor changes its mode of operation into a high gain error amplifier as explained in the previous chapter. The zener diode ZD1, which during the 'start up' condition is starved of current, experiences a progressive increase in current through it from the resistor R1 (15k) and the 10kΩ resistor and diode D1, ensuring a stable reference voltage across it. The V_{out} can be varied by a variable resistor feeding the base of transistor VT1.

Dynamic Trip

In a power supply of this nature, it is essential that there is adequate protection against any current overload i.e. if the BU105 went short circuit or in the event of a tube 'flash-over'.

The collector current through the BD410 is sensed by monitoring the voltage across the resistor R3, this being proportional to it. The waveforms are shown in Photograph 6.

A plastic S.C.R. (TIC46) is used as the dynamic trip. The voltage developed across resistor R3 is fed to the cathode of the TIC46 after the necessary integration of sharp edges. The gate of this device is pre-biassed so that at the desired overload condition value the S.C.R. gate trigger voltage is reached at the end of the BD410 conduction period, as shown in Figure 3. If, therefore, an overload occurs, the TIC46 is triggered before the end of BD410's normal conduction period, shortening the drive pulse to it during subsequent cycles. Since the BD410 collector current waveform has a triangular component, this trip technique is effective. While the S.C.R. is latched 'on' transistor VT3, due to normal monostable action, will switch 'on'. This helps to reset the TIC46 by draining current from it to below the holding current level -- thus switching it 'off'. The S.C.R. is then ready for the next cycle of the dynamic trip action. Waveforms illustrating the circuit operation during a BU105 short circuit are shown in Photographs 9, 10 and 11.

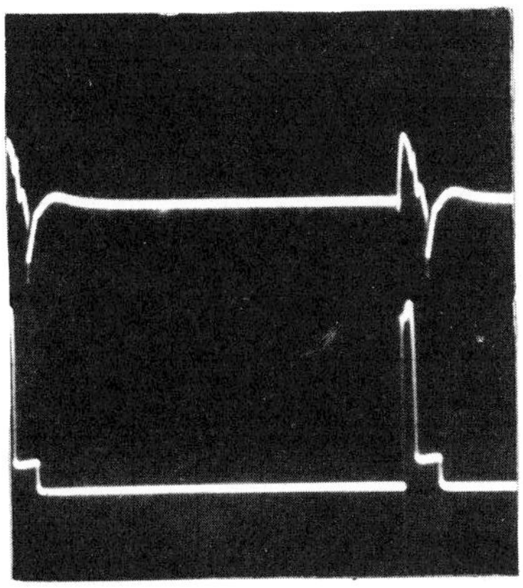

PHOTO. 9

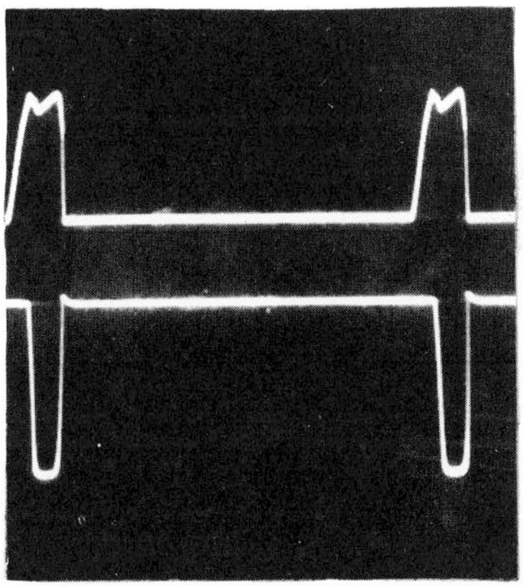

PHOTO. 10.

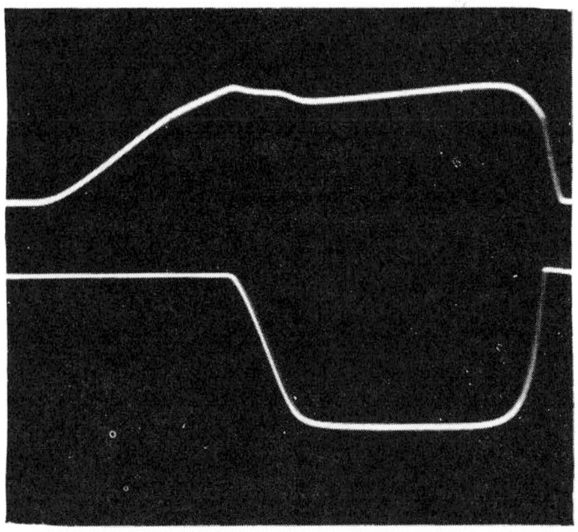

PHOTO.11 (10 EXPANDED)

WAVEFORMS OCCURING WHEN V_{out} IS SHORT CIRCUITED THROUGH 56 OHM

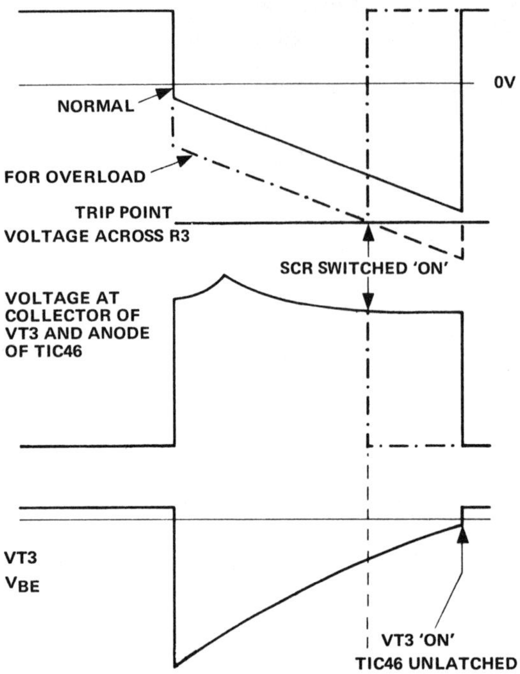

FIGURE 3

An alternative trip circuit approach is to use two silect transistors rather than the SCR as shown in Figure 4. Costs are again kept low by using an NPN–PNP configuration.

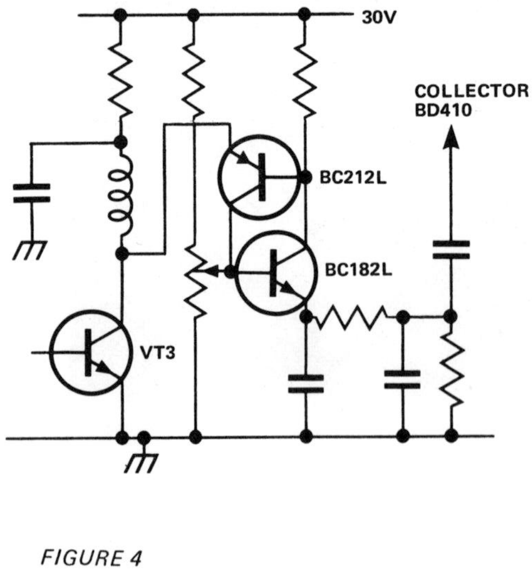

FIGURE 4

Line Output Stage

A turn-off delay of about 10μs has been introduced into the system, as illustrated in Photograph 7, by putting a choke of 40μH (L1) in series with the base of the BU105. This ensures uniform extraction of excess stored charge from the device towards the end of its conduction period. To produce fast switch off times in very high voltage transistors, e.g. the BU105, the base current waveform must have the turn off delay referred to above and shown in Photograph 7. If this is not present the clean and low dissipation BU105 collector current switch off, shown in Photograph 8, would not occur.

RESULTS

Regulation achieved

ΔV_{out} Full load to No load	= 2V
ΔV_{out} Mains ±20%	= 5V
15kHz ripple	= 1.5V
50Hz ripple	= 0.3V

CONSTRUCTION

A printed circuit board layout is shown in Figure 6 and the component positions on this board illustrated in Figure 5.

Transformer Details

BD410 Driver Transformer T1	Filter Choke and Transformer T2
Core = FX2238	Core = FX2241
N_p = 300 turns of 38 SWG	N_p = 400 turns of 34 SWG
N_S = 60 turns of 34 SWG	N_S = 80 turns of 28 SWG
Air Gap = 0.01 in.	$N_s 2$ = 8 turns of 26 SWG
	Air Gap = 0.1in.

CONCLUSIONS

The circuit concept described should be particularly useful for television manufacturers where maximum efficiency is of prime consideration. Simpler circuits available for deriving 135 – 150V supply rails have the disadvantage of approximately 80W power losses. Such losses cannot be tolerated in T.V.s. which have a plastic cabinet and/or polythene back covers. This system with its low losses (≏20W) and hence low temperature rises is useful generally and ideal for sets employing thermoplastic cases, especially where they are intended for use in hot countries.

The circuit described could also be used to produce a regulated 150V 800mA supply. This would be suitable for 90° colour televisions using a single BU108 line output device. The series switch (VT4) in this case would have to have an appreciable gain at 1A collector current. This is beyond the capabilities of the BD410 and a R2010 would have to be used. Increases in loop gain and choke size would also be necessary for the higher power required.

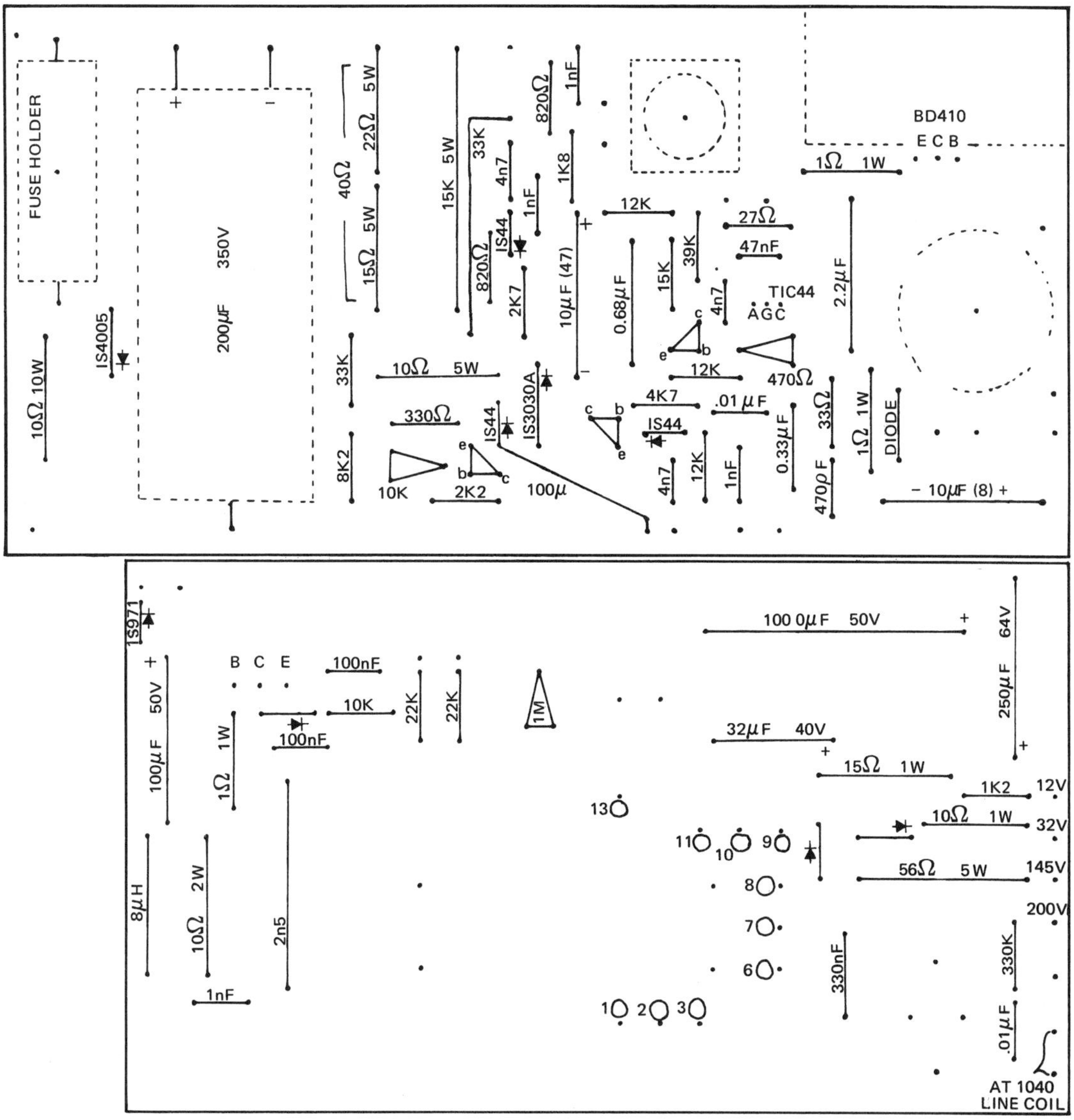

FIGURE 5 Printed Circuit Board Component Layout

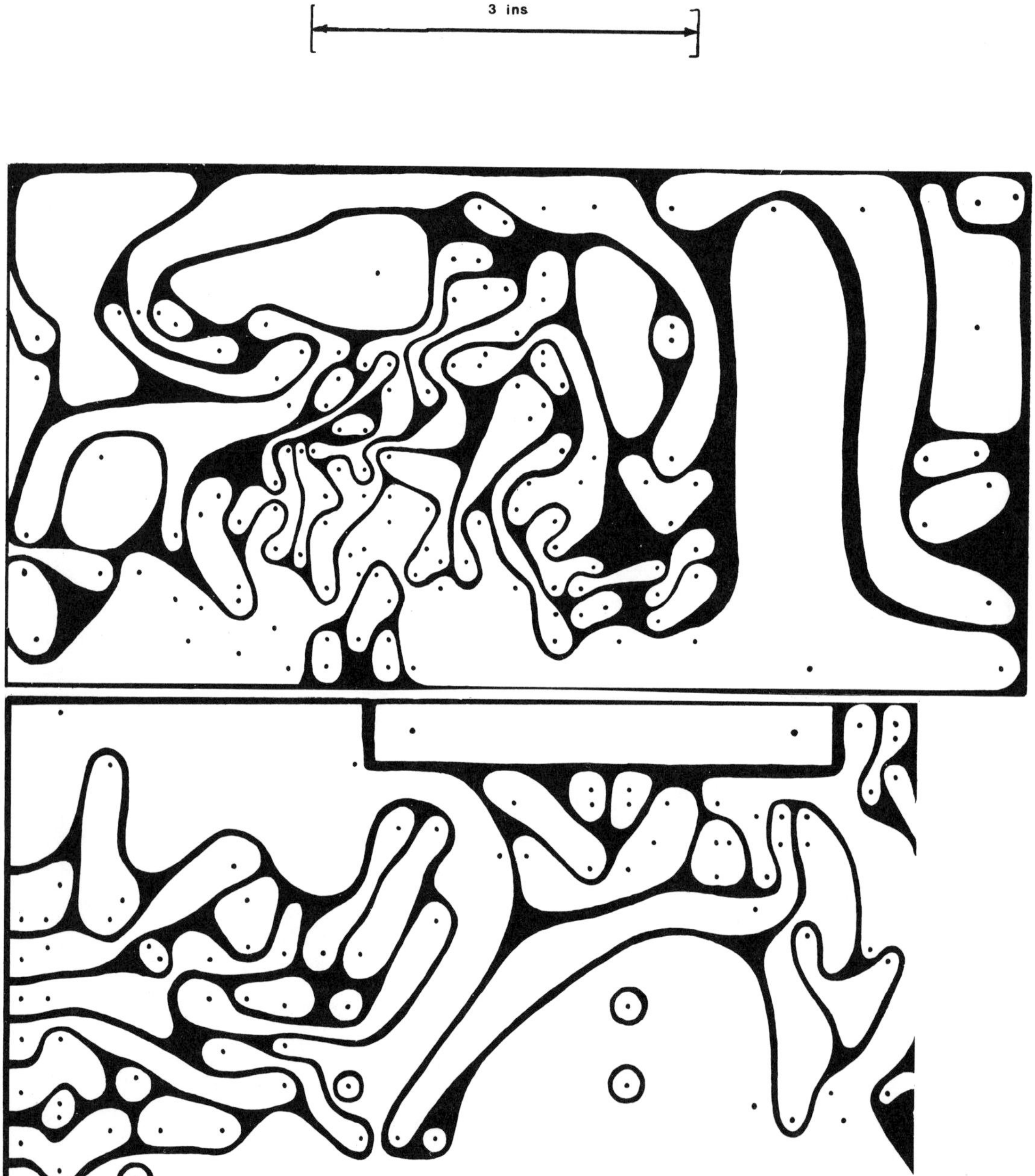

FIGURE 6 Printed Circuit Board Copper Side

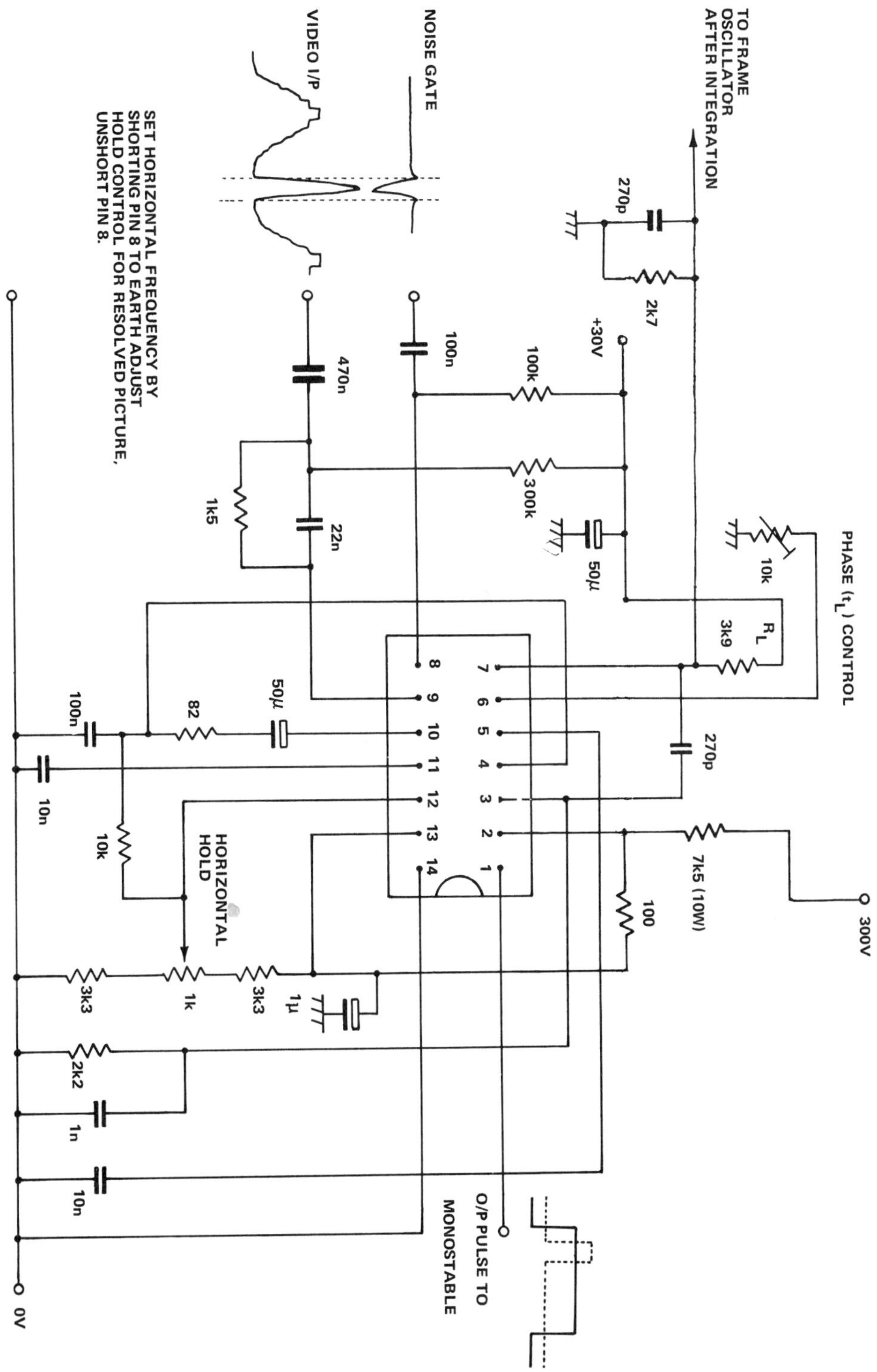

APPLICATION OF A SN76533 AS A LINE OSCILLATOR AND PHASE COMPARATOR

IX CHOPPER POWER SUPPLIES USING HIGH VOLTAGE POWER TRANSISTORS

by Mick Maytum

The BUY69 and BUY70 range of transistors have been designed for high voltage inductive switching applications. Their power switching losses are very small as their maximum switching fall time is 1μs. Hence using them, high power switching mode power supplies, operating at 20kHz and directly from the mains, can be made. Further details of this type of supply are given in Chapters VII, VIII and X.

GENERAL PRINCIPLES

All the designs described in this report are intended to run from rectified mains or D.C. supplies in the 100V to 500V region. Small sized high frequency transformers provide isolation and step-up or step-down of the primary waveform as required. Variations of loading and mains voltage can be reduced by modulating the switching devices conduction period.

SINGLE DEVICE SHUNT CHOPPER POWER SUPPLY

Figure 1 illustrates a simple form of chopper power supply. It is directly equivalent to the usual mains transformer, rectifier and smoothing capacitor arrangement used as the d.c. source for power amplifiers, series stabilisers, etc. For powers above 50VA a switching mode power supply offers lower weight and volume compared with the conventional unit. Its basic output power and voltage characteristics versus output current are shown in Figure 2.

Circuit Operation

The mains supply is full-wave rectified by diodes D1 – 4 and smoothed by capacitor C1. Resistor, R1, 'bleeds' a small current into the base of transistor VT1 to initiate the oscillation of the supply. Transformer, T1, consists of four windings on a ferrite core. Winding N1 provides the base drive for transistor VT1, winding N2 forms the primary and winding N4 forms the output secondary. The purpose of winding N3 is to maintain a constant output voltage of up to the maximum output power. (See Figure 2). This winding may be bifilar wound with winding N2, or preferably sectionalised.

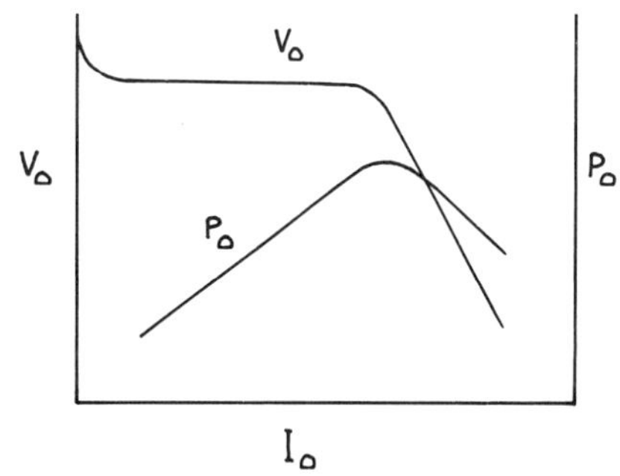

FIGURE 2

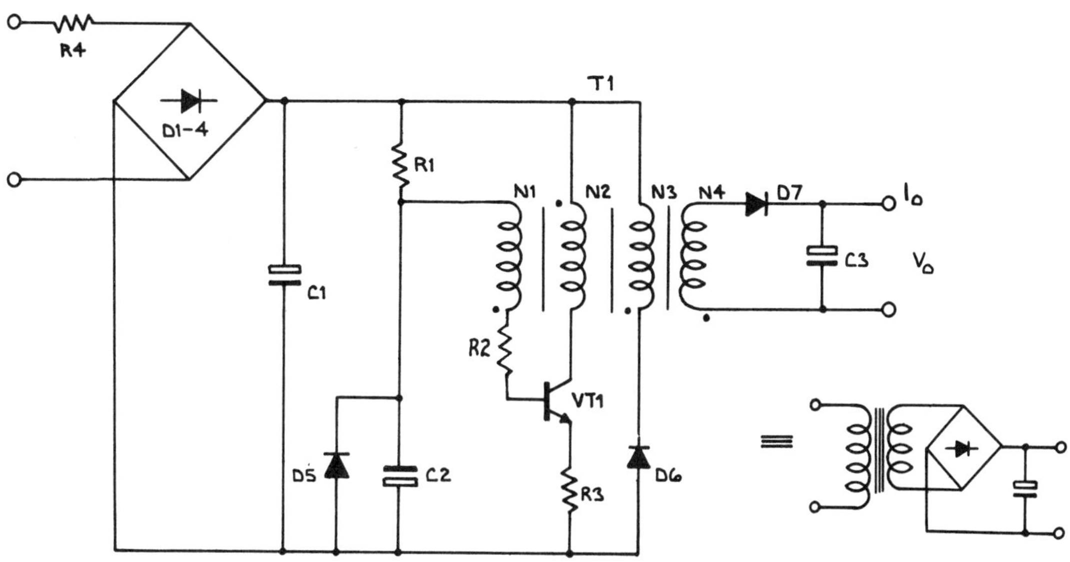

FIGURE 1

The collector current of transistor VT1 (Figure 3(c)) will be a sawtooth whose slope ($\Delta I_C/\Delta t$) will be determined by the inductance L2 of winding N2 and the supply voltage V_s

$$\Delta I_C / \Delta t = V_s/L2 \qquad A/s$$

The peak value of collector current is limited to:—

$$I_{C(max)} \approx (V_s . N1/N2 - V_{BE}) / R3$$

This is the condition for zero base current, i.e. the point in the operation where the device starts to switch off. The device in starting to switch off (i.e. V_{CE} rising) reduces the base drive voltage and regenerative action gives a fast device switch-off.

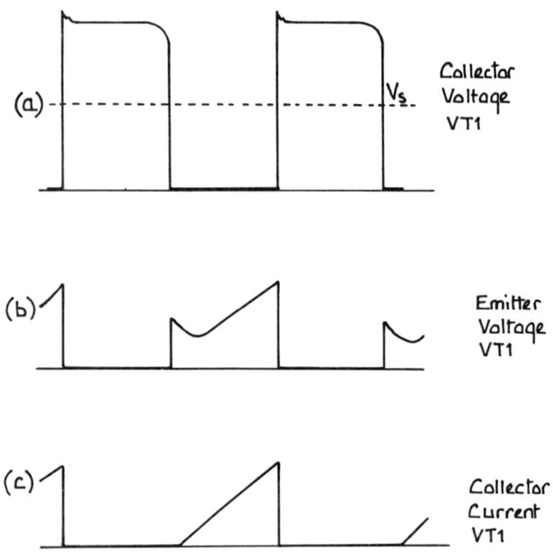

FIGURE 3 Circuit Waveforms

Energy, stored in the transformers core during the devices conduction period has two alternative 'sinks'. It can be returned to the supply V_s via winding N3 and diode D6 or transferred to the secondary load. In normal operation some energy is returned to the supply and some to the load. The regulation of the secondary voltage starts to fall when all the energy is transferred to the load. This type of supply, if correctly designed, can tolerate open circuit and short circuit load conditions. If the windings N2 and N3 have equal turns the peak collector voltage will be 2Vs and the collector voltage waveform will be a square wave (Figure 3 (a)). As windings N2, N3 and N4 may handle equal powers, equal winding spaces must be allotted to these three windings. The power supply's output voltage will vary by the same percentage as the input mains but the frequency of operation will remain sensibly constant.

Design Considerations

The BUY70B device is specified at I_C = 4A and has a $BV_{CER} > 400V$ and a $BV_{CBO} > 800V$. From a rectified supply of 300V an equal turns ratio for windings N2 and N3 will produce a peak collector voltage of 600. Working to a peak collector current of 4A, the maximum output power will be 300. 4. 0,5. 0,5 = 300W assuming a 50% duty cycle and neglecting transformer and rectification losses. Use of a BUY69B device working to a 8A peak collector current would give 600W.

Extensions to the Basic Circuit

The supply described will have the same percentage output voltage change as the input supply. It is possible to stabilise the output voltage against supply variations, and one form the additional circuits could take is shown in Figure 4.

Here a unijunction transistor oscillator defines the oscillation fequency. A fast thyristor turns transistor VT1 'off', the thyristors firing point being determined by an error amplifier and opto coupling link from the secondary.

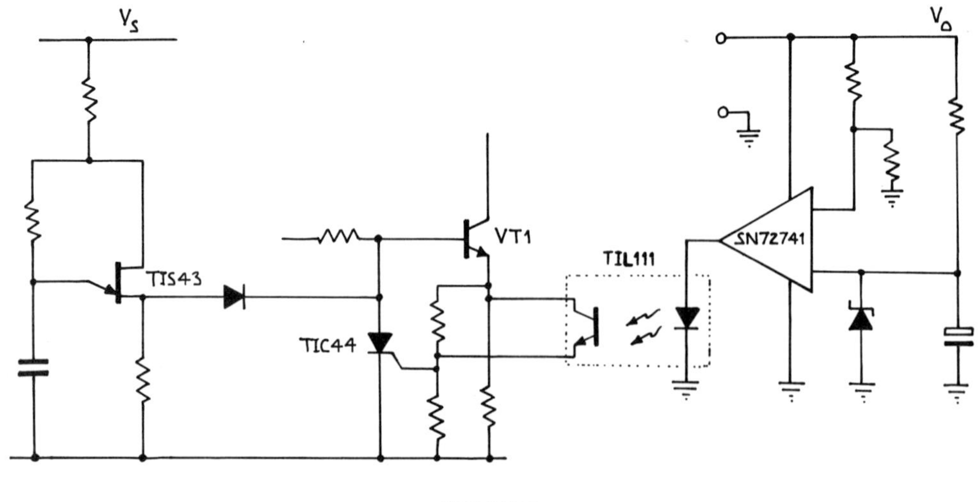

FIGURE 4

MODIFIED CIRCUIT TO INCREASE OUTPUT POWER

The power output of a single device shunt chopper may be increased by reducing the slope of the collector current of transistor VT1 and switching the device 'on' before transformer T1 has delivered all its energy to the secondary load. Typical waveforms are shown in Figure 5.

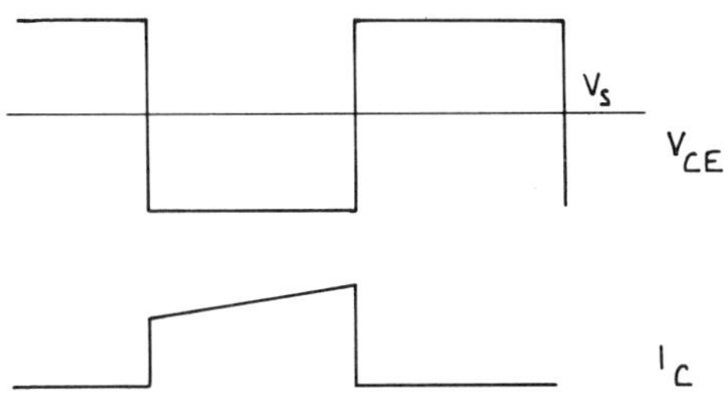

FIGURE 5

Taking the previous BUY70B example and specifying the collector current rising from 3A to 4A the power output becomes (3 + 4) x 0.5 x 0.5 x 300 = 525W. This output power for a BUY70B device operating at a collector current up to 4A is very nearly equal to the 600W obtained from a BUY69B device operating at 8A collector current when the initial collector current is zero. Thus increasing the transformer size to produce a smaller ramp on the collector current pulse, makes better use of the devices peak current by giving more output power.

The circuitry to drive VT1 of necessity increases in complexity. A block diagram of this is shown in Figure 6.

Circuit Operation

Transformer T1 acts as an energy store, and when the transistor VT1 conducts the transformers stored energy is increased. This energy increase is transferred to the output load by diode D2 after transistor VT1 has switched 'off' and the transformer's voltage reversed. Suitable selection of the output decoupling capacitor, C2, reduces ripple and noise to an acceptable level. The value of capacitor C2 will be relatively small due to the high 'chopping' frequency.

An error amplifier, A1, which may be fabricated from any common integrated circuit differential amplifier (e.g. SN72741P) monitors the power supply's output voltage compared with a fixed voltage reference diode, D3 (e.g. a Y95AS). The signal from amplifier A1 controls the pulse width of the modulator, hence holding the power supply output voltage constant. Linking the control circuitry and the output transistor VT1 is a driver stage consisting of transistor VT2 and transformer T2. It is usual to arrange the windings of transformer T2 so that when transistor VT2 is conducting transistor VT1 is 'off' and vice-versa. This form of operation gives a strong and defined turn-off drive for transistor VT1.

The current limit circuit, if activated, overrides the output of the amplifier A1 and reduces the output device's (VT1) conduction time to limit its peak current.

Waveforms for the circuit are shown in Figure 7.

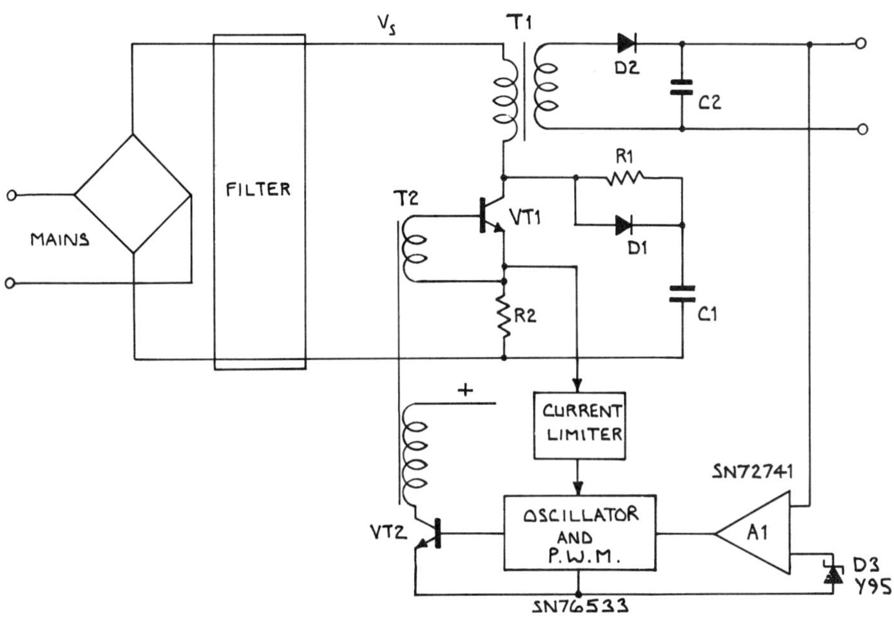

FIGURE 6

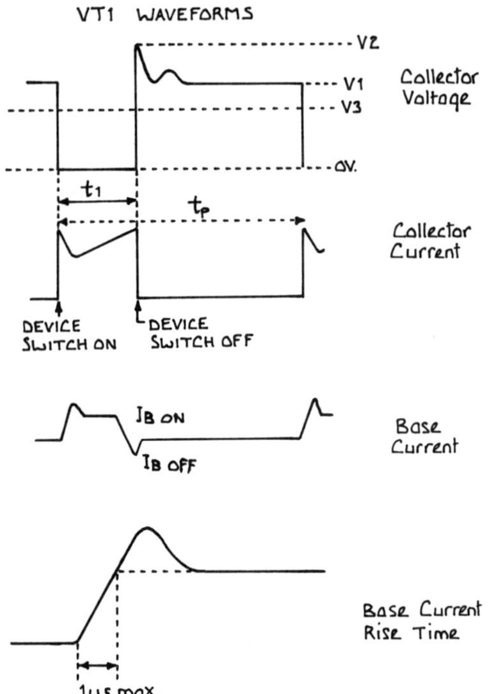

FIGURE 7

Switch-off Conditions of Transistor VT1

To ensure reliable operation of the transistor VT1 it is desirable that the device's collector current should have stopped flowing before its collector voltage exceeds the device's BV_{CER} rating. Capacitive damping applied to the primary of transformer T1 will slow the rate of rise of collector voltage at switch-off of transistor VT1. Resistor R1, capacitor C1, and diode D1 slow the voltage rise to enable the collector current to fall to zero before the collector voltage has risen above the supply voltage V_S (Figure 8). Typical values of capacitor C1 are 1.5nF per Amp. of collector current.

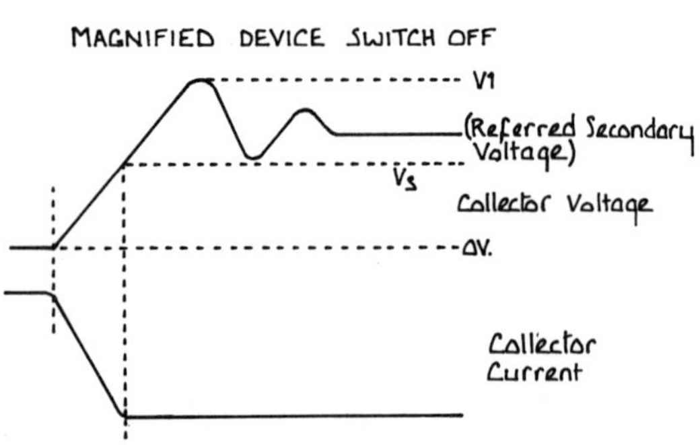

FIGURE 8

Supply Start Up

During the initial supply start-up the conduction time of transistor VT1 should be as short as possible. The reasons for this are as follows:–

(i) A small pulse width at the supply start-up gives a gentle rise of the output voltage with no overshoot.

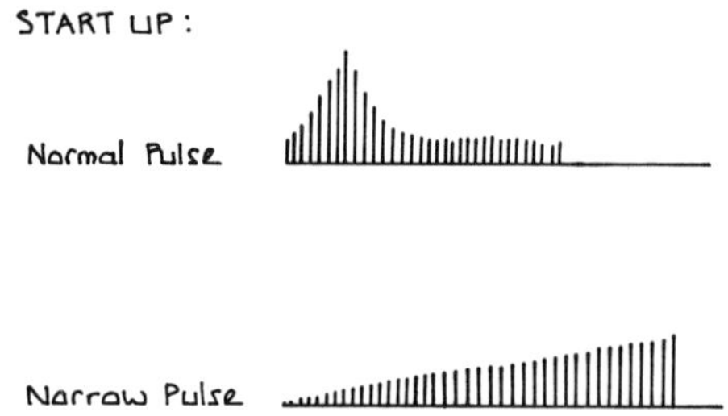

FIGURE 9 Collector Current at Start Up

(ii) The peak current that transistor VT1 experiences is very much reduced. (Figure 9).

(iii) Drive and voltage levels in the control circuits are given time to stabilise.

Large pulse width 'start-ups' could cause the device VT1 to come out of saturation and also possibly cause saturation of transformer T1.

An alternative is to rely on the current limit circuit. The voltage developed across resistor R2 provides collector current sensing. If this current exceeds a preset value the pulse width can be made to reduce itself and hence the peak collector current (dynamic current limiting). The maximum modulation capability of the pulse width modulator should be of the order of 50% of the total period. This again reduces possible stress on the transistor VT1.

Selection of Transistor VT1

(a) The device could be operating (i.e. conducting) when $V_{CE} = V_s$ and hence $BV_{CER} > V_s$(maximum).

(b) The leakage inductance of transformer T1 causes a voltage overswing of V'. This voltage V' may be reduced by clamping, damping or improved transformer design. It is necessary that $BV_{CEX} > V'$. I_{CEX} is usually quoted in manufacturers published data, X being the reverse base-emitter voltage which is normally $-2V > X > -8V$.

(c) The forward base drive current should be in the order of $I_{B(on)} = I_{C(max)}/4$ and have a rise time $< 1\mu s$. The reverse base drive current is normally arranged to be $I_{B(off)} = 0.75\ I_{B(on)}$. Under these conditions the collector current fall time will be less than $1\mu s$.

PUSH–PULL SHUNT CHOPPER POWER SUPPLIES

In this type of inverter, width modulated pulses are averaged out in a filter to produce a mean D.C. Any variations in the required D.C. output voltage are compensated by increasing or decreasing the pulse widths. A block diagram of this type of power unit is shown in Figure 10 and waveforms are given in Figure 11.

To avoid large currents at device switch on, Filter 2 is usually a 'choke-input' type. This is a useful configuration as the transformer's leakage inductance can be considered as part of the filtering system. Hence a very low leakage inductance output tranformer is not required as in the case of the single transistor chopper transformer.

Design Considerations

Bridge rectifying the mains will produce an on-load voltage of about 300. Assuming a BUY69B device is to be used running at 8A peak collector current, the following procedure enables an estimate of the output power to be made. Due to the transformer magnetising current some slope on the collector current is inevitable (Figure 11). Assuming this slope to be 2A gives an effective collector current during the device conduction time of 8 – 2 x 0.5 = 7A. As the two power devices cannot be allowed to conduct simultaneously a guard space must be left between their conduction periods. If the guard spaces and width modulation occupy 20% of the available working time the mean effective primary power will be 300 x 5.6 ≏ 1.7kW. Transformer and secondary rectifier losses will reduce the power output to about 1.5kW.

Hence two BUY69B transistors are capable of generating power outputs of 1.5kW and similarly two BUY70B devices could produce 800W.

The major problem with this type of inverter is its load variations. If the load current drops to a very low value the storage time of the switching transistors becomes very long. In the case of the BUY69 2.5A of base drive is necessary to ensure a low saturation voltage at the maximum expected collector current of 8A. When the collector current is low, say 1A, the 2.5A base drive will heavily overdrive, the device and the storage time can increase to as much as 10μs.

The effect of this pulse lengthening is to cause the secondary rectified output voltage to rise. Several methods are available to overcome this problem. Base current variation with loading, pre-loading the power supply, and emitter current switching are obvious solutions.

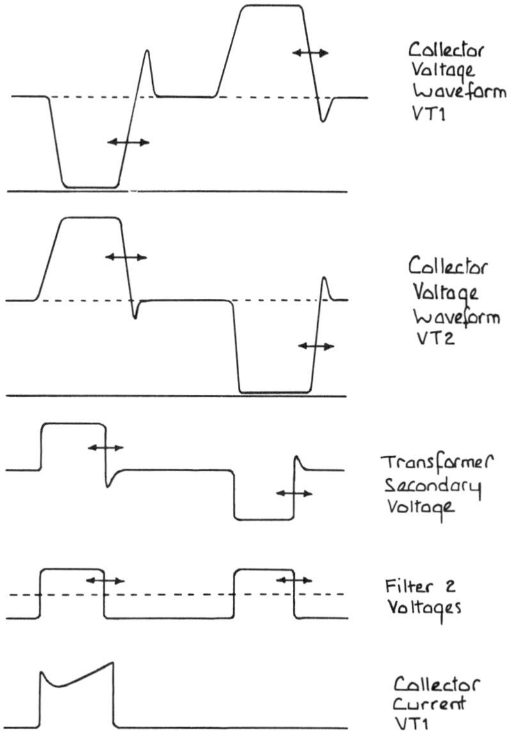

FIGURE 11 Circuit Waveforms

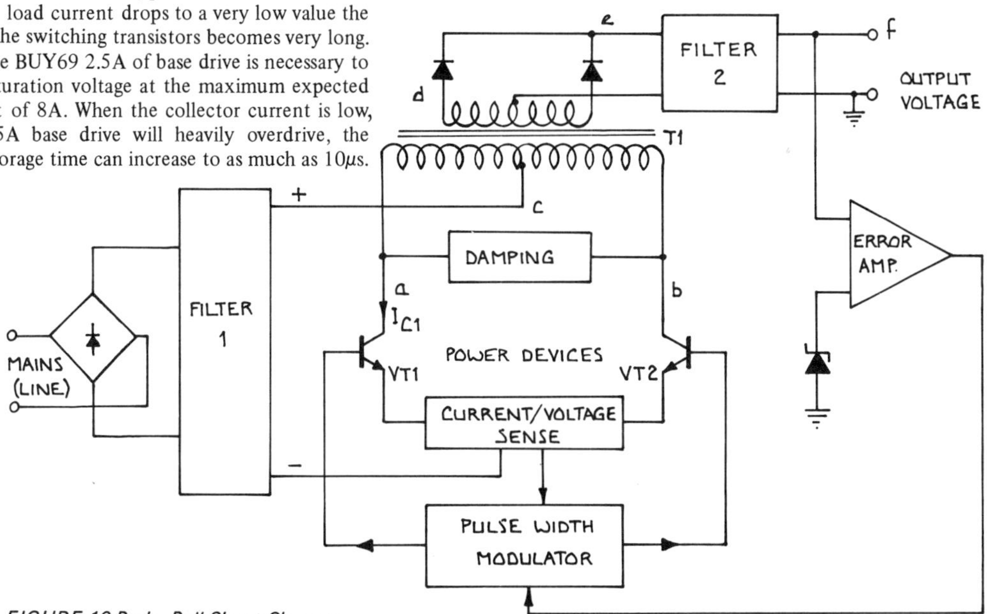

FIGURE 10 Push–Pull Shunt Chopper

CONCLUSIONS

The following table summarises the possible output powers from BUY69 and BUY70 transistors operating in chopper power supplies driven directly from rectified mains.

	BUY69 (B) W	BUY70 (B) W
Simple regulated power supply unit	600	300
Regulated and stabilised power supply unit	1000	500
Push-Pull	1500	800

X INVERTERS

by Mick Maytum

Initially developed to overcome the problems caused by mobile electronic equipment having to be operated from a single d.c. power source, inverters and converters are now being increasingly used in mains operated equipment reducing space, weight and heat losses.[1,2]

An inverter produces an a.c. waveform from a d.c. power source, perhaps the most well known application is for the fluorescent lighting units in public vehicles. Converters are merely inverters whose output is rectified to produce d.c. again usually at a different voltage level to the input voltage, (see Figure 1).

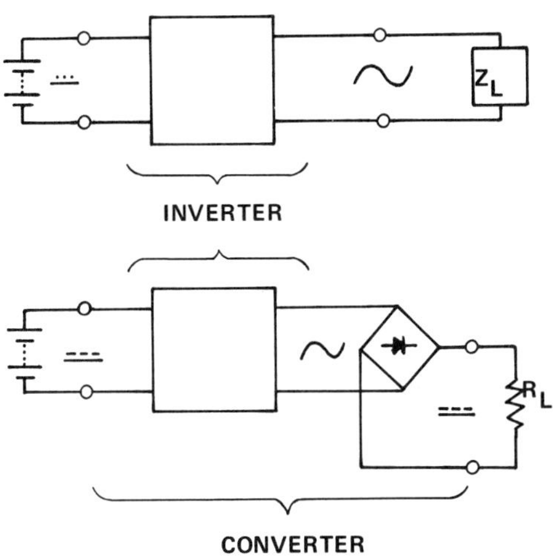

FIGURE 1

Taken as a two port network the ability of the unit to transfer power at the input terminals (P_i) to power at the output terminals (P_o) is called the efficiency η

Figure 2 (a) shows a typical efficiency curve, the maximum efficiency occurring at the maximum rated loading. As P_o decreases fixed losses in the unit cause the efficiency to drop.

Higher powers cause overloading, usually the switching devices become unsaturated and efficiency drops. A typical converter regulation curve is shown in Figure 2 (b). In producing a multiple of the input voltage at the output most units will have the same output voltage stability as the input.

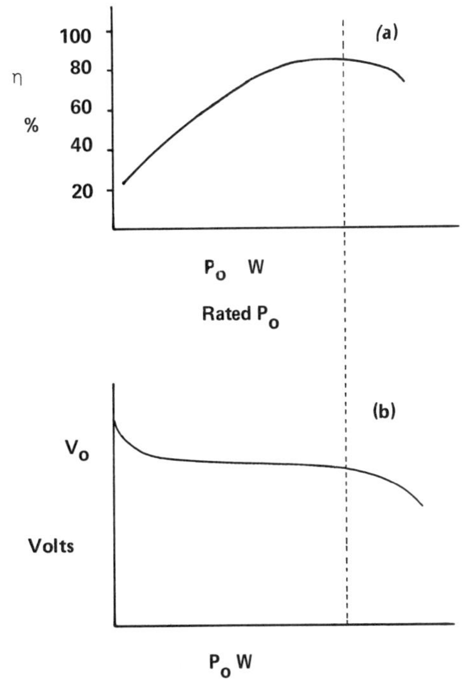

FIGURE 2 Converter Performance

Efficient units are desirable from the point of view of conserving power from a limited capacity source and to reduce the cooling necessary. Efficiency must also be optimised against size, weight and cost. The inversion frequency also has some influence in the cases where a definite running frequency is not required.

Figure 3 (a) shows a typical push-pull type inverter circuit. The transistors are alternatively switched on and off at a certain frequency by the base drive waveforms and Figure 3 (b) shows the collector voltage waveform. A square wave is the easiest and most efficient wave form to generate. Using a transformer of turns ratio n will give a secondary voltage of almost $\pm nV_S$. Most inverters are made self oscillating by positive feedback rather than driven as in Figure 3. In each inverter some form of timing mechanism will be present which, when it runs out, removes the base drive from the conducting transistor and causes the inverter to switch over. This mechanism will often influence the inverters overall efficiency and the frequency stability with loading.

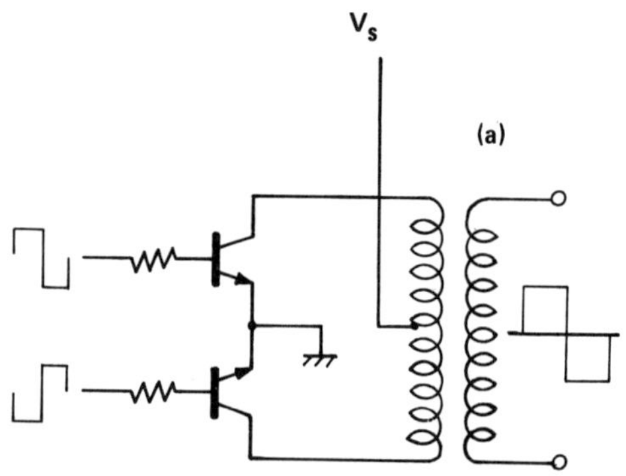

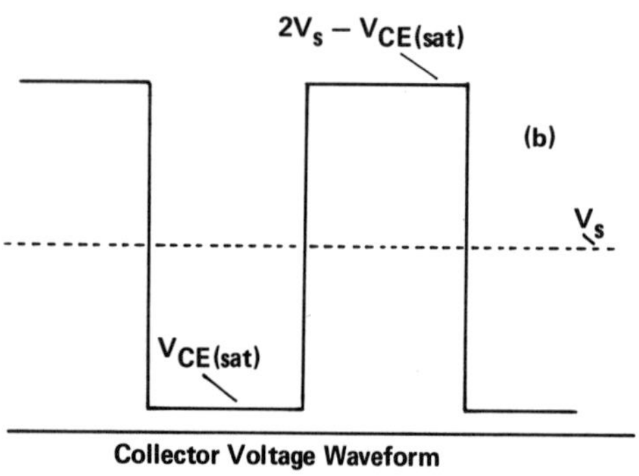

FIGURE 3

If the power source is any distance from the inverter it is desirable to decouple the inverter centre tap to earth to provide a low impedance supply at the instant of switch over. Radiation will also be reduced and twisted supply leads also helps in this direction.

Wet tantalum capacitors make excellent decoupling capacitors, but are somewhat expensive. A typical decoupling combination is 0.47μF in parallel with a 50μF electrolytic. Capacitors used in this service can have their ripple current ratings exceeded and this should be checked.

POWER SOURCES

In a report of this size it is not possible to discuss all the different types of power source available. Rather than describe many different types the particular case of the lead-acid battery is taken to show the facts which should be known about the intended power source.

Lead-Acid Batteries

Fostered by the motor vehicle the lead-acid battery is the most common and economic secondary (rechargeable) battery of today. As a result most inverters, including those in this report, are designed to run from 12V or 24V supplies. An awareness of battery characteristics is most important to the designer in defining worst case conditions and to the user to take account of output variations with battery condition.

A 12V battery will consist of six series connected nominal 2V cells and twelve for a 24V battery. Complete batteries, in good condition and at the same state of charge, may be connected in series to give higher output voltages or parallel to give increased capacity.

Before continuing the terms capacity and final voltage must be defined.

Final Voltage

When the cell voltage falls to this value the cell is deemed discharged. For automotive batteries this voltage, which is a compromise between battery life and voltage consistency, is taken as 1.75V. Thus a 12V battery is considered fully discharged when the terminal voltage on load falls to 6 x 1.75 = 10.5V.

In situations where relatively high currents are taken and the output voltage is not of prime importance, a final voltage of 1.33V per cell is adopted (8V for a six cell battery).

Capacity

Strictly, this is the total charge output in ampere-hours (Ah) in a certain time at a given battery temperature.

A specific example will make this clear.

A 12V 40 Ah automotive type battery which is rated at the 20 hour 25°C condition will supply 40/20 = 2A for a minimum of 20 hours before the on load voltage falls below 10.5V. By the same reasoning a 60 Ah will supply 3A and 100Ah will supply 5A.

Not many people will want to use an inverter for exactly twenty hours nor the battery at its twenty hour discharge current rate. Currents above the twenty-hour rate and battery temperatures below 25°C will reduce the Ah capacity as shown in Figure 4 and Figure 5. Conversely, lower currents, higher temperatures (not above 40°C for battery longevity) and intermittent use will increase the capacity.

Consider an 80% efficient inverter running an 80W piece of equipment. The power demand from the battery will be 80/0.8 or 100W. Assuming a mean battery voltage of 12 the current drawn will be 100/12A, roughly 8.5A. This is 8.5/2 = 4.25 times the 20 hour discharge rate for the previously referred to 40Ah battery. Figure 4 shows the Ah capacity reduced to 80% by operating at this current. Thus the effective Ah capacity becomes 0.8 x 40 = 32Ah running the inverter for 32/8.5 - 3.8 hours. From Figure 1 the mean voltage during this time will be 6 x 1.93 = 11.6V. In cold weather, say 10°C, the capacity will be further reduced (Figure 5) to 32 x 0.85 = 27 Ah giving a 27/8.5 = 3.2 hour running time. This should be compared with the figure of 40/8.5 = 4.7 hrs. which might have been expected. Incidentally, if the battery forms part of a vehicle starting system it should not be discharged more than 60% to ensure restarting the engine.

To bring out several points about battery operation the above example was made deliberately unfair by operating a relatively high power inverter with a low capacity battery.

If the inverter is operated when the vehicle is mobile the Ah rating is no longer a problem since the vehicle charging system will keep the battery 'topped up' so to speak and for most of the time supply the inverter direct. Voltage variation is a problem in this case though. Figure 5 shows the band of maximum voltage settings for a voltage regulator normally fitted to a small car. A maximum cell voltage of 2.7V might seem high but this is explained in Figure 6.

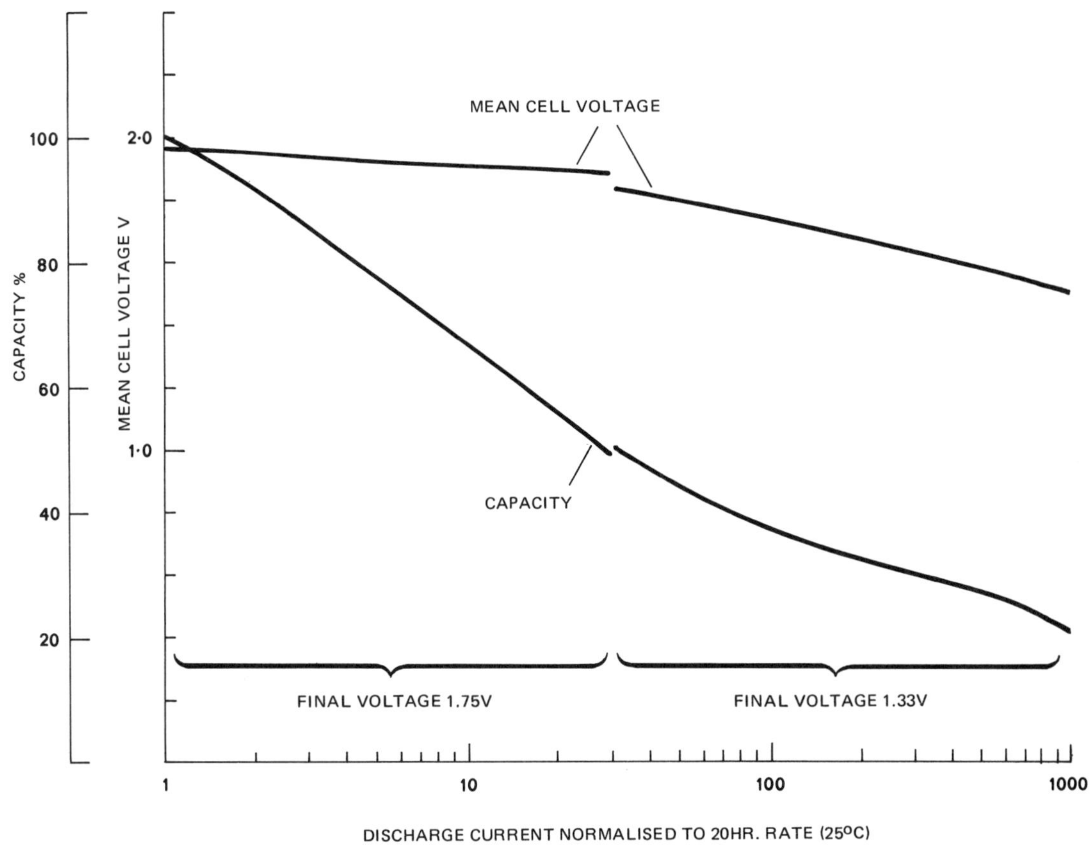

FIGURE 4

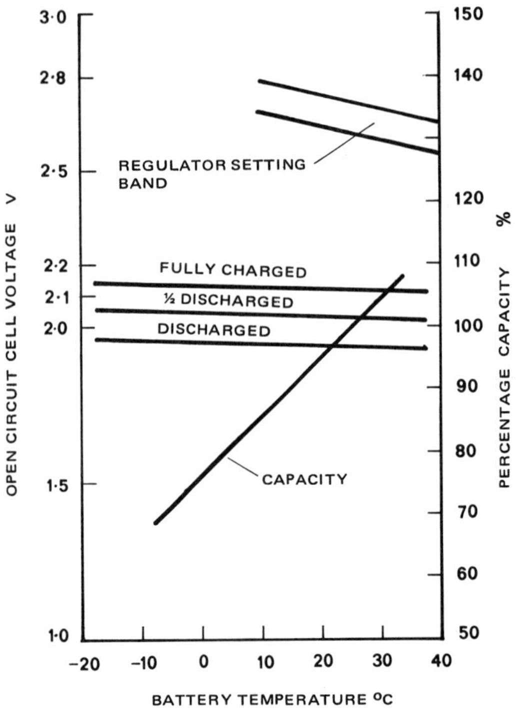

FIGURE 5

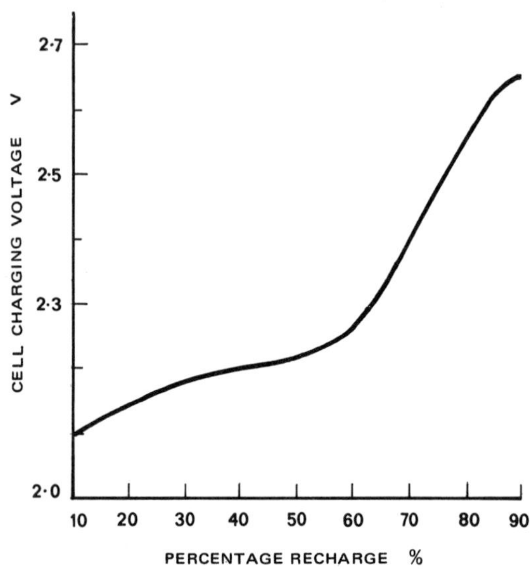

FIGURE 6

To fully recharge a battery the cell voltage must go up to at least 2.65V. When the cell voltage reaches 2.4 gassing starts to take place and the charging rate should be reduced to 7% of the battery Ah capacity. The cell open circuit voltage should not be regarded as an indication of the charge state unless measured 12 hours after charging or one hour after discharge.

To summarise, batteries separate from a charging system will show a variation of 2.1 to 1.75V per cell (12.6 to 10.5 for a six cell battery) on discharge. There is a possibility of the initial cell voltage being as high as 2.65V (15.9V for six cells) if a freshly charged battery is connected. Inverters for this application are classified as 12V input type. Mobile equipment should cater for variations of 2.7 to 2.1V per cell (16 to 12.6V on 6 cells) and 2.1 to 1.8 per cell (12.6 to 10.5 on 6 cells) when stationary. Inverters used in case are termed 14V types. Very large systems, such as aircraft, can often have voltage spikes many times the nominal battery voltage on the supply rails resulting in voltage rating required being greatly increased.

MAGNETIC PROPERTIES

Fundamental Magnetic Relationships[3,4]

The magneto motive force (m.m.f.) applied to a toroid of magnetic material wound with N turns is:

$$\text{m.m.f} = N.I \qquad \text{A}$$

where I is the current passing through the N turns. If the mean path length of the toroid is L′ in metres the magnetic field strength, H, will be:

$$H = N.I/L' \qquad \text{A/m} \quad . \; . \; . \; . \quad 3.1$$

Corresponding to H there will be a flux density, B in Teslas, given by:

$$B = \mu.H \qquad \text{T}$$

where μ (= $\mu r.\mu_o$) is the materials permeability
μ_r is the materials relative permeability
μ_o (=4.π. 10^{7} H/m) is the permeability of free space

Summing B over the whole of the toroids cross sectional area, A, in square metres, gives the total flux, Φ Webers.

$$\Phi = B.A \qquad \text{Wb.} \quad . \; . \; . \; . \quad 3.2$$

Inductance, L is defined as the flux linkage per unit current. In terms of the toroids parameters this is:

$$L = dN\Phi/dI \qquad H$$

Substituting for Φ from the previous equations gives:

$$L = \mu N^2 . A/L' \qquad H$$

The Law of induction states that

$$e = -dN.\ \Phi/dt \qquad V$$

where e = mean induced winding voltage
dt = time in which e occurs

Substituting from the previous equations gives two addition forms:

$$e = -\ L.\ dI/dt \qquad V \ .\ .\ .\ .\ 3.3$$

$$e = -NA.dB/dt \qquad V \ .\ .\ .\ .\ 3.4$$

A special case of the last equation is when the magnetic material is driven from one saturated limit ($B_{(sat)}$ to the other in time t_1 by a voltage V_1.

$$\therefore\ V_1 = 2.N.A.B_{(sat)}/t_1 \qquad V$$

or

$$t_1 = 2.\ N.A.B._{(sat)}/V_1 \qquad s\ .\ .\ .\ .\ 3.5$$

Equation 3.5 shows how core saturation may be used as a timing mechanism. A core starting at one $B_{(sat)}$ limit will take exactly t_1 seconds to reach the other $B_{(sat)}$ limit under the influence of voltage V_1.

A linear inductance, L, after the time t_1 would be carrying a current I_1 of:

$$I_1 = V_1 .\ t_1/L \qquad A\ (\text{from } 3.3)\ .\ 3.6$$

Current I_1, is the peak magnetising current of the inductor. Linear transformers are often designed with magnetising currents below one tenth of the referred load current. Equation 3.6 enables an estimate to be made of the required minimum primary inductance. The linear transformers working flux density, B_w, must be calculated to ensure core saturation is not approached and the core is not overdissipated. Using equation 3.4 gives:

$$B_w = V_1 .\ t_1/N.A. \qquad T$$

Transistor Operation with Saturable Cores

Transistor, VT1, shown in figure 7 has an h_{FE} of 50. Its base can be shorted to earth or supplied with 0.1A from resistor, R1, depending on the position of switch, Sw. The maximum collector current, I_C, the transistor can pass is 50 x 0.1 = 5A.

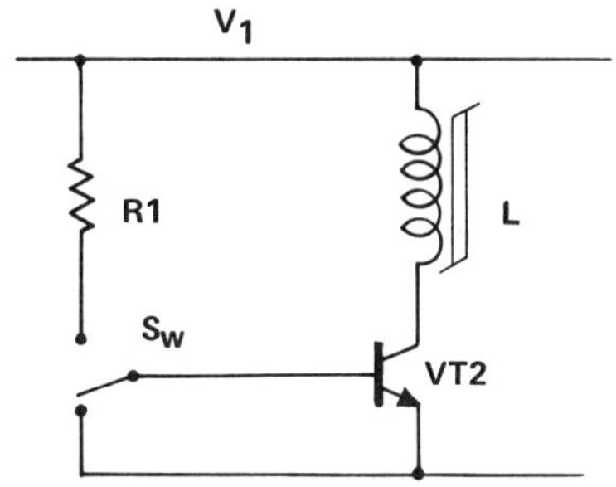

FIGURE 7

However, the initial collector current will be limited by the impedance of inductor L the collector load, The inductor core is initially at 'a' on its BH loop, figure 8.

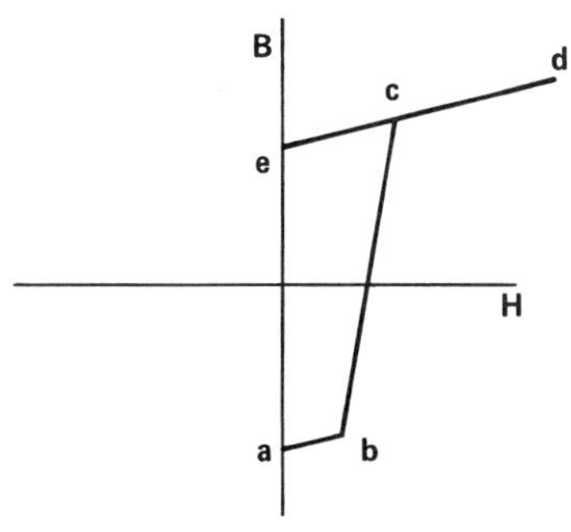

FIGURE 8

Switching Sw to resistor R1 turns transistor VT1 'on' applying voltage V_1 across inductor L (neglecting the $V_{CE(sat)}$of VT1). The instantaneous collector current will be zero and will gradually increase with time. Reference to equation 3.4 shows that for a constant applied voltage V_1, dB/dt must also be constant. Hence the B axis of figure 8 is transversed at a constant rate. The H axis reflects the amount of collector current needed to produce H at any given time. Hence the collector current, figure 9(a) after

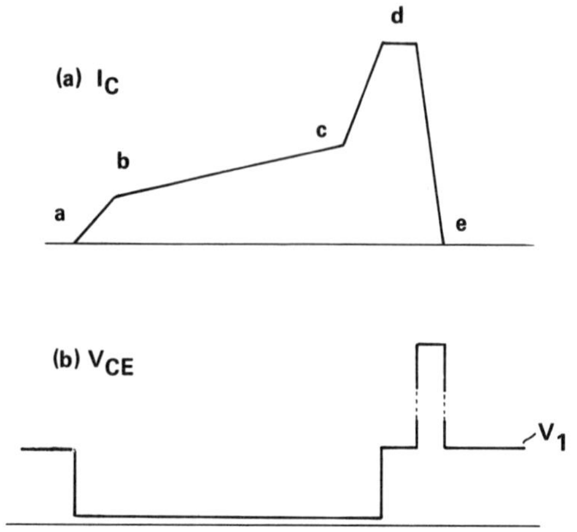

FIGURE 9 Collector Current and Voltage against time

switch on, will rise rapidly a to b and then increase slowly, b to c. At point c on the hysteresis loop the core saturates. To maintain a constant dB/dt the collector current must increase rapidly. When the collector current reaches 5A, point d, the transistor comes out of saturation. As the collector current is then constant, dI_C/dt is zero and the voltage across L falls to zero.

If the base of VT1 is then shorted to earth by Sw a large voltage spike will be generated as I_C falls to zero. As a practical example consider VT1 has a fall time, t_f of 1μs, and the saturated inductance value of L along path d e is 50μH. Using equation 3.3 gives a spike voltage of 50 x 5/1 = 250V showing how easily excessive voltages can be generated in inductive circuits.

DEVICE SELECTION

Three main device technologies are employed at present by Texas Instruments to produce its power transistors. These are in order of decreasing device robustness, epitaxial base single diffused, triple diffused and epitaxial planar.

Epitaxial Base Single Diffused (EBD)

Most of the Silicon Power Transistors used today are of the single diffused type. Low cost, wide availability, general device robustness and complementary types have contributed to this situation. In its basic form the single diffused is somewhat slow in terms of switching speed. Use of an epitaxial base gives a very useful increase in speed and pushes the normal 0.7MHz f_T to about 10MHz. Epitaxial base units are suitable for inverter circuits up to about 30kHz although this figure is somewhat dependent on the inverter circuit used. Present metal-can (TO3) products are capable of up to 30A continuous I_C and with up to 100V BV_{CEO} ratings in NPN and PNP. The modern trend is to develop and extend the range of plastic encapsulated power device (TIP) range. Complementary NPN and PNP devices in 1, 3, 6, 10 and 25A ratings with 40, 60, 80 and 100 BV_{CEO} ratings makes this range extremely comprehensive.

Triple Diffused Mesa (3DM)

This technology has been the vehicle for producing high voltage NPN power transistors like the BUY23A which has 10A I_C, 300V BV_{CEO} and 700V BV_{CEX} ratings. With ratings like this operation from rectified 230V a.c. mains is possible. Continuing improvements in high voltage transistor manufacture will produce devices whose BV_{CER}s are in the 1 – 2kV region. In speed a high voltage triple diffused type is about twice as fast as its equivalent epitaxial base single diffused type. However, with the BUY51 type device triple diffusion has been used to greatly increase the switching speed while still giving a moderate (100V) BV_{CEO} rating. Its ability to switch off 30A in 150ns rivals many epitaxial planar units but its robustness and safe area of operation is far superior.

Epitaxial Planar (EP)

Very fast switching speeds can be achieved with these types typically 5 – 10 times faster than an epitaxial base single diffused. Planar devices feature a high and reasonably constant h_{FE} over a wide range of collector current. This reduces the current drive needed for the base at maximum collector current compared with other types and to a certain extent the drive power.

For the same current rating an epitaxial planar power transistor uses a much smaller chip than other types. This, together with the very thin base region gives a device with a relatively poor safe operating area. Some form of distributed emitter resistance is often used to improve the safe area equalising the current distribution in the chip.[5,6] Any increase in emitter resistance will also increase $V_{BE(sat)}$

and $V_{CE(sat)}$ which is one of the trade offs made with this kind of unit. Additionally being mainly second breakdown limited, which does not derate very much with increasing T_j, the safe area is often given for T_{case} V 100°C. Planar devices are very often flattered in terms of safe operating area by giving the maximum allowable device power against V_{CE} rather than the more conventional I_C against V_{CE}. Appreciable amounts of low voltage power may be dissipated even at 100°C case, since the lower leakage currents of the planar construction permit junction temperatures of 200°C.

The use of a smaller chip means the bonding wires must be very fine and the peak current rating is made much more real in the sense of wire fusing rather than the fall off in current gain used in other units.

Robustness

Device reliability, which is a measure of the stability of its characteristics, should not be confused with robustness, which is a qualitative statement of the devices ability to withstand power overloads and be thermally limited, rather than second breakdown limited. (See Application Report B167, Second breakdown and power transistor area of operation). Hence while the epitaxial planar power transistor is the most reliable of the three types mentioned it is generally the least robust.

Switching Speeds

As power device f_Ts are normally quoted at relatively low values of I_C they should only be regarded as general indication of the devices switching performance at high current levels.

Figure 10 shows the typical switching waveforms which occur and how the device switching times are defined. To give an idea of the relative switching speeds of the different device technologies several types were measured operating in a two transformer converter. Table 4.1 is the result.

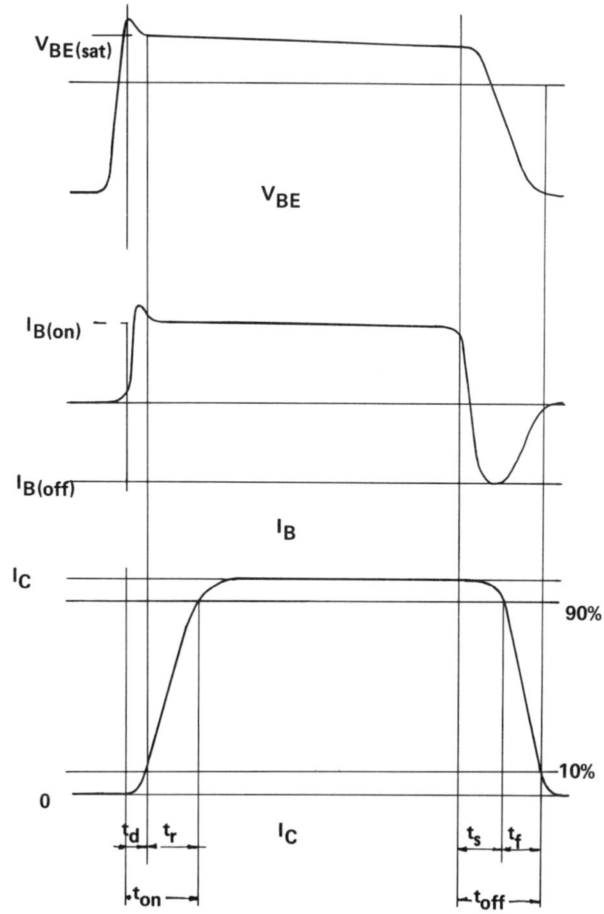

FIGURE 10

A BLY63, a very fast R.F. power device, was used as the control device to indicate how much the converter circuit itself was limiting the switching times. The last two columns show that epitaxial planar devices still switch off quickly without any strong reverse base drive. (i.e. the device is always operating in its forward biassed safe area).

Table 4.1

Device	ns: I_C = 5A $I_{B(on)}$ = $I_{B(off)}$ = 0.5A			I_C = 5A $I_{B(on)}$ = 0.5A $I_{B(off)}$ = 0 R_{BE} = 470	
	t_r	t_s	t_f	t_s	t_f
Single Diffused	3700	2600	3500	2300	21000
Epitaxial base single diffused TIP33	650	680	840	1250	4200
Triple diffused BUY20	800	1800	350	1300	1100
Epitaxial Planar	500	450	200	800	340
Epitaxial Planar BLY63	350	360	120	290	100

Although Data Sheet switching specifications usually only provide t_{on} and t_{off}, these can be used to gain some idea of t_s, and t_f the most important parameters for inverter transistors. If t_{on} and t_{off} are measured at a moderate value of I_C, t_d is usually small compared with t_r, therefore $t_{on} \triangleq t_r$ and as $t_r \triangleq f$.

$$t_f \simeq t_{on}$$

$$t_s \simeq t_{off} - t_{on}$$

Higher currents will increase t_r and t_f and decrease t_s and vice-versa for lower currents.

Selection Summary

To arrive at a short list of devices suitable for a particular inverter configuration the designer must know or decide on P_o, η, $V_{s(max)}$ and the operating frequency f_o.

If a two transformer inverter, Figure 26, is assumed, the devices should be able to pass at least

$$I_C > P_o / \eta . V_s \quad \text{A}$$

$$> 4P_o / 3V_s \qquad \text{for } \eta = 75\%$$

Allowing for 20% collector voltage overshoot gives:

$$BV_{CEO} > 1{,}2 \times 2V_s$$

$$> 2.4V_s$$

f_T gives some indication of the device switching rate and a good rule of thumb is to select

$$f_T > 100f_o$$

Inverters rarely work devices up to their maximum power dissipating limits so this parameter is of little importance. It is merely a problem of finding the smallest heat sink which will keep T_j below its maximum value. Knowing the device power loss P_1, the maximum junction temperature $T_{j(max)}$ and ambient temperature $T_{a(max)}$ the maximum junction to ambient thermal resistance θ_{ja} can be calculated:

$$\theta_{ja} = (T_{j\,max} - T_{a\,max}) / P_1 \quad {}^{o}C/W$$

The data sheet will give a value of θ_{jc} and the case to sink mounting θ_{cs} will be of the order of 0.5°C/W thus:

$$\theta_{sa} = \theta_{ja} - \theta_{jc} - \theta_{cs}$$

If a square heat sink is used of side length Lmm its thermal resistance is:

$$\theta_{sa} \simeq \frac{560}{\bar{L}} \quad {}^{o}C/W$$

Provided the heat sink is not made too long and thin the required heat sink area $\bar{L}^2mm^2$ can be formed in any shape.

High efficiency (η) is the criterion on which most inverters are judged and device selection often reduces to finding the device with the lowest power losses.

If the inverter switching times are small compared with the operating frequency transistor d.c. losses can be put at:

$$\hat{V}_{BE(sat)} I_{B1} + \hat{V}_{CE(sat)} . I_{C1} + 2V_s . \hat{I}_{CBO}$$

to keep the transistor in saturation

$$I_{B1} \geq I_{C1} / \check{h}_{FE} \qquad \text{where}$$

^ has been used to show a maximum value
ˇ " " " " " minimum "

Strictly speaking h_{FE} should be measured with the device at the edge of saturation particularly for 3DM and E.P. types whose I_C against V_{CE} 'knees' are not very sharp (i.e. h_{FE} drops as V_{CE} approaches $V_{CE(sat)}$).

The available base drive voltage from the transformer should be in the order of $2V_{BE(sat)}$, and $V_{BE(sat)}$ being dropped by the base current limiting resistor and the starting circuit. As the feedback resistor, R_F, is usually dimensioned to drop half the collector to collector voltage the total base drive power is:

$$4\hat{V}_{BE(sat)} . I_{C1} / \check{h}_{FE}$$

Total switching losses for this type of inverter will be:

$$2.V_s . I_{C1} . t_f . f_o/3 \qquad \text{(Figure 66)}$$

So the total amount of power loss caused by the transistors will be:—

$$P_1 = \hat{V}_{CE(sat)} . I_{C1} + 2V_s . \hat{I}_{CBO}$$
$$+ 4.\hat{V}_{BE(sat)} . I_{C1}/\check{h}_{FE} + 2.V_s . I_{C1} . t_f . f_o/3$$

$$= I_{C1} . (\hat{V}_{CE(sat)} + 4\hat{V}_{BE(sat)}/\check{h}_{FE})$$
$$+ 2V_s . (\hat{I}_{CBO} + I_{C1} . t_f . f_o/3)$$

Some manipulation of the first term is possible by increasing the base current above its minimum value I_{C1}/h_{FE}, (effectively reducing h_{FE} in the power loss expression) giving a decrease in $V_{CE(sat)}$ and a slight increase in $V_{BE(sat)}$. Similar analysis can be performed on the other types of inverter described in this report. Very often a transistor type is found whose typical units would result in a highly efficient inverter. However, in designing the inverter to take full spread in device parameters a good deal of the efficiency can be lost. In situations where consistent high efficiency is required for large numbers of the same inverter type the designer could decide to pay a small premium and have a selection on the standard device type.

SWITCHING REGULATORS

High speed high power switching transistors have produced an upsurge of interest in switching mode regulators[7] When compared with the conventional type of regulators a considerable reduction in bulk can be achieved by operating these units at high frequencies. Further, unlike conventional convertors, a regulated output voltage can be produced in spite of wide variations in input voltage. Regulators of this type use four main components, see figure 11; a transistor to act as a switch, an inductor to act

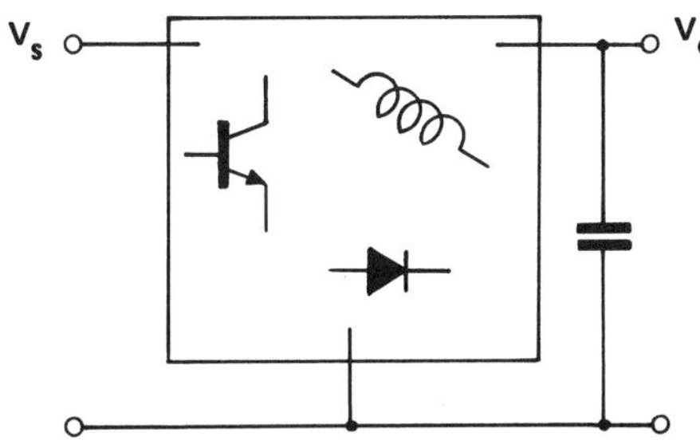

FIGURE 11

as an energy transfer device, a diode to reclaim stored energy from the inductor after the transistor has switched off, and a capacitor which is connected across the output terminals to act as an energy storage device and to reduce the ripple voltage[8] Several arrangements of these four basic components are possible between the input, output and common line of the supplies. Two broad groups are formed

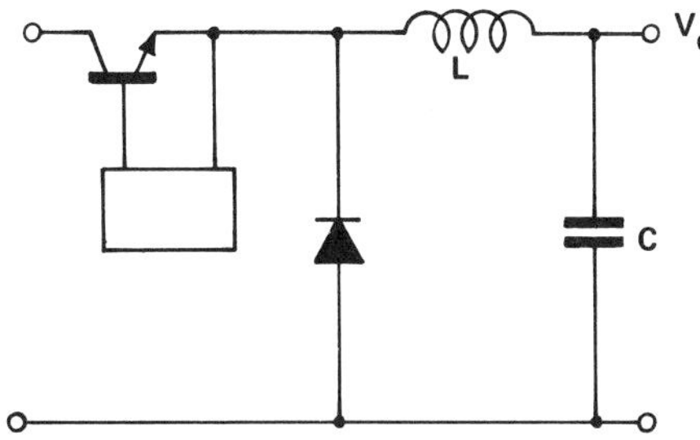

FIGURE 12 Basic Series Chopper Circuit

– series and shunt choppers[9] In the series chopper, Figure 12 the inductor is connected between the input and output whenever the transistor is conducting. In the case of the shunt or parallel chopper the inductor is connected across the input supply whenever the transistor is conducting. By varying the transistor's on to off time it is possible for both types of convertor to nullify the effects of load and input voltage variations and so maintain a stable output voltage. This is discussed in detail in Chapter VII.

Two modes of inductor operation are possible for both series and shunt choppers. The first mode is when the inductor current does not fall to zero at any time. This is the usual operating condition for a series chopper and for a shunt chopper during the initial start up. High voltage shunt type choppers usually operate in the second mode, that is when the inductor current falls to zero before the transistor conducts again. Series choppers will also operate in this mode under light loading conditions.

The Series Chopper

With this type of unit the output voltage will always be less than the input voltage. This is shown in Figure 13, where

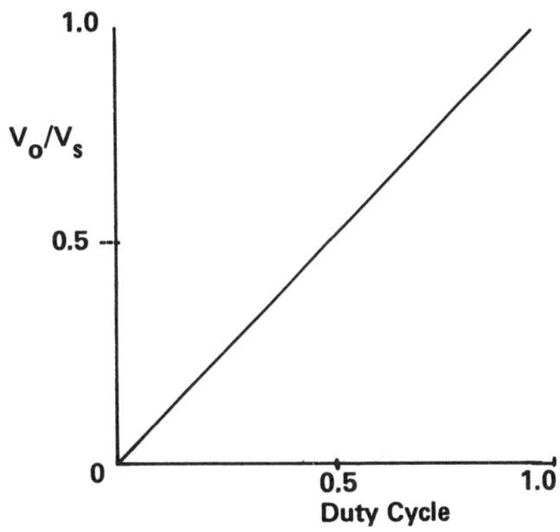

FIGURE 13

the output/input voltage ratio (V_O/V_S) is plotted against transistor conduction duty cycle. This circuit is very tolerant in terms of choke resistance. A choke resistance of 1% of the load does not cause any discernible change to figure 13. Devices chosen for this service must be able to withstand the supply of voltage V_S and also be capable of carrying the maximum inductor current which is equal to the load current plus an additional amount caused by the choke's finite inductance. Generally this will be about 10% of the maximum load current. The absolute values of L and C will be decided by the system requirements. These could be (a) minimum ripple voltage at the output (b) minimum overshoot of output voltage when the load is suddenly decreased (c) minimum undershoot of output voltage when the load is increased (d) a minimum volume solution (e) a minimum cost solution. So, for a 5V 8A supply generated from a 28V source a suitable device would be a 10A 40V transistor. Assuming an overall efficiency of 80% 50W would be drawn from the source. This is a mean current of 50/28 or roughly 1.8A from the supply. As the transistor draws this current in a series of 8A pulses good input decoupling for this type of supply is essential. Although chopper supplies are capable of giving efficiencies of >90% the example given has an inherent limitation in that as the supply voltage is much higher than the output voltage the transistor will be off for most of the time. This means that the diode will be carrying the load current almost continuously. Since the diode voltage drop can be as high as 1V it is evident that 1/6th of the total output power will be lost in the diode, limiting the maximum circuit efficiency of the circuit to about 80%. Losses like this can be reduced by operating high current diodes far below their maximum rating.

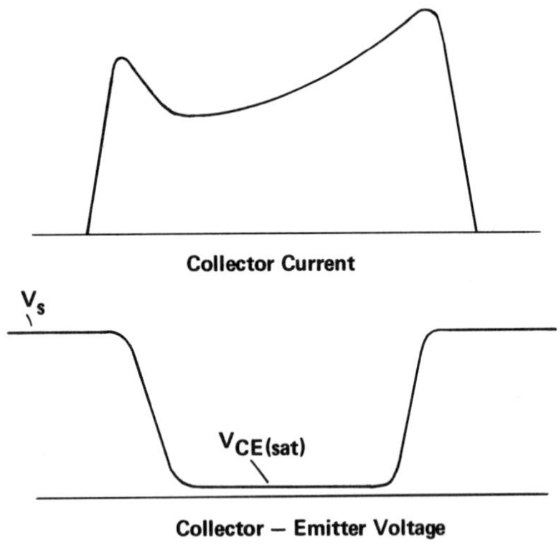

FIGURE 15 (a)

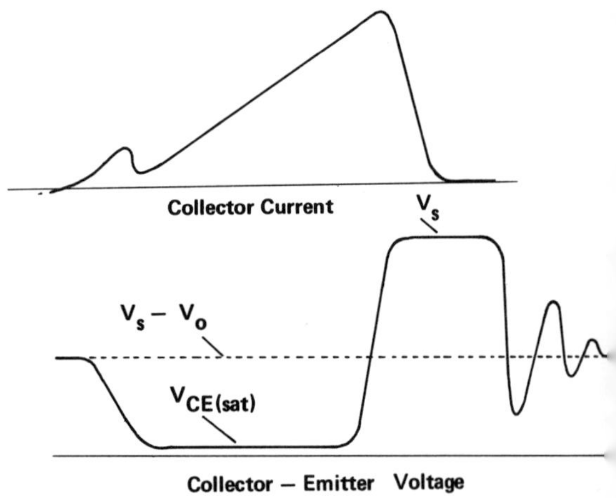

FIGURE 15 (b)

Special attention must be paid to the chopper transistor operating area shown in Figure 14. In figure 15(a), which is for the continuous choke current case, it can be seen that two separate operations take place each time the chopper transistor switches on and off. First, the transistor must take over the decaying choke current from the diode and it must neutralise any stored charge in the diode before the voltage across the transistor can start to fall. The diode stored charge is extracted in the form of a spike which can be seen at the start of the collector current. After this the collector voltage can fall to $V_{CE(sat)}$, there then follows a period of constant voltage across the choke which results in a linearly increasing choke current. To switch off the voltage across the transistor must increase sufficiently to turn the diode on. In other words V_S plus the instantaneous

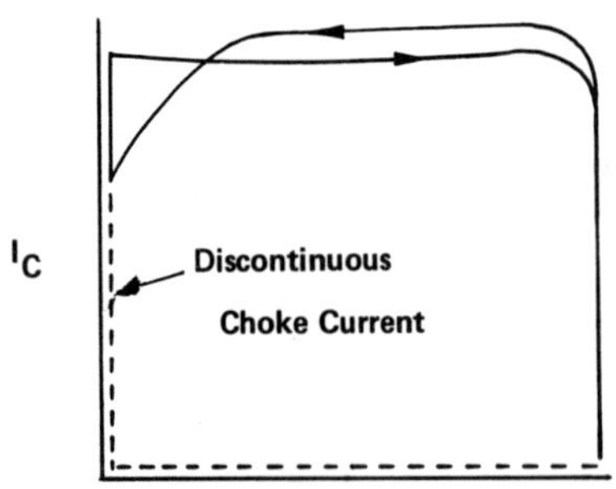

FIGURE 14

forward drop of the diode. The collector current can then fall to zero. It can be seen that this form of switching results in very high transient dissipation in the device. To maintain a low average switching power loss the device switching times should be short compared with the total device conduction time. The turn on period does not impose any stringent problems to the device providing that it is turning on from a voltage below BV_{CEO} and it is within its safe operating area for forward bias conditions (where the device is at its most robust). On switch off, however, the device operating area should be examined for any possibility of reverse second breakdown occurring. With discontinuous choke current, figure 15(b) and shown dotted in figure 14, the device switch-on is not critical since the transistor only has to charge the diode capacitance. Switch off, however, is exactly the same as in figure 15(a).

The Shunt Chopper

Shunt switching regulators are able to produce some magnification in output voltage with the input voltage. Two types are shown in figures 16 and 18. The output/input ratios against normalised transistor conduction periods are shown in figures 17 and 19. The effect of using a choke with 1% of the load resistance is shown dotted in both cases. It will be noticed that the output voltage starts then to decrease at very long conduction periods, limiting the maximum magnification that can be achieved. The opposite polarity shunt chopper, figure 18, can be considered a variant on figure 16 with the choke and transistor interchanged. This type of converter can be considered as a pulse width modulated version of the ringing choke converter. By using a transformer instead of a choke and arranging the secondary rectifier diodes to conductor during the transistor off time, a very flexible form of converter is produced with the ability to isolate and stabilise any value of output voltage and the use of a transformer permits multiple supply rails to be generated from other windings. Again, determining the device operating area is of the utmost importance, especially so in that the collector voltage can swing well in excess of the input supply voltage. A critical examination of the converter start up process should be conducted to check for safe device operation during this period. Restricting the devices maximum conduction period during start up is one of the best ways of tailoring the V_{CE} versus I_C locus. Switch on is not a problem as only the diode junction capacity needs to be charged. By using a suitable fast diode types such as the 1S070 series for up to 1A 1N3874 series for up to 6A, and the 1N3889 series for 12A, very low rectification losses can be achieved to beyond 100kHz. It is well worth degrading the device switch-on to ease the switch-off conditions, see Figures 20, 21. This can be done by placing a small capacitor in parallel with the transistor. At switch off the transistor will have stopped conducting before the voltage swings above the supply rail. The current

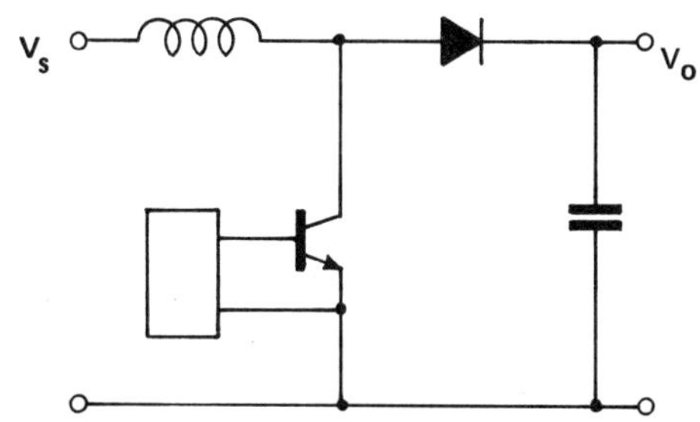

FIGURE 16 Basic Shunt Chopper Circuit

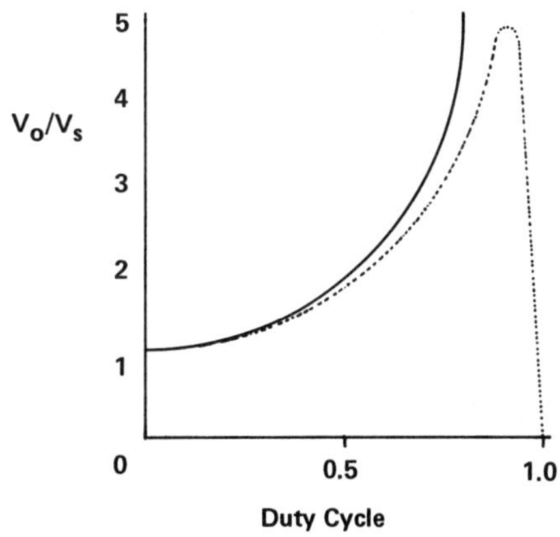

FIGURE 17

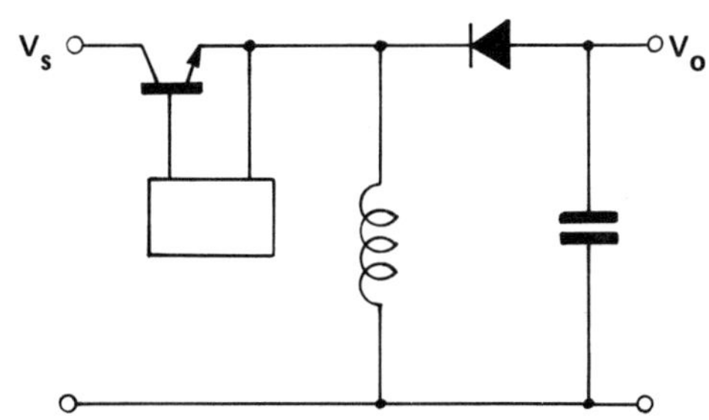

FIGURE 18 Basic Shunt Chopper Circuit (negative output)

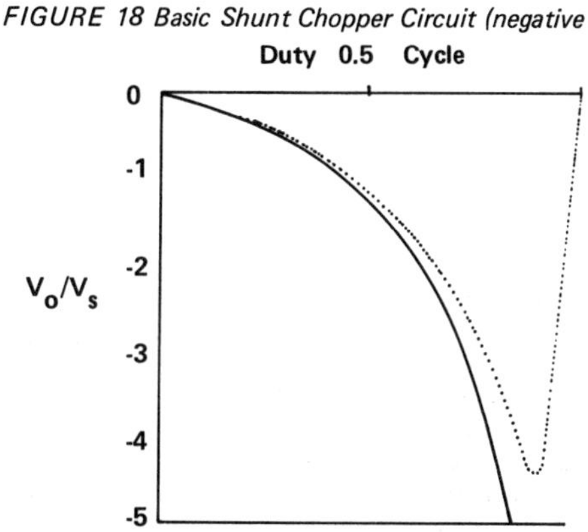

FIGURE 19

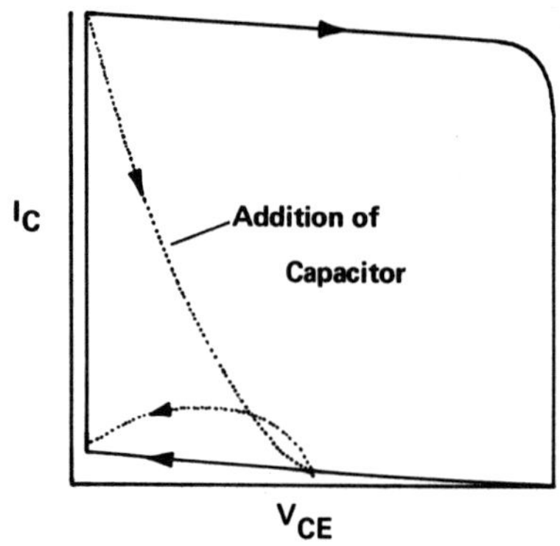

FIGURE 20

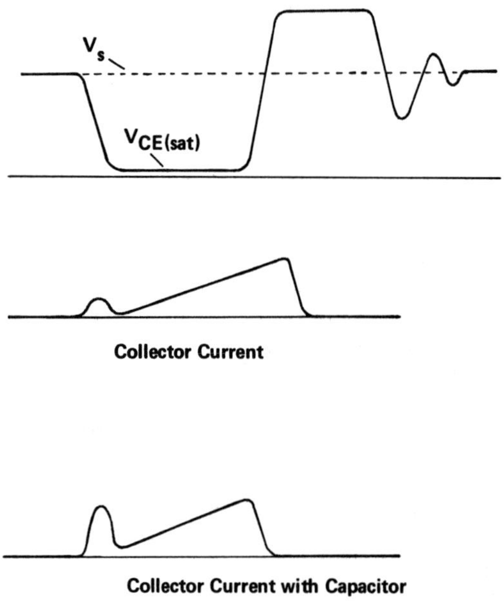

FIGURE 21

is maintained to the inductor by the capacitor over the remainder of the switch-off period. At switch on the transistor must discharge the capacitor which by then is charged to the supply rail voltage. However, as pointed out before, the device is most robust under forward bias conditions, and this causes no problems except for a very slight loss in conversion efficiency. Artifices like this are particularly useful when used with devices such as the BUY23A which has a BV_{CEO} rating of 300V, but under inductive switching conditions, is allowed to swing to 600V providing the collector current has stopped by the time the BV_{CEO} rating is reached. So, with this device and a supply rail of 300 volts, the peak collector voltage can be allowed to swing as high as 600. For a given power output and frequency the use of high voltage transistor circuitry produces a reduction in primary smoothing component bulk compared with low voltage transistor circuitry, as the current is correspondingly less.

One particularly useful application of the shunt chopper is to maintain a constant supply voltage from a battery. Batteries, as mentioned, can vary widely in output voltage depending on the conditions of use. One of these regulators could be used, for instance, to stabilise a nominal 28V battery voltage to, say, 30V, and this output voltage would be maintained constant for battery variations from 20V to 30V.

SINGLE SATURATING TRANSFORMER CONVERTER

Operation

A typical circuit for this type of converter is shown in Figure 22 and the collector voltage and current waveforms in Figure 23.[10] Timing is by core saturation. Just after the converter has switched over the core will be at $-B_{(sat)}$ and heading towards $B_{(sat)}$ under the influence of $V_s - V_{CE(sat)}$ applied half the primary. After a time equal to the inverter half period t_1,

where $t_1 = 2.B_{(sat)}.A.N/(V_s - V_{CE(sat)})$ from equation 3.5.

the core will saturate allowing the collector current to increase to $h_{FE}I_B$. At this point dI_C/dt is zero and the collector voltage rises to V_S; also the base drive voltage which was maintaining I_B falls to zero. This unstable situation is terminated after the transistors storage time by the collector current falling to zero. This induces the turn-on drive for the other transistor which then drives the core from $+B_{(sat)}$ to $-B_{(sat)}$. Neglecting the transistor storage time the half period time can be expressed in terms of the oscillation frequency f_o as:

$$1/2f_o = 2.B_{(sat)}.AN/(V_s - V_{CE(sat)})$$

$$\therefore\ N = (V_s - V_{CE(sat)})/4.B_{(sat)}.Af_o \quad . \quad . \quad . \quad 6.1$$

For a particular core and core material the number of primary turns completely defines the operating frequency.

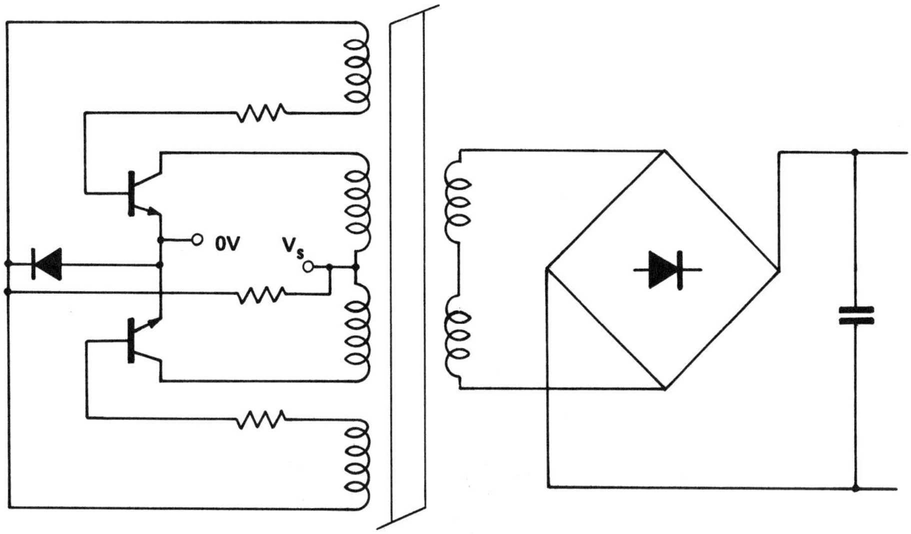

FIGURE 22 Single Saturating Transformer Converter

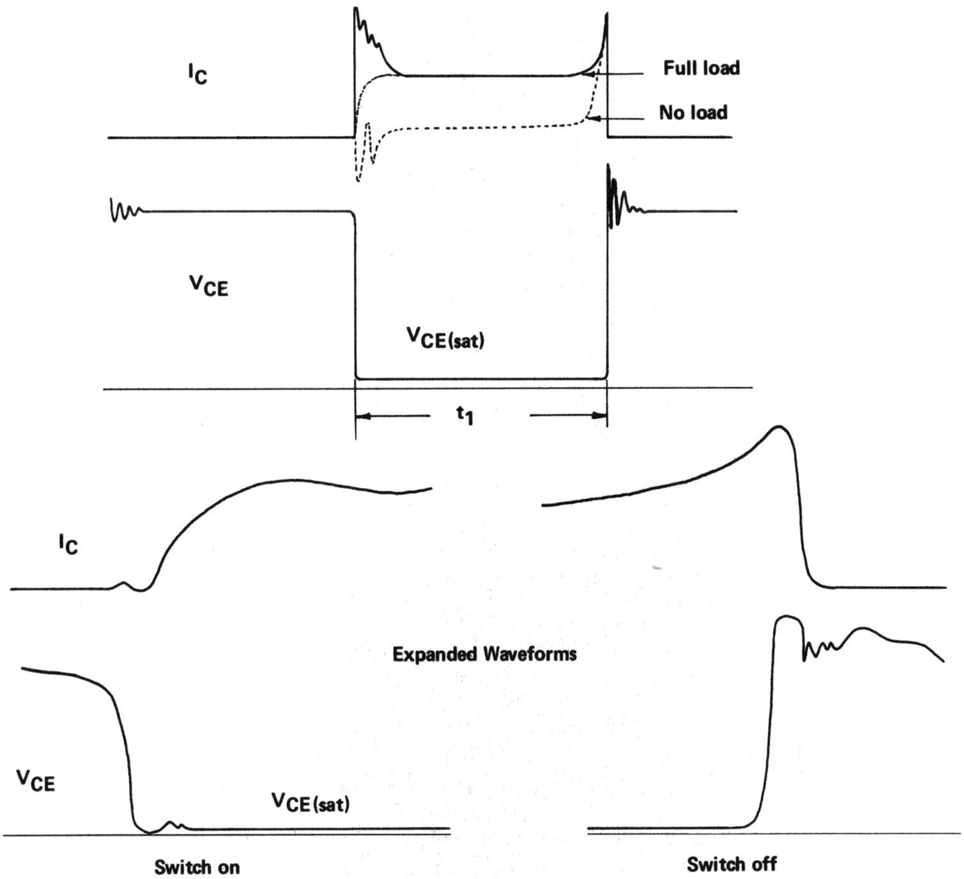

FIGURE 23

Two good features of this circuit are: it is simple requiring only one transformer and the running frequency is independent of loading.

Turn-on dissipation is low but the turn-off dissipation is extremely high and independent of loading, see Figure 24.

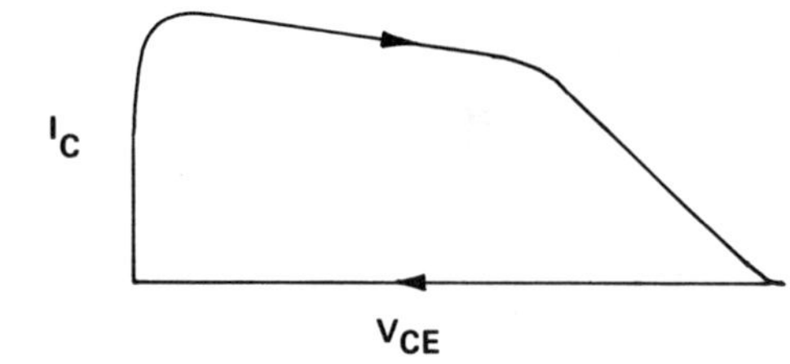

FIGURE 24

Losses should be calculated for square loop switching in time t_{off}. Safe area checks are most important for this type of inverter and good supply decoupling is required otherwise the high peak collector currents will cause R.F. interference from the power supply leads. The lowest switching losses occur with materials like HCR and mumetal which have very sharp saturation characteristics. High power versions of this circuit will require large amounts of these expensive core materials. Also as the core volume increases its area does not increase at the same rate. This results in a higher surface temperature, as the loss per unit volume is constant since the core is taken into saturation. So at higher powers the inverter is limited by the maximum core temperature rise rather than its winding space. In practical terms this means there is a maximum frequency (which decreases with increasing core size) at which a certain core may be used without suffering excessive temperatures. The rapidly increasing cycling losses with frequency of conventional laminated materials (which can be offset somewhat by using very thin laminations) has led to the almost exclusive use of ferrites and permalloys above 20kHz for this type if inverter. Ferrites are to be preferred for their faster switching times but they are often not available in such large piece parts as the permalloys. As a ferrite $B_{(sat)}$ is in the 300 – 400mT region, compared with the 1.5T of HCR, the lower loss ferrite core must be larger than the HCR one it replaces since $B_{(sat)}$ AN has to be constant at a given frequency. Slow rectifiers and core eddy current losses cause a peak at the beginning of the collector current (Figure 23). If this 'bath-tub' effect becomes too large and is comparable with $h_{FE}.\,I_B$ the inverter will switch over at the first current peak. As the inverter will then be operating higher than its designed frequency switching losses increase possibly causing device failure due to over-dissipation. The requirements that the reflected load current should be below $h_{FE}.\,I_B$ for the lowest gain device under worst case conditions means high gain units will produce collector currents far in excess of the load current. All these factors increase the peak current ratings for the transistor and subjects high gain units, which tend to have longer storage times, to large switching powers.

Base Drive Power

There is an optimum value of base winding voltage which will provide the highest $V_{BE(sat)}$ unit with the minimum required base current $I_B(=4P_o/3h_{FE}.V_s)$ and use a minimal amount of drive power for a typical $V_{BE(sat)}$ unit. Figure 25 shows the equivalent base drive circuit. Its assumed $V_{CE(sat)}$ variations cause little change in the drive voltage.

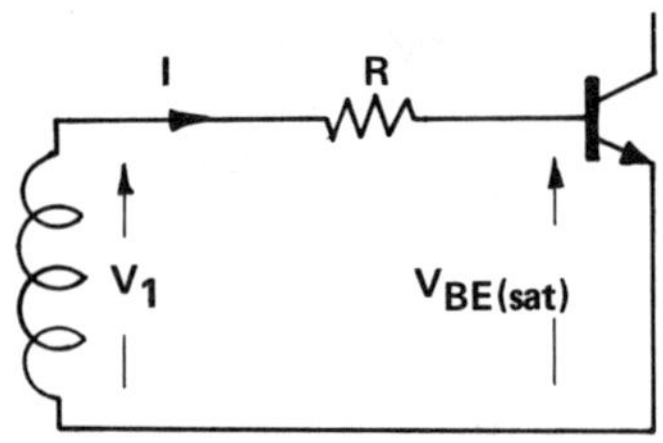

FIGURE 25

V_1. The starting diode maximum and typical forward voltage drops can be added to the $V_{BE(sat)}$ values. For worst case conditions.

$$V_1 = \check{I}_B.R + \hat{V}_{BE(sat)} \quad . \; . \; . \qquad 6.2$$

where R is the base circuit series resistance.

With typical devices drive power, P, is

$$P = (V_1 - V_{BE(sat)}).V_1/R \quad . \; . \qquad 6.3$$

Substituting for V_1 from 6.2 in 6.3 gives

$$P = (\check{I}_B.R + \hat{V}_{BE(sat)} - V_{BE(sat)}).(\check{I}_B.R + V_{BE(sat)})/R \qquad 6.4$$

To find minimum power differentiate 6.4 with regard to R and rearranging gives

$$dP/dR = \check{I}_B.^2 - V_{BE(sat)}.(\hat{V}_{BE(sat)} - V_{BE(sat)})/R^2 \ldots 6.5$$

$$0 = \check{I}_B^2 - V_{BE(sat)}^2.(\hat{V}_{BE(sat)}/V_{BE(sat)} - 1)/R_o^2$$

$$R_o = V_{BE(sat)}\,(\hat{V}_{BE(sat)}/V_{BE(sat)} - 1)^{1/2}/\check{I}_B$$

The optimum drive voltage V_{1o} will be

$$V_{1o} = \hat{V}_{BE(sat)} + V_{BE(sat)}(\hat{V}_{BE(sat)}/V_{BE(sat)} - 1)^{1/2}$$

This analysis has neglected the fact that $V_{BE(sat)}$ will be larger for the typical unit since a current higher than I_B is being forced into the base. Differences in V_{1o} caused by this are usually slight.

TWO TRANSFORMER (SATURATING) INVERTERS

Operation

Figure 26 illustrates a typical two transformer circuit, together with its operating waveforms shown in Figure 27.

The absence of the single saturating transformers collector current turn-off spike is very noticeable. Timing is defined by the saturation of the driver transformer, T1. Across the primary of the output transformer, T2, there is a square

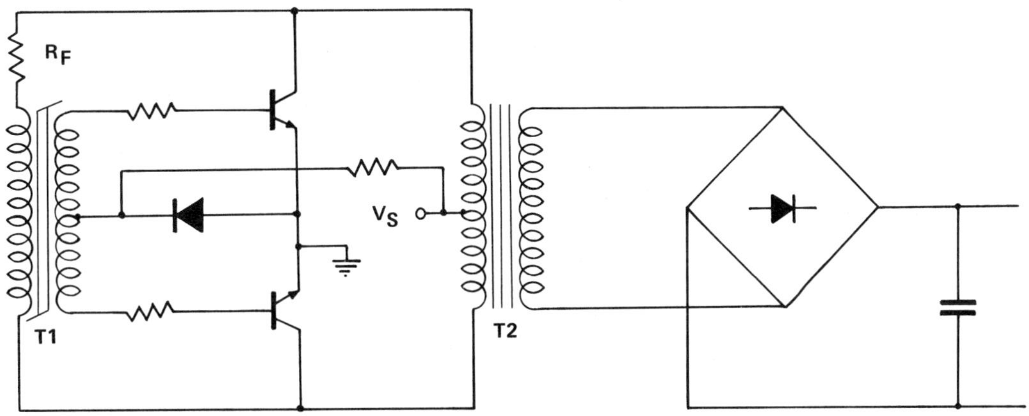

FIGURE 26 Two Transformer Converter

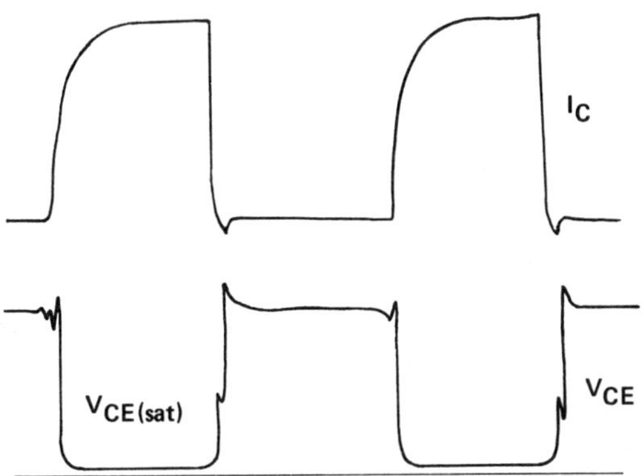

wave of $\pm 2(V_s - V_{CE(sat)})$. Resistor R_F is usually chosen to drop half of this leaving $\pm(V_s - V_{CE(sat)})$ across the unsaturated T1. Exactly the same formula used for the single transformer inverter applies, the oscillation ferquency, f_o, being:

$$f_o = (V_s - V_{CE(sat)})/4.B_{(sat)}.A.N_p$$

where N_p is number of primary turns of transformer T1. (This only applies when R_F drops $V_s - V_{CE(sat)}$ volts.)

Once a particular core is chosen the ability of R_F to allow sufficient current to saturate the core should be checked. That the transformers magnetising current is less than 10% of the referred base current should also be checked. From

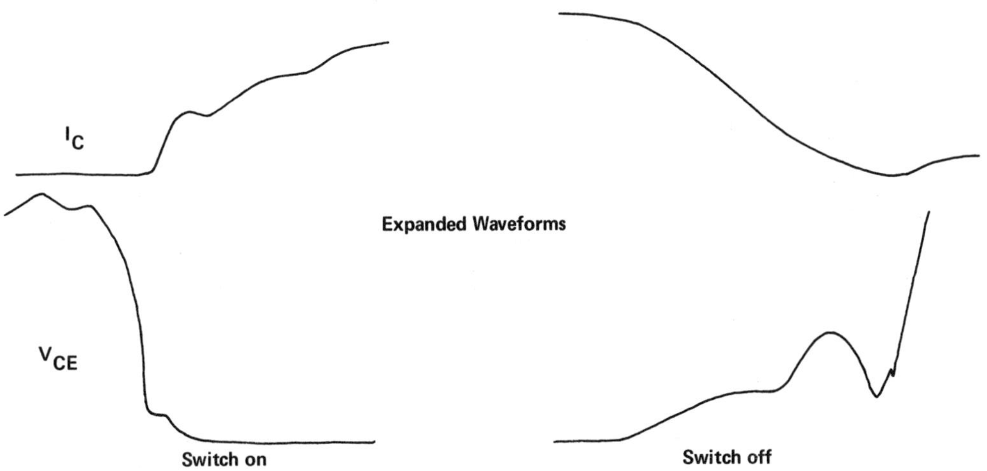

FIGURE 27

the core material data H_O for $B_{(sat)}$ can be found. Thus the minimum current, I_O, R_F should supply is:–

$$I_o = H_o.L'/N_p < 2V_s/R_F$$

After a precise number of volt seconds from the beginning of a device conduction time, the driver core will saturate dropping the transformer base drive voltage to zero. A negative base current will flow out of the transistor. Except for a slight increase in current, due to resistor R_F being connected directly across the collectors, as far as the collector circuit is concerned nothing has happened until the transistor switches off at the end of its storage time, t_s. The magnetising current of the linear output transformer T2, after the transistor has stopped supplying current, reverses the voltage across T2 primary desaturating T1 and driving the other transistor on to start a new cycle. This sequence of core saturation, t_s drift, switch-off (t_f) results in very low switch of losses and a tight I_C versus V_{CE} locus, see Figure 28 – especially under no load conditions. Capacitors across the base resistors, to speed up the switch-off, can be harmful with fast devices as the dotted portion of Figure 28 shows. The operating frequency will be load independent, but somewhat dependent on supply voltage. It is possible to use a cheap core material for the linear output transformer so that costly low loss material is only required for the relatively small driver transformer. This makes the high power two transformer inverter type more economical than the single saturating one. Again, conventional core materials are limited to about 20kHz. Also as the transistors are not subject to high peak current their ratings can be lower. Used with suitable transistors 100kHz operation and above is perfectly feasible, but core losses should be taken into account. Fine timing of the inverter frequency can be done by varying resistor R_F. The disadvantages of this type of inverter are few, probably the major one being two transformers are needed.

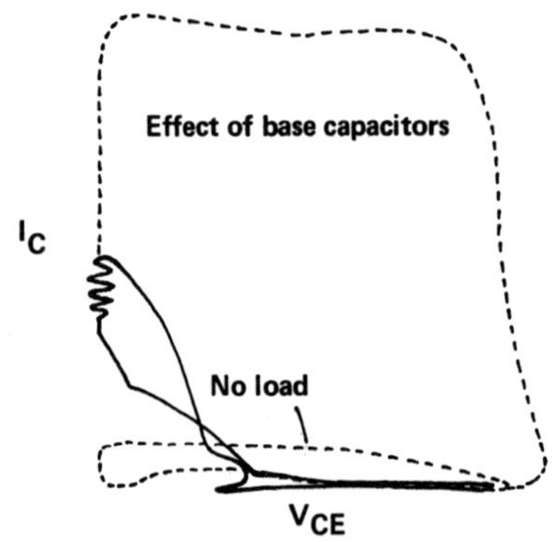

FIGURE 28

Current Feedback Version

By feeding back the collector current rather than voltage as shown in Figure 29 a new type of two transformer inverter is formed. This type of inverter performs well into low impedance loads as the more collector current that flows the more base drive current is produced. The driver transformer turns ratio defines the forced h_{FE} for the devices and base drive power losses are relatively small. $V_{CE(sat)}$ is increased by the reflected base drive voltage but this increase is not likely to be more than 20%. Two or more starting diodes in series will help stabilise the mark space ratio and frequency at the expense of collector circuit voltage losses. With this type of drive the transistors will almost certainly 'die' if the load is short circuited. The ability to strongly start into very low loads makes the circuit ideal for converters where the output capacitor must be charged up at switch on. However, starting is very difficult under light loading (low I_C) conditions. Even when the inverter is running a large reduction in its power output, causing very low values of I_C, could result in failure to oscillate.

To overcome some of the starting and running problems a combination of voltage and current feedback may be used to the driver transformer.

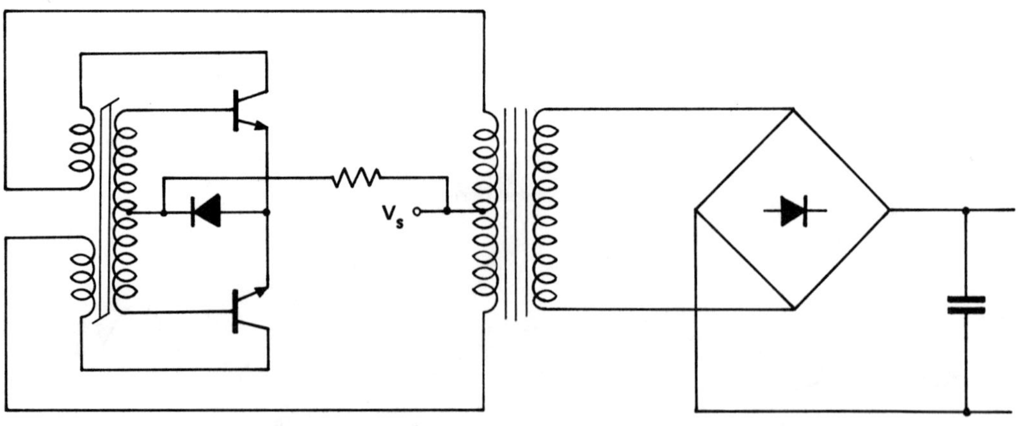

FIGURE 29 Current Feedback Two Transformer Converter

Calculation of Resistor R_F

Assuming the inverter is 75% efficient the base drive must support a minimum collector current of:

$$\check{I}_C = 4.P_o/3(V_s - V_{CE(sat)})$$

With the lowest gain device (h_{FE}) the minimum base current requirement is:

$$\check{I}_B = \check{I}_C/\check{h}_{FE}$$

Some power must be lost in this kind of circuit to stabilise the base drive voltage. If not, unequal mark to space ratios occur (in the same ratios as the $V_{BE(sat)}$s) and the overall frequency is strongly dependent on the summation of the two device V_{BE}s. Referring R_F to the secondary of the driver transformer gives Figure 30 where V_F is the starting diode forward voltage drop, R is the series base resistance, n is the turns ratio of the base drive transformer.

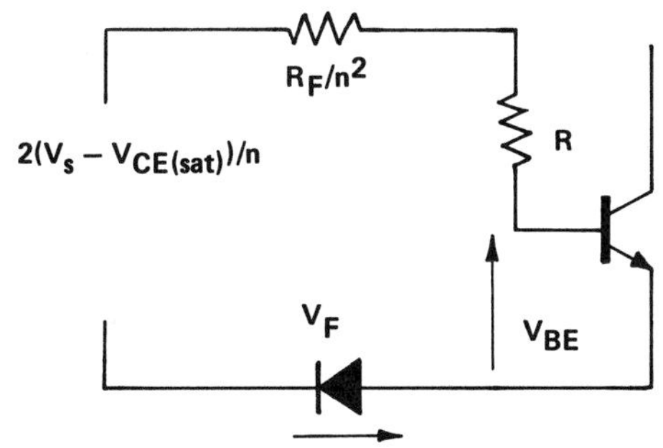

FIGURE 30

Substituting values gives an easily solved quadratic in n. Knowing n, the value of resistor R_F may be calculated and should be rounded off to the next lowest preferred value, as this analysis neglected the additional magnetising current taken by the primary of transformer T1.

CR TIMED INVERTERS[11,12]

Operation

These types of inverter use the exponential decay of voltage or current in a CR circuit to define the inverter timing. The conventional multivibrator could be used as an inverter with the collector loads replaced by the transformer primary. However, its efficiency would not be as good as the CR timed type described here. The reasons for this are (i) a high dissipation switch over takes place by one device turning on with $2V_s$ across it and forcing the other device off. (ii) quite large amounts of power can be lost to the base resistors.

The CR timed inverter described by Norwicki, Figure 31 uses an exponentially decreasing base current drive for the 'on' transistor. Unlatching will occur when $I_B = I_C/h_{FE}$. At turn-off the device exhibits very little storage time and t_f is the switch-off time, thus giving low dissipation switch-offs. Typical waveforms are shown in Figure 32.

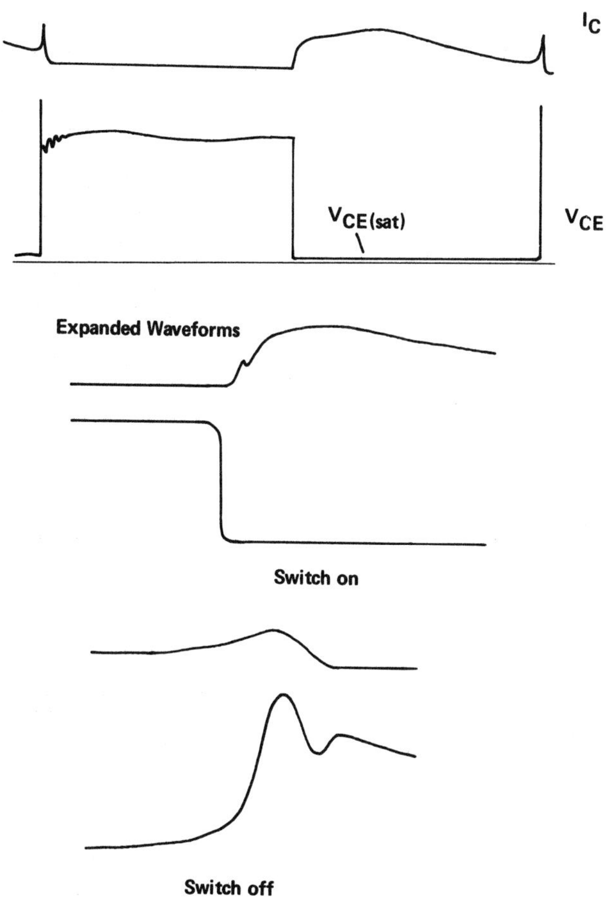

FIGURE 32

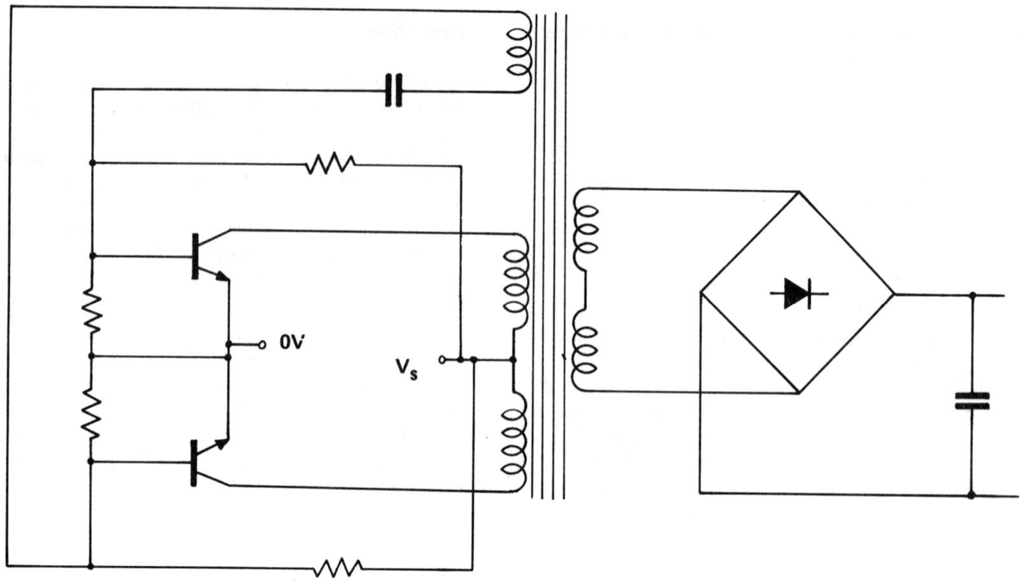

FIGURE 31 CR Timed Converter

At switch-over when the timing capacitor, C, is almost completely charged in one direction the feedback voltage changes polarity and adds to the voltage stored across the capacitor. This drives a very large initial current round the base circuit which exponentially decreases with time as the capacitor voltage falls to zero and recharges in the opposite direction. Just before the capacitor is completely charged the "on" transistor switches off because of lack of base drive and the (linear) transformers magnetising current swings the inverter over, turning on the other device.

Quite a large number of base drive circuits can be derived but they can all be simplified to the basic circuit of Figure 33.

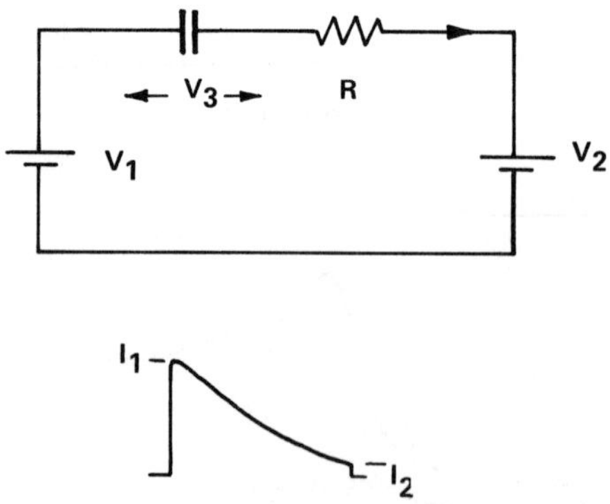

FIGURE 33

Here

V1 = Feedback winding voltage

V2 = Equivalent opposing voltage due to V_{BE}s and diode V_Fs.

V3 = Initial voltage on capacitor C

I1 = peak capacitor current

I2 = minimum capacitor current to keep device in saturation, including starting resistor 'bleeds'.

Q = I1/I2

R = Equivalent series charging resistance

The time t_x for the current to decrease from I1 to I2 is:

$$t_x = C.R.\log_\varepsilon Q$$

In terms of frequency f_o; $(=1/2t_x)$

$$f_o = 1/2.\ C.R.\log_\varepsilon Q$$

Norwicki uses Q = 5 giving

$$C = 1/3.2.f_o R$$

The value of resistor R is given by V1 and Q to give a peak current of I1.

i.e. R = 2.Q. (V1 – V2)/[I1 (Q + 1)]

= 2 (V1 – V2)/[I2 (Q + 1)]

for Q = 5

R = (V1 – V2(3I2)

Resistor R will dissipate a little less than

4.C. (V1 – V2)2. f_O . . . (Q = ∞ case)

Capacitor C will suffer peak voltages of

± (V1 – V2).(Q – 1)/(Q + 1)

or for Q = 5

±2.(V1 – V2)/3

Voltage V1 is usually chosen to be 4 to 10 times V2.

In reducing base drive circuits to the form of figure 33 the slope resistances of conducting base emitter junctions and diodes should be included.

Base Drive Circuits

Figure 31 shows the simplest form of base drive circuit. Its main disadvantage is that nearly 2.V1 can be impressed on the off transistors base emitter junction which places a limit of $BV_{EBO}/2$ on V1. This can be incompatible with the requirement V1 > 4V2. Adding an extra resistor in series with capacitor C will 'pot down' the voltage applied to the off transistor to reasonable limits.

Planar units as explained in Device section are best not reverse biassed under inductive switching conditions. Table 4.1 shows that there is very little penalty paid in t_f increase for not driving the base negative, e.g. as in Figure 34. When the base drive of transistor VT1 changes from forward to reverse, fast diode D1 shuts off leaving resistor R1 to discharge the base region. Slow diodes act like capacitors and can drag large negative base currents out of the device.

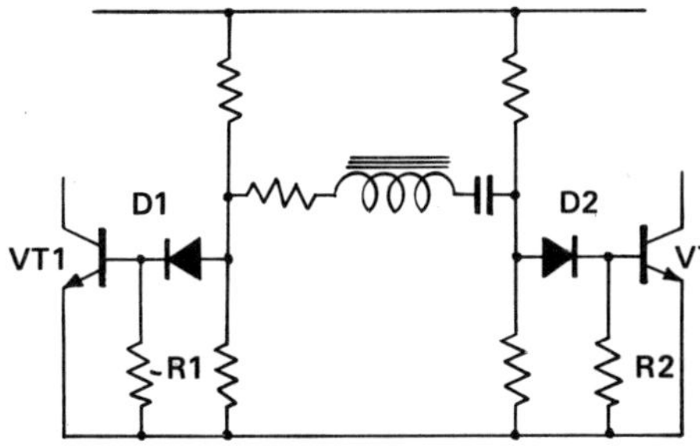

FIGURE 34

Shunt diodes, as shown in Figure 35 can be useful. Here R1 and R2, the starting resistors, can be considerably increased in value increasing the overall efficiency. Negative undershoots on the collector voltage will be clipped by the series combination of the diode and collector base junction.

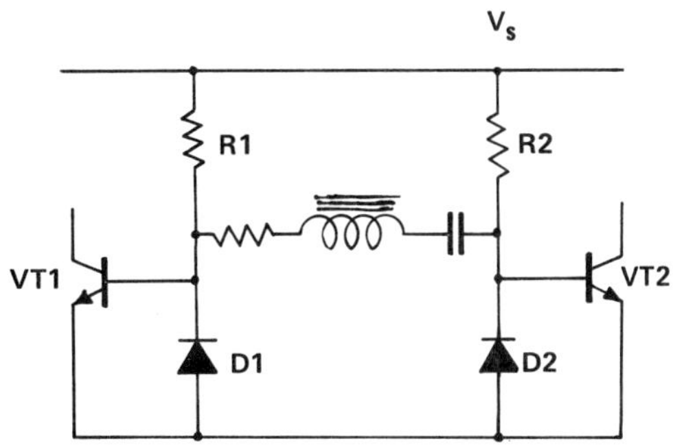

FIGURE 35

Characteristics

As the base current is relatively large just after switch over extra collector current is available to counteract any sudden drops in the load impedance as could be caused by slow rectifiers or small smoothing capacitors. This circuit is highly suitable for high frequency operation since the single linear transformer eases stability problems and manageable values of capacitor C occur.

Unfortunately, the running frequency is strongly dependent on the load (I_C) and h_{FE}. A mild form of compensation takes place for mis-matched h_{FE} units in the same inverter. The highest gain unit will allow the capacitor to charge far longer giving an increased discharge time with the low gain unit.

Decreasing the load from maximum will cause the frequency to decrease until a point is reached when the frequency stays constant for further decreases of load. Two things can stabilise the frequency at low loading. One is the transformer cone saturates with the larger number of volt seconds applied to it, so the circuit becomes a saturating core inverter. Typically the transformer will support at least three times its normal working volt second rating before saturating. Thus saturated operation will only occur below 30% of the full load rating. A second form of frequency stabilisation takes place by the starting bias· 'bleed' overcoming the capacitor current and turning the off device on. This gives a voltage rather than a current timed inverter which tends to result in high dissipation switch offs. Thus excessive starting bleeds should not be used as it can throw out the timing. For a constant resistive load the operating frequency is relatively constant with supply voltage change, making this type of inverter a suitable base drive unit for a high power driven inverter.

SINEWAVE, BRIDGE AND DRIVEN INVERTERS

Sinewave Inverters

Many different circuit configurations can be used to produce a sinewave output inverter. A class B push-pull circuit with a tuned collector to collector load is one form the inverter can take, Figure 36. As the conducting transistor is always 'making good' the difference between the supply voltage and the generated sinewave the overall inverter efficiency is rather low.

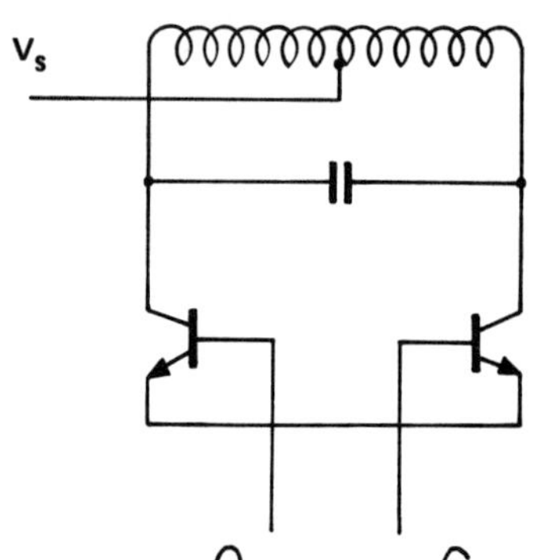

FIGURE 36

This circuit can be improved in terms of efficiency by biassing the transistors off into class 'C' operation so they only conduct at the sinewave peaks when the collector voltage is low. As the transistor is now conducting for a shorter time the peak collector current must increase quite dramatically and this is one of the drawbacks with class C operation necessitating a high current transistor. One major problem with self oscillating sinewave generators of this type is that the base drive is a sinewave giving a slow device turn on and an indefinite off-drive. With class C types special attention must be given to safe-guarding the base-emitter junction against the occurring high reverse drive voltages.

Two elegant and very efficient forms of sinewave generator called class D types have been described by Baxandall.[13]

In the voltage switching type of class D inverter, Figure 37 the collector voltage is a square wave. The series connected resonant LC circuit between the square wave voltage and the resistive load passes only the square wave fundamental frequency. Hence devices operating in this type of inverter have a square wave of collector voltage and when the device is on (saturated) the collector current is a half sinusoid.

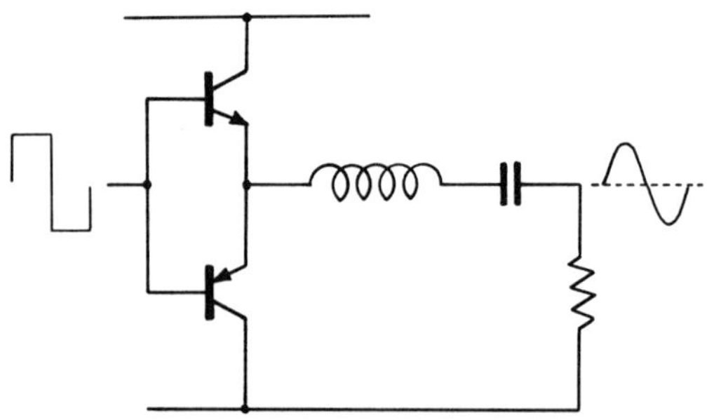

FIGURE 37

This type of inverter is particularly suitable for R.F. output stages in the H.F. range.

The current switching class D inverter, Figure 38, is the dual of the voltage switching type. Devices in the current switching inverter pass a square wave of collector current when in the on (saturated) condition and have a half sinusoid of voltage on the collector in the off condition. A constant current feed to the inverter centre tap is achieved by using inductor L. This current is alternately switched between the two transistors at the running frequency of the inverter. The parallel tuned circuit between the two collectors, resonant at the inverters operating frequency, by-passes the higher harmonics of the square current pulse leaving only the fundamental component generating a sine wave across the circuit.

Each end of the tuned circuit is alternately 'held' by the transistors as they conduct so the voltage wave at the centre tap consists of a series of half sinusoids. If resistive drops in the current feed inductor are negligible the mean voltage across it must be zero. Simple a.c. theory shows that the peak voltage from zero at the centre tap must be: $\pi V_S/2$. Hence the peak transistor collector voltage is πV_S (neglecting $V_{CE(sat)}$) and the peak to peak sinewave voltage across the whole transformer primary is $2\pi.V_S$.

This type of inverter is particularly suitable for E.H.T. generation.[14] A high turns ratio between primary and secondary is typical for this application giving relatively poor coupling a high secondary winding capacitance. Rather than trying to overcome the secondary's natural resonance, the inverter is allowed to operate using the secondary as its tuned circuit. It is often necessary to introduce an air gap into the core to bring the inverters operation above the audible range.

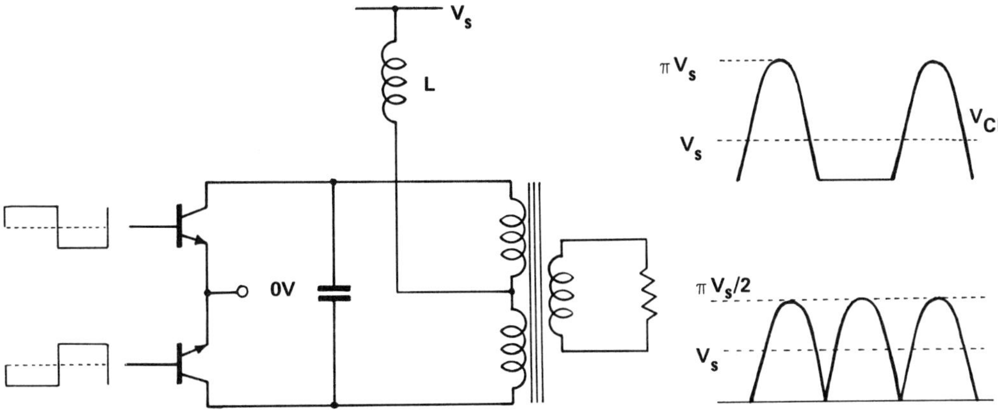

FIGURE 38

Centre Tap Voltage

A particularly interesting form of the current switching inverter has been described by Ridgers.[15] By pulse-width modulating the inverter base drive it is possible to control the amount of energy fed to the tuned circuit which in turn controls the a.c. voltage. Inverters using this principle can nullify the effects of changing loads and input voltages.

50Hz Inverters

Mobile use of normally mains operated equipment presents several compatibility problems.[16] True sinewave inverters for 50Hz are not really practical since the tuned circuit elements assume such very large values. If the equipment contains rectifiers which take large current pulses at the peak of the sinewave, the tuned circuit Q requirement becomes very stringent to maintain a reasonable sinewave during this period. The use of a square wave voltage to drive the equipment gives long rectifier conduction times and greatly reduces the peak currents. This solution, while easing the inverter loading, creates additional problems. One of these is deciding on the square wave amplitude. If it is not near the rms sine wave value, lamps and valve heaters will not be operated under the correct conditions. Unfortunately, making the square wave rms compatible with the mains usually results in low d.c. outputs from the rectifier circuits. A solution would be to employ high rectifier surge limiting resistors (which must dissipate large amounts of power under sine wave operation) to make the sine and squarewave d.c. outputs nearly equal. Alternatively the inverter complexity could be increased by programming it (or several inverters) from logic circuits to synthesize and approximate sinewave. Light filtering of the stepped wave form results in a low distortion sinewave. Some form of mild filtering is always desirable on 50Hz square wave inverters as the sharp wave transients tend to break through into the equipment circuits.[17]

Finally, if the inverter is to drive a changing mechanical or electrical load its operation from the initial switch-on to normal running should be checked.[18] It is possible that an inverter can 'latch up' at an extremely high frequency at switch on. This does not cause any change in the load and hence the inverter does not return to its designed frequency. One method of overcoming 'latch up' is to use a driven inverter.

Bridge Inverters

In a bridge inverter the load is connected between the centre points of two bridge arms across the input supply. Figure 39 shows a two transformer (saturating driver) version. Driver transformer T2 has its base windings phased so that diagonally opposite transistors conduct together. Figure 40 shows how the conducting pairs of transistors alternately reverse the supply voltage across the load. The maximum load voltage will be the supply voltage minus the $V_{CE(sat)}$s of diagonally opposite transistors.

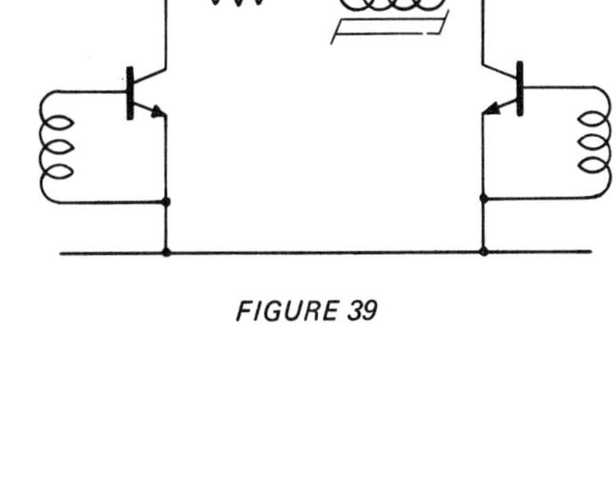

FIGURE 39

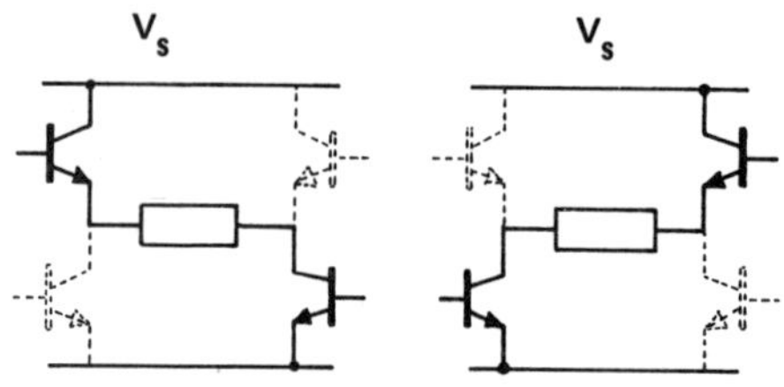

FIGURE 40

This arrangement is useful in cases where the input supply voltage is very high and devices with a twice supply voltage rating for a normal push-pull inverter cannot be obtained. As there is a distinct possibility that devices in a bridge inverter will be forward biassed with the full supply voltage across them (at the device turn-on) the voltage rating BV_{CEO} must be greater than V_s. With current devices simple bridge inverters can give about 3kW using a BUY23A and 300V supply. Using four power transistors in one inverter is costly and disastrous if one should fail. A two transistor variant of the bridge circuit is shown in Figure 41. Here one active arm has been replaced with a passive one.

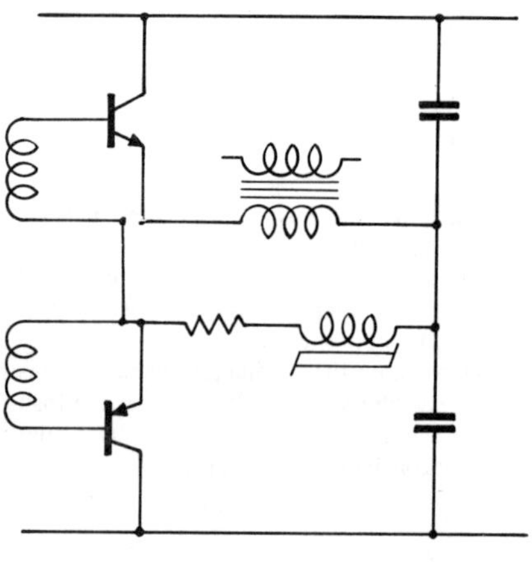

FIGURE 41

Ideally the capacitive arm centre point should be at $V_s/2$. However, component tolerances usually cause the centre point voltage to drift. By using diodes and additional windings on the output transformer the centre point voltage can be stabilised. In this form of the inverter the load will only experience a voltage of $\pm V_s/2$ but the transistors can easily have the full supply voltage across collector to emitter.

Collector current pulses for bridge inverter transistors can often have twin spikes on them. Figure 42. As the transistor turns on the first spike is stored charge extraction from the off going transistor. At switch-off the second spike is the device's stored charge being pulled out by the newly conducting transistor.

FIGURE 42

To achieve a better torque inverter operated motors can be shunted by a capacitor resulting in a collector current such as shown in Figure 43.

FIGURE 43

Driven Inverters

A simple driven inverter really consists of two inverters in one, Figure 44. The first relatively low power inverter concentrates on producing a good base drive for the output pair and getting the timing right. Freed of any influences on the timing, the output stage becomes relatively easy to design. Switching dissipations at high frequencies can be very high in this type of inverter since as one transistor is being driven off the other is being driven on. A twin spike collector current can occur as mentioned in the bridge inverter section.

Ideally the two base drives should be 'skewed' by the respective transistor storage times so that the one transistor is not driven on until after the storage time of the other transistor.

By stabilising the voltage supply to the low power driver inverter, good frequency stability against supply voltage variations can be achieved. Further increases in frequency stability may be obtained by triggering or counting down from an external reference, such as a crystal oscillator.

Compared with the normal two transformer inverter the starting is better since the base drive is independent of loading. The overall efficiency increases as neither the starting 'bleed' for the output transistors of the series diode in the base circuit are necessary. The most dangerous thing that can happen to this type of inverter is a short circuit; As fuses are often too slow to save the output devices, a fast acting current trip to shut off the driver inverter is preferred.

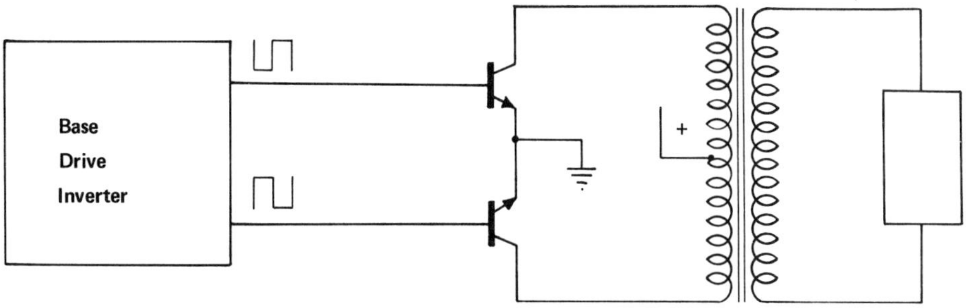

FIGURE 44 Driver Inverter

TRANSFORMER DESIGN

This section is intended to broadly outline design techniques and to indicate the most important parameters for the different types of transformer. Linear and saturating transformers are considered at both high and low frequencies.

Physical Restraints

All of the available space on a core cannot be usefully used in making a transformer. Concessions have to be made for insulation and mechanical stability.[19]

Cores

Laminated cores by virtue of the interlamination insulation have a smaller magnetic area (A) than their physical cross sectional area (A_{cs}). The ratio of these two area is called the stacking factor S.

$$S = \frac{\text{magnetic cross sectional area}}{\text{physical '' '' ''}} = \frac{A}{A_{cs}}$$

Typically S will be about 0.9 but it can drop as low as 0.5 for very thin laminations such as 0.0005 inch HCR. Even a very small air gap in the cores magnetic path can drastically alter the overall magnetic properties. This problem can be attacked three ways:

(a) A single piece part core can be used; but this restricts the core material to one which can be moulded, i.e. ferrites.

(b) Two piece part core with carefully matched mating faces can reduce air gap effects to acceptable limits, i.e. 'C' cores, pot cores.

(c) The laminations can be chosen or assembled in such a manner as to reduce the air gap as much as possible. Toroids can be tape wound and E and 1 laminations can be alternated as in Figure 45

FIGURE 45 Air Gap Reduction by Alternating E and I Combinations

Windings

Two types of winding space will be considered, that of an EI core Figure 46 and a toroid figure 47. Not all of the E cores window area A_W can be used for winding. Coil formers, insulation and lead-out wires all reduce the available winding space. A factor K_e is defined which is the ratio of total wire area A_n to window area A_W.

$$K_e = A_n/A_W$$

In most cases K_e will lie between 0.6 and 0.9.

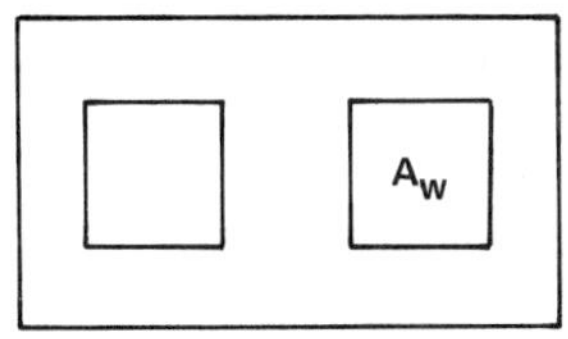

FIGURE 46

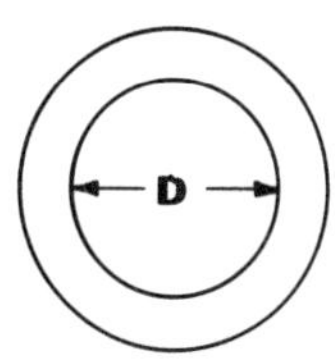

FIGURE 47

Although a toroid does not need a former a certain amount of winding space is lost in core protection and the extreme difficult in winding the toroid right to the centre. The space factor K_t this time is:

$$K_t = A_n/D^2$$

where D = inside diameter of toroid.

In most cases K_t will lie between 0.1 and 0.4

Calculations involving the wire area A_n assume the wires lay as in Figure 48 and not as in Figure 49. The total area

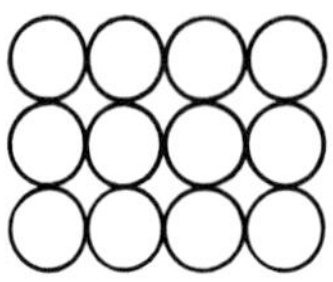

FIGURE 48

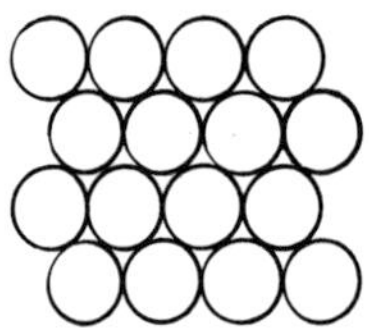

FIGURE 49

occupied by N_1 turns of d_1 mm diameter wire laying as Figure 48 would be:

$$N_1 . d_1^2 \qquad mm^2$$

The factor d_1^2 is often given in wire tables. In non-metric tables d^2 is the wire diameter in thousanths of an inch squared and given in units of circular mils (cmil) to distinguish this area from the wires true cross sectional area.

It is usual to divide up the total wire area for the windings in proportion to the powers which they handle. A 20W transformer with 15W and 5W secondaries will have its wire area split as 20 : 15 : 5 or 0.5 for the primary, 0.375 for the 15W secondary and 0.125 for the 5W secondary. Wire current densities of 1.5A/mm^2 are typical, 2.2A/mm^2 being used for 50% duty cycle windings. Transformers with mixed 50% and 100% duty cycle windings have a slightly different method of winding area calculation as shown later.

Core Size

Using all the above information a rough estimate of the minimum core size can be made.

Starting with equation 3.7

$$V = 4.B.A.N.\ f_o$$

substituting for A in terms of A_{cs} mm^2

$$V = 4.B.S.A._{cs}N.f_o.10^{-6} \qquad 10.1$$

or

$$A_{cs} = 10^{+6}.V/(4B.S.N.f_o)\ mm^2 \qquad 10.2$$

If the transformer is handling power P_w the primary current I_p will be:

$$I_p = P/V \qquad A \qquad 10.3$$

Assuming the primary and secondary occupy equal areas and both are 100% duty cycle the required window area, A_w mm^2, is

$$A_w = 2.NI_p./(1.5K) \qquad mm^2 \qquad 10.4$$

Substituting from equation 10.3 for I_p

$$A_w = 4.N.P./(3.K.V.) \qquad mm^2 \qquad 10.5$$

combining 10.2 and 10.3 gives

$$A_{cs}.A_w = P.10^{+6}/(3.B.K.S.f_o) \quad mm^4 \quad 10.6$$

For transformers with mixed windings a better value of A_w can be obtained from:

$$A_w = [N/(K.V)]\ \Sigma\ (P_1/1.5 + P_2/1.5 + P_a/2.2 + P_b/2.2) \qquad mm^2 \qquad 10.7$$

where P_1, P_2 etc are the powers handled by the normal windings.

P_a, P_b, etc. are the powers handled by the 50% duty cycle windings

combinging 10.2 and 10.7 gives:

$$A_{cs}\ A_w = \Sigma\ (P_1/1.5 + P_2/1.5\ ...\ P_a/2.2 + P_b/2.2).10^6 /(4.B.K.S.f_o) \qquad mm^4 \qquad 10.8$$

It is desirable that the core to be used should have a A_{cs}. A_w product slightly larger than that of equation 10.8. In the case of a saturating transformer the flux density used will be B_{sat}.

Transformer Core Losses

Hysteresis and eddy current losses are the two major losses to be taken account of in power transformer design.[20,21]

Hysteresis loss represents the energy used to magnetise and demagnetise the core material as its hysteresis loop is traversed. Each traversal of the hysteresis loop causes a fixed amount of energy to be lost. Hence the average power lost will be proportional to the frequency at which the loop is cycled. A generalised equation for this loss is

$$P_h = C_h.\ f_o.\ \hat{B}^n \qquad W/m^3$$

where f_o = frequency of operation

$\hat{B}$ = peak flux density

n = Steinmetz exponent, normally $1 < n < 3$

C_h = constant

Eddy currents are caused in a conducting magnetic material by the varying magnetic field. These currents absorb extra power giving a loss, P_e of

$$P_e = C_e . \hat{B}^2 . f_o^2 \qquad W/m^3$$

where

C_e = constant

The total normalised core loss, P_t, is thus

$$P_t = P_h + P_e \qquad W/m^3$$
$$= f_o . B. (C_h . B^{n-1} + C_e f_o . B) \qquad W/m^3$$

Figure 50 shows the expected variation of the total normalised core loss against frequency for a constant peak flux density. Obviously core loss increases rapidly with frequency when the eddy current loss term becomes significant. The eddy current loss at a given frequency may be reduced by decreasing the lamination thickness or in the case of ferrites choosing a higher resistivity grade material.

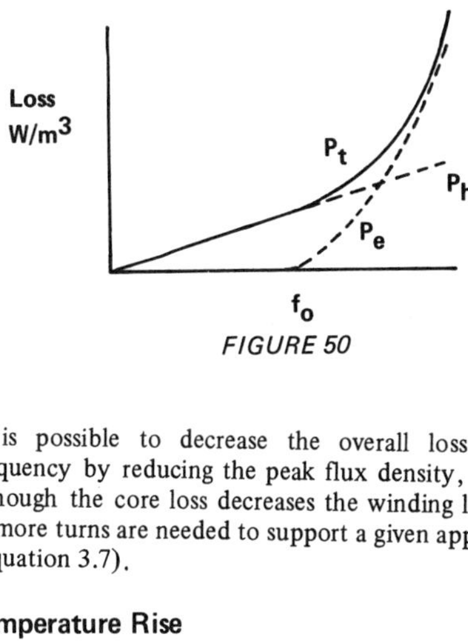

FIGURE 50

It is possible to decrease the overall loss at a given frequency by reducing the peak flux density, B. However, although the core loss decreases the winding loss increases as more turns are needed to support a given applied voltage. (Equation 3.7).

Temperature Rise

Most of the transformers losses will appear as heat. This heat will be lost to the surroundings by radiation, convection and conduction. The references given show how these losses may be calculated. A widely used rule of thumb is to allow $0.2W/in^2$ of surface area ($300\mu W/mm^2$) which results in a transformer surface temperature rise of 30 – 40°C above ambient.

Loss Distribution

Minimum transformer total loss usually occurs when the winding loss and the core loss are equal. With linear transformers at low frequencies this condition cannot be obtained as the peak flux density would have to approach B_{sat} to make the losses equal. Additionally low frequency transformers use large numbers of turns which can cause output voltage regulation problems. So a design might be optimised for voltage regulation rather than efficiency.

At high frequencies the two losses can be equalised. Regulation requirements and the high frequency skin effect in the conductors can make it undesirable to equalise the losses. Careful arrangement of the windings can improve the regulation.

High Frequency Transformer Core Size

Choosing a transformer core size can be difficult unless based on past experience. Similar circuits at the same frequency would have nearly equal power/weight ratios provided the extrapolation was not too large. Frequency scaling is more difficult. Ferrites in the 1 to 20kHz region tend to have a constant power/frequency ratio. So the same core used at twice its original frequency would give twice as much power output.

A very elegant method of core selection has been given by E.C. Snelling.[18] Some designers will be forced to choose from a preferred range of cores. In this case, a fairly rapid selection procedure would be as follows:–

(a) Calculate the cores effective surface area for heat loss (mm^2)

(b) Multiply this area by $300\mu W/mm^2$ to arrive at the cores dissipation (use half the area for pot cores). If this power is more than 10% of the total power to be handled the core is too large.

(c) Divide this power by the cores volume to find the core loss in $\mu W/mm^3$.

(d) Reference tc the core materials curves will show the flux density corresponding to this loss at the inverters frequency of operation. Saturating transformers will require a minimum flux density of B_{sat}.

(e) Knowing B the minimum $A_{cs}.A_W$ product required can be calculated and compared with the cores actual value.

Cores passing (b) and (e) requirements can then be examined in greater detail such as efficiency, regulation, and size to arrive at a final choice.

POWER DIODES

Diodes are used to convert the inverter produced a.c. to d.c. Rectification losses must be small to keep the efficiency high.

Diode Power Losses

Forward Loss

Assuming the diode characteristic can be approximated as Figure 51 the forward loss, P_F, for 50% duty cycle is:

$$P_F = (V_{FR} \cdot \bar{I}_M + I^2R)/2$$

where I_M = mean forward current during conduction

$\bar{I}$ = rms forward current during conduction and

R = the diode slope resistance

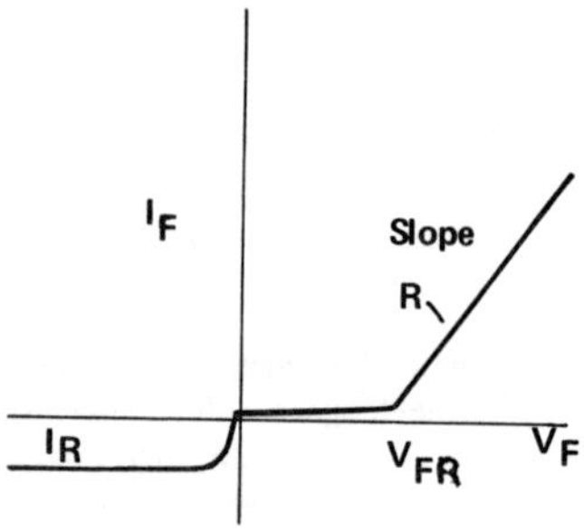

FIGURE 51

V_F will decrease at about 2.5mV/°C but at high currents the semiconductor bulk resistance can predominate giving an overall positive temperature coefficient for P_F.

Reverse (Blocking) Loss

For square wave operation the reverse power loss, P_R, will be:

$$P_R = I_R \cdot V_R/2$$

where V_R = reverse voltage applied to the diode. This loss will approximately double every 10°C and should be taken into account at high temperatures.

Transient Losses

Transient losses occur as the diode is switching from one state to another and can be the diodes major power loss.

Forward Transient Loss

When the diode is quickly driven from reverse bias to forward conduction the instantaneous forward voltage is higher than the steady state value as shown in Figure 52. Usually this loss is negligible compared to the reverse transient loss.[22]

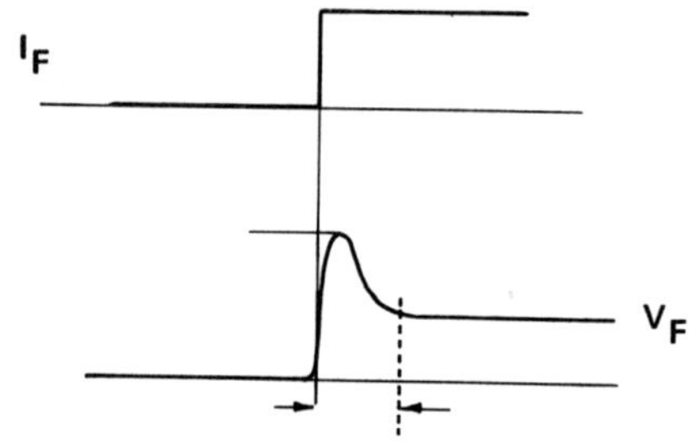

FIGURE 52

Reverse Transient Loss

The waveforms associated with switching a conducting rectifier off are given in Figure 53. First its stored charge must be removed and secondly its junction capacitance charged to voltage V_R.

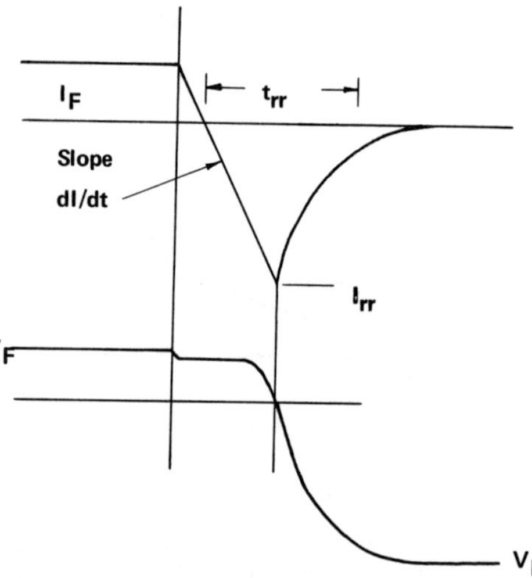

FIGURE 53

The principle power loss occurs as the diode current is falling to zero and the reverse voltage is building up.

An approximate value for this power loss, P_{RT}, in fast rectifier diodes is

$$P_{RT} = V_R \cdot I_{rr} \cdot t_{rr} \cdot f/4$$

where f is the square wave frequency.

I_{rr} will rapidly increase with increasing dI/dt as the diode stored charged has less time to recombine. Doubling V_R or f will cause double the power loss. This type of loss is reasonably constant with temperature.

Rectifier Circuits

Bridge

A bridge circuit (Figure 54) makes maximum use of the transformer winding space (unlike the biphase circuit) since

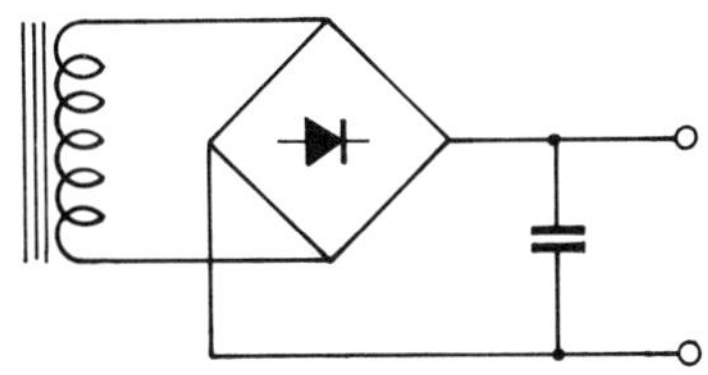

FIGURE 54

the single winding is continuously supplying the output current. Compared with the bi-phase circuit only one half the voltage rating is required for the diodes. Slow diodes can cause problems by 'locking up' charge in the slowest device which is suddenly thrown back into the inverter as the output voltage goes through zero. Mistiming at certain output currents can occur as a result of this switchover being the most critical phase of an inverters operation. 50Hz (sinewave) bridges sometimes cause this mistiming problem at 400Hz (square wave) operation.

Bi-phase

Only two rectifiers are required for this circuit, Figure 55, which can save costs if the rectifiers are expensive. Below about 20V output this configuration has lower rectification losses than the bridge since only one diode drop occurs between the input and output. Also, if the bridge winding resistance is r, the regulation resistance is r + 2R. For the biphase, however, it will be 2r + R which is usually 40% less.

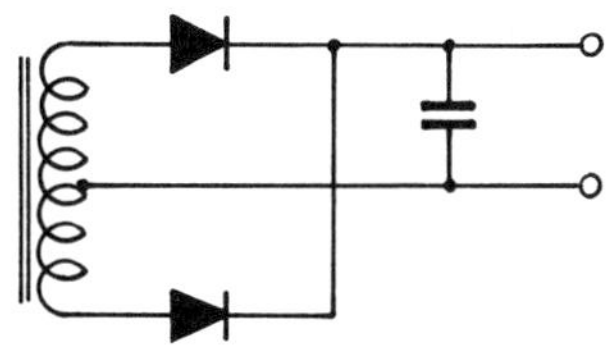

FIGURE 55

Voltage Multiplier

The ideal n stage voltage multiplier, Figure 56, driven by $\pm V_0$ would produce an output voltage from B2 of $+2V_0.n$ volts. Reversing the output chain, i.e. connecting A2 to A and B2 to B will give $-2V_0.n$ at B1, a useful feature for dual polarity supplies. Each capacitor and diode needs to be rated to at least $2V_0$ and the optimum output impedance for a fixed total capacitance is given by:

$$C1a = C1b = 2.C2a = 2C2b = \ldots\ldots 2^{n-1}.\ Cnb$$

Often it is mechanically easier to make all the capacitors equal. High frequency operation and high speed rectifiers will produce a very small efficient unit. Output impedance and capacitor value formulae have been given by M.C. Jackson.[23] Multipliers are mostly used to generate voltages above 1kV, a region where transformer output waveform purity and secondary insulation is becoming a problem.[24]

Smoothing Capacitor

Not only must the smoothing capacitor supply the load current during the inverter switch-over period, but it must supply the diodes transient loss currents. A charge of $I_{rr}.t_{rr}/2$ will be lost by a diode. This can be added to the load current multiplied by the transition time giving the total charge lost. Dividing the latter by the required ripple voltage gives a starting value for the smoothing capacitor. In the 100V to 400V range a 1μF is often sufficient while below 100V 25μF is a typical value. It is most important to check that the ripple current ratings of the capacitors are not exceeded.

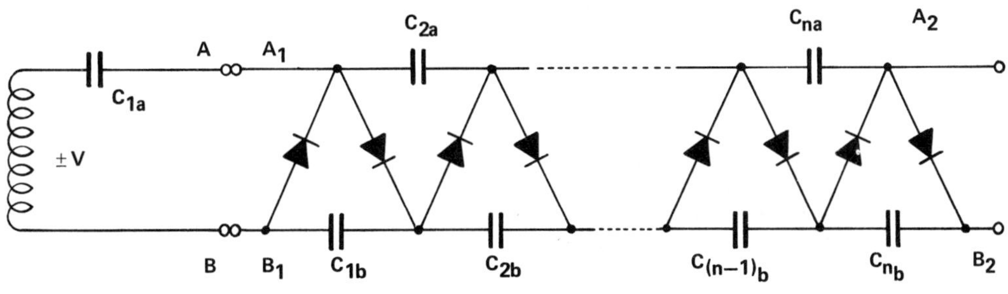

FIGURE 56 Voltage Multiplier

SPIKES AND SWITCHING LOSSES

Positive Spike

Positive overshoots on the collector waveforms of push-pull square wave inverters, Figure 57, can cause transistor over-voltage and possible failure. Device selection should have allowed some 20% extra on voltage rating to cover this situation. If however the prototype inverter shows spikes very much in excess of this, the designer has three alternatives (a) Change the device specification i.e. to increase the device voltage rating to cover peak voltage imposed by the overshoot or to choose a slower unit (b) Improve the transformer winding, (c) Employ some form of network to damp or clip the spike.

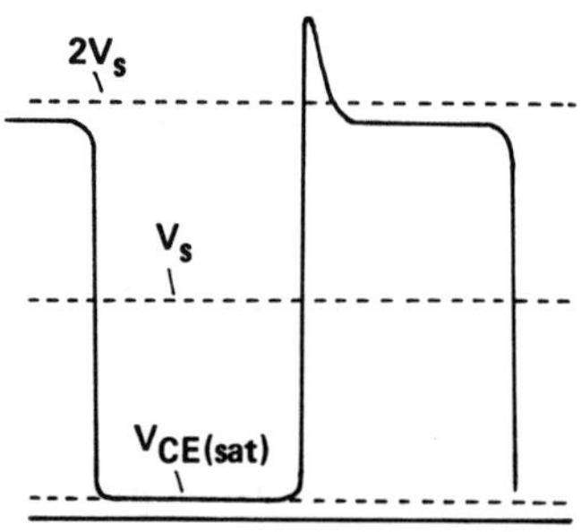

FIGURE 57

When prototypes are tested it is advisable to increase the power supply voltage to the circuit slowly, and also connect power zeners, with zener voltage slightly below the devices maximum rated voltage, between the collector emitter terminals. This action is useful for checking phasing and to show if spikes are likely to be troublesome. In most cases no-load conditions on the converter give the largest spikes. These spikes are caused by the transformers leakage inductance circuit capacitance and resistance interacting with the falling collector current. Obviously the faster the device turns off (t_f smaller) the larger dI_C/dt will become and the greater the spike voltage. Slow devices will certainly give no spike problems but switching losses are higher and the efficiency usually suffers. Bifilar winding of the primary is certainly recommended to reduce the leakage inductance to a minimum.

Figure 58 shows two examples of damping circuits. Collector-collector damping is the most polular form. However, the independent damping circuit may be preferred for cases where the transformer saturates and the leakage inductance is then comparable to the saturated magnetising inductance. Sometimes it may be found that the spike is only present on one collector – then one R.C. network to earth is adequate.

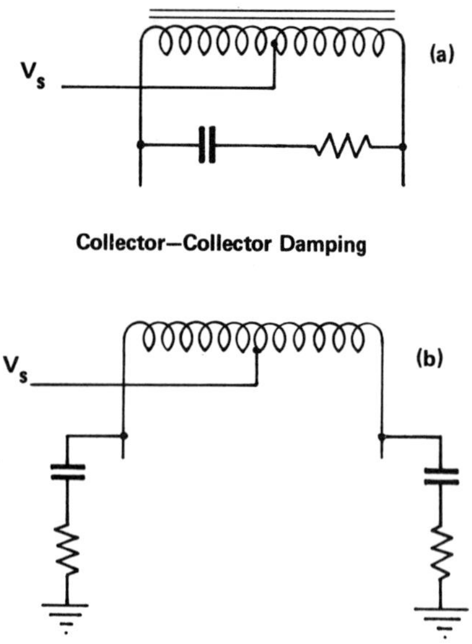

FIGURE 58

Based on the argument that the inverter load is effectively isolated from the collectors by the leakage inductance (and by the rectifier system for a converter); under switch over conditions, R1 (Figure 58 (a)) is made equal to the collector to the collector load, i.e.

$$R1 = 4.V_s^2/P_o$$

Capacitor C1 is then chosen to give a C1.R1 time constant equal to the spikes.

Another approach used when efficiency is a prime consideration is to decide how much power can be lost in damping. To charge a capacitor from zero to a voltage V via a resistor extracts CV^2 of energy from the charging source 50% of which is lost in the resistor. Since C1 in figure 58(a) is charged from $+2V_s$ to $-2V_s$ and back to $+2V_s$ each inverter cycle the damping power lost, P_d, is:

$$P_d = C.(4V_s)^2.2.f_o$$

$$= 32.C.f_o.V_s^2 \qquad (W/kHz.V^2)$$

Further, if k% of the power out (P_o) is allotted to damping, the value of C becomes:

$$C = 3.k.P_o/10.f_o.V_s^2 \quad \ldots\ldots\ldots \mu F$$

So if 3% loss is allowed for a 16V, 25W, 5kHz inverter, C1 has a maximum value of 22nF. R1 will have to dissipate $0{\cdot}5 . P_o.k/100$ = 0.375W so a ½W 39Ω resistor will be suitable. Values of R1 and C1 suggested here are initial ones only, practical work on the inverter will optimise these.

Figure 59 shows how C.R. damping modifies the transistor switch-off locus to a safer region of the transistor characteristics.

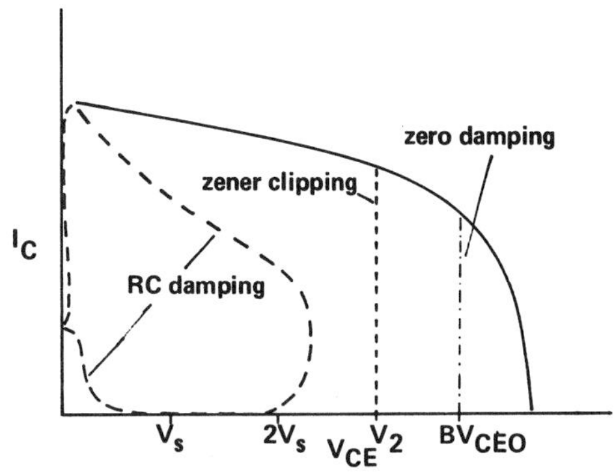

FIGURE 59

CR damping tends to degrade the device turn on loss and can give quite a high initial spike to the collector current. Some designers equate R1 to zero and just use a capacitor between the collectors. High device turn-on currents and losses can result from this and starting is made more critical as the loop gain, instead of being modified with a step network is rapidly reduced at higher frequencies. One useful thing the single capacitor circuit does, is to provide some extra decoupling against supply line transients.

The actual increase in supply power by using a CR network is not as great as calculated above since some of the energy used to charge the capacitor is already present in the circuit. This is in the form of $L_L . I_C^2/2$ the energy contained in the leakage inductance, L_L, which causes the spikes.

Zener diode clipping, Figure 60, is the most efficient way of reducing spikes. With the peak primary voltage now limited to V_Z the leakage inductance will have $V_Z - 2V_s$ across it. Hence the zener will be conducting for:

$$L_L . I_C/(V_Z - 2V_s)$$

seconds carrying a sawtooth of current. Each zener will dissipate a mean power P_Z, of:

$$P_Z = L_L . I_C^2 . f_o \cdot V_Z/2.(V_Z - 2V_s)$$

A plot of $V_Z/(V_Z - 2V_s)$ against V_Z/V_s shows it desirable to choose V_Z as close to the device BV_{CEO} as tolerancing will allow as this reduces the zener dissipation. It is assumed that the spikes are much shorter than the zener thermal time constant. The zener in the circuit shown in Figure 60(b) requires twice the power dissipation of those as in Figure 60(a). Also two fast diodes are required for the single zener case which has the further disadvantage that negative undershoots are not clipped. Where the transistors are mounted some distance from the transformer and earth planes (extra leakage inductance), the zeners should be mounted close to the transistors preferably directly between the collector and emitter terminals. Care should be taken to ensure that the inverter supply voltage cannot approach or exceed $V_Z/2$ otherwise the clipping zeners will over dissipate and become short circuit.

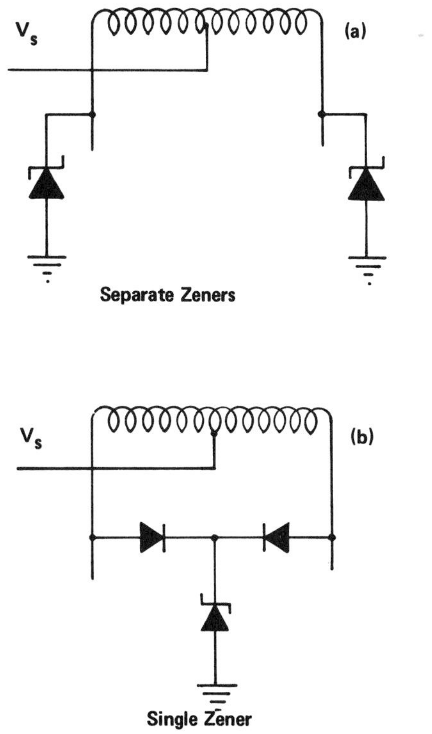

FIGURE 60

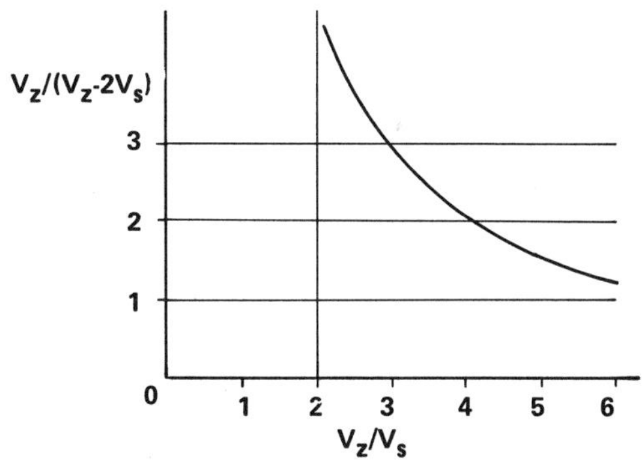

FIGURE 61

Even if the transistor on the primary side of a converter is not being 'over voltaged' by the spikes it is often necessary to reduce those occurring at the secondary. Figure 62 shows a typical converter regulation characteristic.

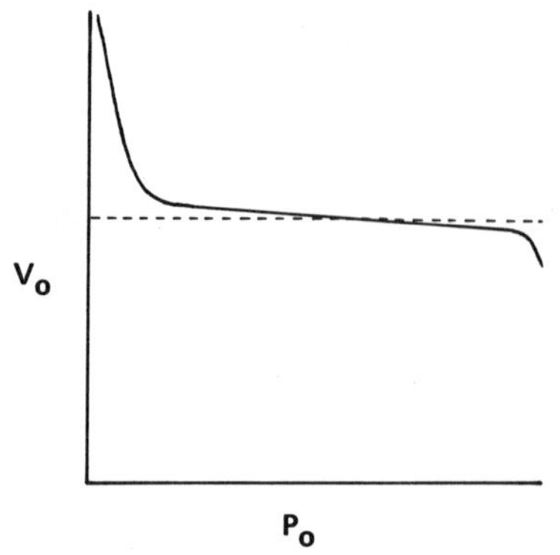

FIGURE 62

At low power outputs, spike rectification causes a rapid increase in V_O and may increase the required P.I.V. ratings of the rectifiers by 100%. Apart from this, the rapid increase in voltage could damage the circuits which the converter supplies. One solution is to pre-load the converter by a bleed resistor which reduces the full load output and efficiency. Another would be to use a power zener rather than a resistor to maintain the full load output and efficiency. C.R. damping networks could also be used to reduce the spikes on the secondary winding.

Undershoots on the collector voltage, as shown in Figure 63, are quite common for unloaded converters and inductively loaded inverters. Predominantly inductive loads

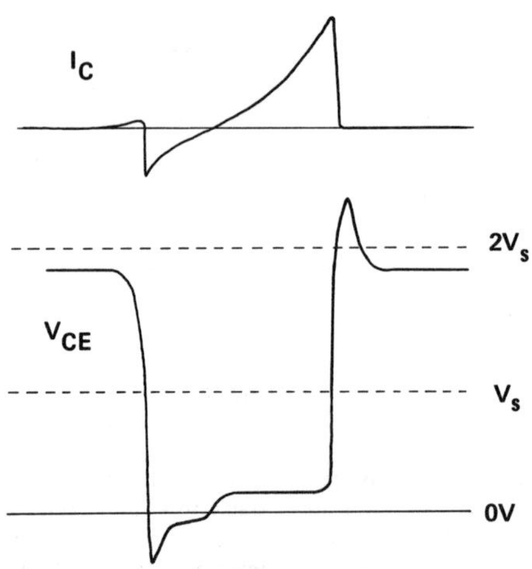

FIGURE 63

cause the load current to lag a little behind the load voltage. Hence when the transistor (NPN) is first turned on it finds the load trying to pull current out of the collector.

To accept this negative current the collector-base junction becomes forward biassed resulting in a negative collector base voltage. If the base circuit impedance to earth is very small, all the negative collector current will come from the base drive circuit, into the base and out of the collector. The collector voltage will be clamped at or below earth depending on the V_F of the collector-base diode and the exact value of the base drive impedance. Higher base circuit impedances cause a different form of clamping to take place. Shown in Figure 64 are the typical curves obtained by operating a silicon power transistor in the inverse mode, i.e. with the collector base diode forward biassed and the emitter base diode reverse biassed. Two things are

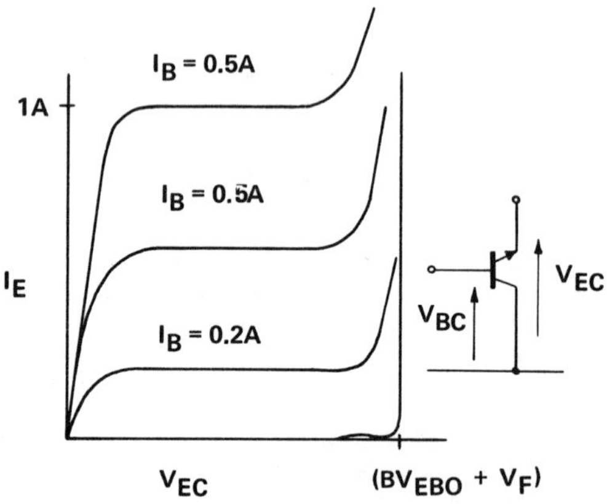

FIGURE 64

immediately noticeable about these characteristics, (a) the inverse current gain (the ratio of emitter to base current) is low typically 5 to 50% of the normal forward gain and (b) when the breakdown voltage BV_{EBO} is reached across the base emitter junction it avalanches, the current flowing out of the transistor through the forward biassed base collector junction. Hence a transistor of 10V BV_{EBO} would clamp a negative current of the collector to about -11V or less if the base impedance was high. Planar types will often show 'switch back' on their breakdown characteristic.

Transistor Limits

Provided the base and emitter bonding wires are of the same size there is no reason why the peak negative base current under inverter operation should not be the same as the peak collector current rating, since the base region has to pass I_C under normal operation. However, operating under inverted emitter collector breakdown conditions is not recommended for two reasons: (a) appreciable amounts of power will be dissipated in the very thin base emitter region (b) the low current gain (important for starting) can be degraded particularly with epitaxial planar units.

Transistor Losses

Generally the transistor switching-on loss can be neglected compared with the switching-off loss. Figures 65 and 66 show two forms the switch off can take. Switch-off power losses quoted are for one device only and it is assumed the maximum collector voltage is $2V_S$.

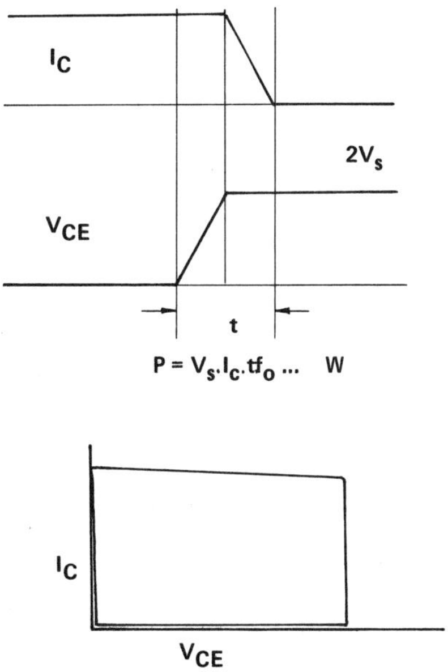

FIGURE 65 Square Loop Switching

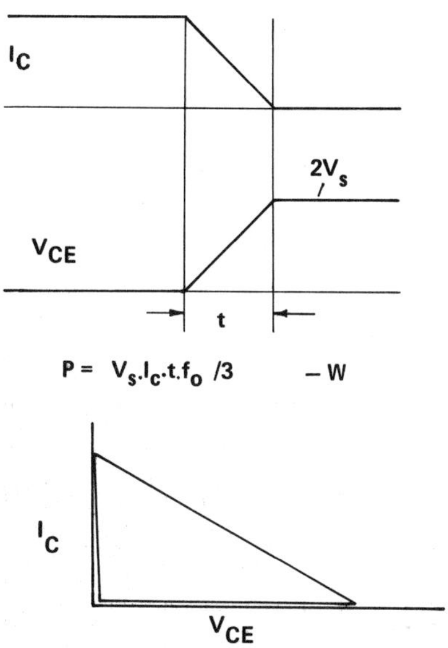

FIGURE 66 Triangular Switching

STARTING CIRCUITS

Operation

A self oscillating inverter as distinct from a driven one, must have some additional components to induce oscillation on the application of the supply voltage. For the inverter to start one device must be biassed above a certain critical value of I_C. This value of I_C is found by equating the inverters loop gain (written in terms. of the device transconductance being $40I_E$ mS) to 1. Converters require higher starting currents because of the reflected smoothing capacitance.

If the output is likely to be short-circuited, which will stop the inverter, it is as well to check that thermal runaway caused by the standing current induced by the starting circuit, does not occur. Inverters using modern high f_T silicon devices start more easily than the older germanium types as the loop gain is maintained to a higher frequency.

Configurations

Figure 67 shows the inverter base drive winding. The centre tap is taken to the supply via a resistor R1 which defines the transistor base current. The initial base current at switch on is almost as large as the normal operating base current. One obvious drawback is that R1 dissipates a large amount of power as it carries the full base current. Another not so obvious drawback is that when the core saturates and the winding voltage falls to zero both transistors switch on. Adding a small capacitor from the centre tap to earth overcomes this problem by holding the centre tap voltage below earth for a short time, thus giving the off going transistor chance to switch.

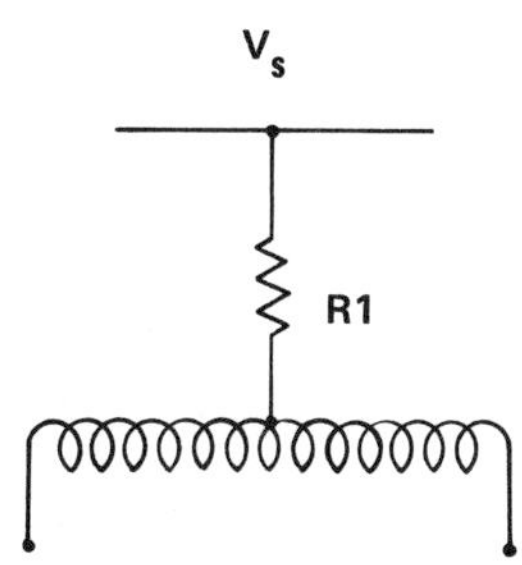

FIGURE 67

A more efficient circuit is shown in Figure 68. Resistor R1 and R2 form a potential divider across the supply. Low V_S, temperature and the maximum load form the worst case starting condition. R2 must carry the maximum base current so its value should not be too high or the BV_{EBO} rating of the off device could be exceeded, Capacitors and diodes are often added across R2 to ease this problem.

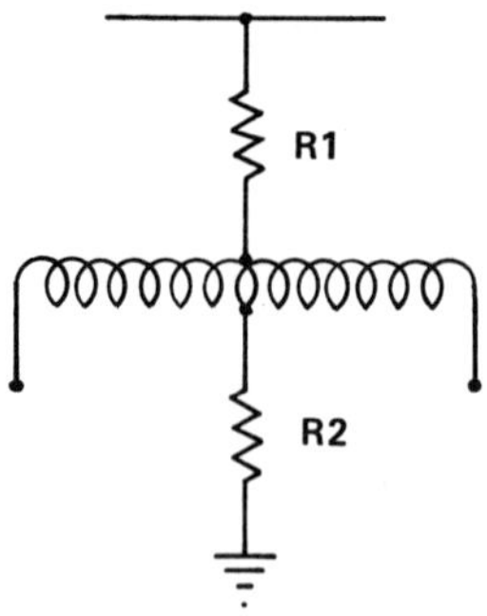

FIGURE 68

Figure 69(a) shows perhaps the most common and efficient starting circuit. All the current in R1 must flow into the transistor bases giving a good starting action. When the inverter is running, the centre tap is held at 0.7V negative by the diode conducting the base current of the on transistor. Figure 69 (b) shows a variant on this, saving one power resistor, but the base winding should not provide more than $\frac{1}{2}BV_{EBO} + V_{BE}$. The diode need not be fast and, for reasons given previously, it is desirable to add some capacitance across the diode.

Even the above circuit may consume large amounts of power when used on high power inverters, Figure 70 shows another approach to the problem.[25] When the supply is switched on, the UJT oscillator starts to work giving current pulses into the tertiary of the driver transformer of the two transformer inverter. As a result of these periodic 'kicks' the inverter starts to switch. Obviously it is undesirable to leave the UJT running, so components D1, D2, D3, R4 and C2 are included to shut it off once the inverter is switching.

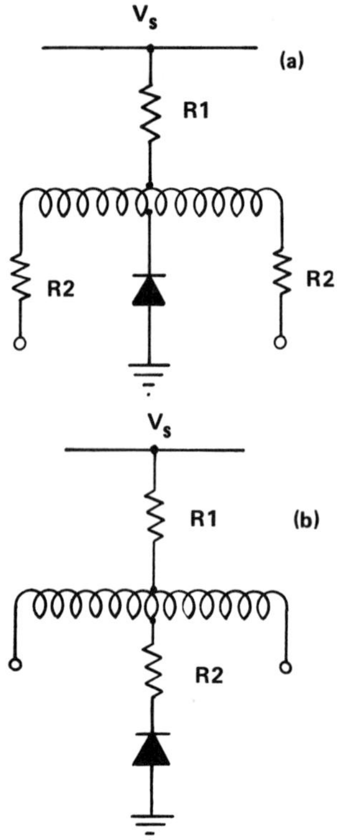

FIGURE 69

Initially the UJT runs from about half the supply voltage (potted down by R1 and R2). As the inverter starts up D1 and D2 charge up C2 to nearly the supply voltage. D3 becomes reverse biassed and the peak point voltage of the UJT is then above half the supply voltage and it stops firing.

FIGURE 70

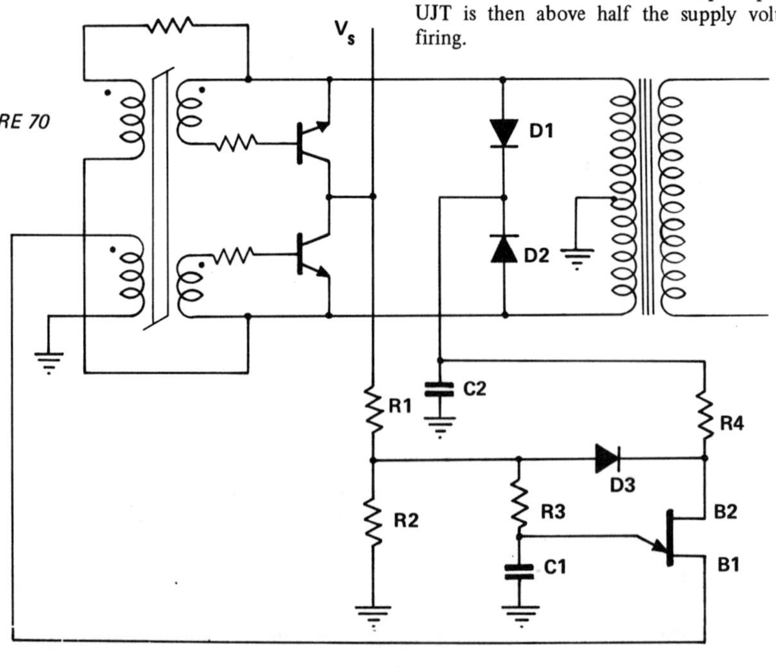

(a)

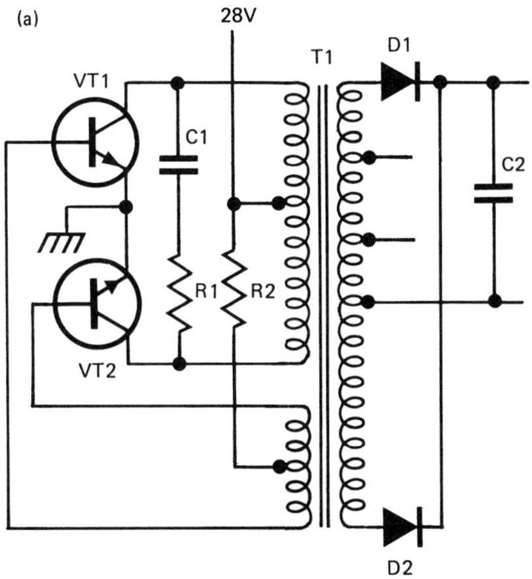

R1	–	22Ω	0.5W
R2	–	680Ω	2W
C1	–	10nF	100V
C2	–	10μF	25V
D1, D2	–	1SX170	
VT1, VT2		BFY50	

T1 – MT351* (12-0-12V secondary)

*Gardners Transformers Ltd.

Power lost per transistor = 570mW

(b) (c)

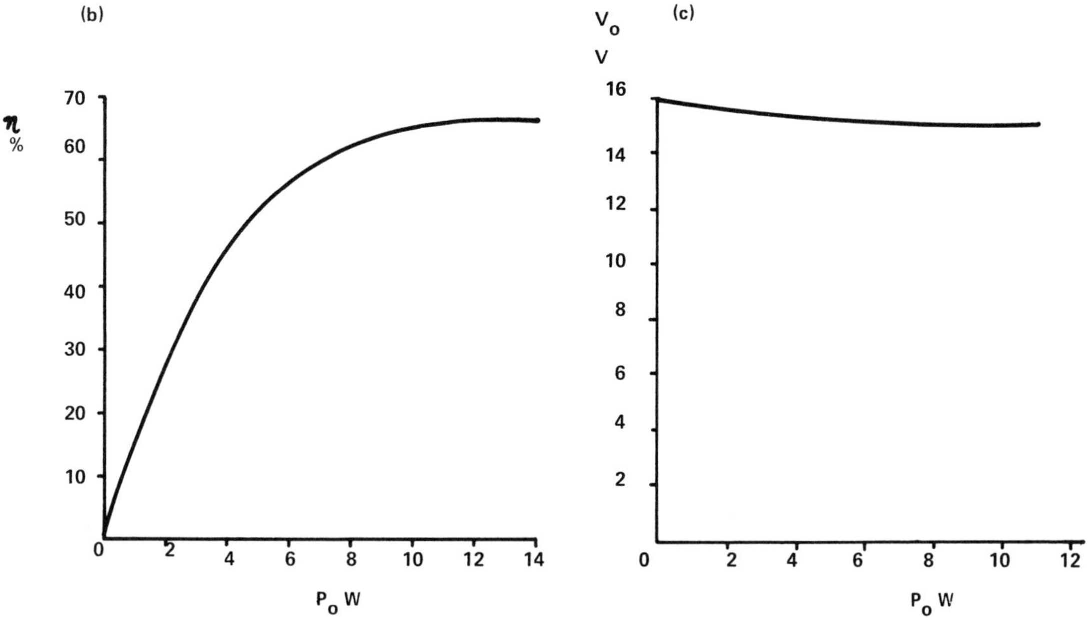

FIGURE 71 10W 20kHz 28V Single Saturating Transformer Converter

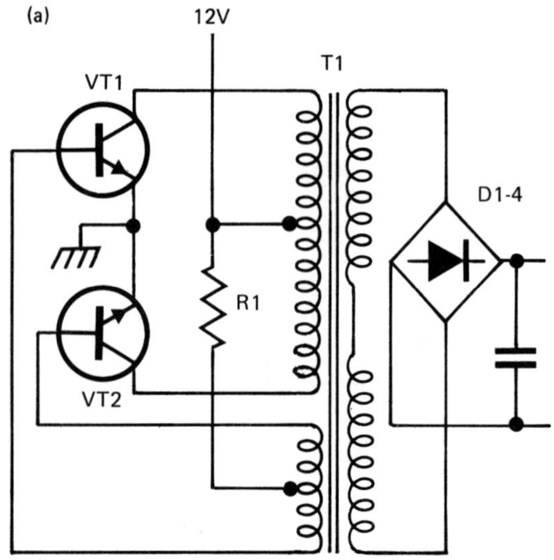

R1	–	270Ω	1W
D1-4	–	1SX172	
VT1, VT2		TIP31A	
T1	–	MT345* (2 x 50V secondary)	
		MT347* (250V secondary)	

*Gardners Transformers Ltd.

Power lost per transistor = 1.2W

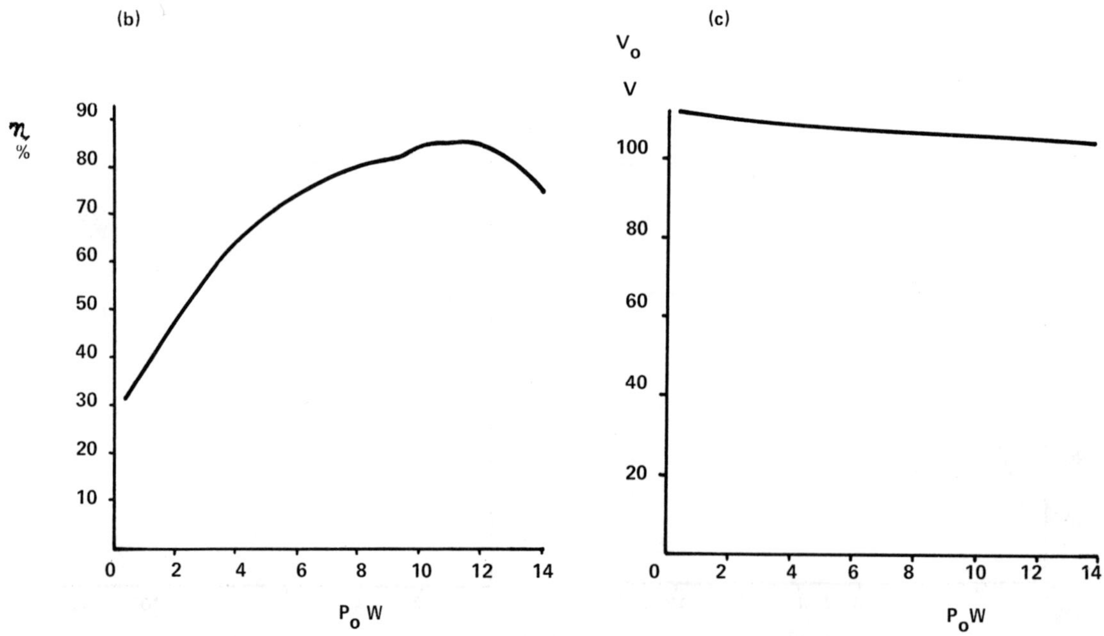

FIGURE 72 10W 20kHz 12V Single Saturating Transformer Converter

(a)

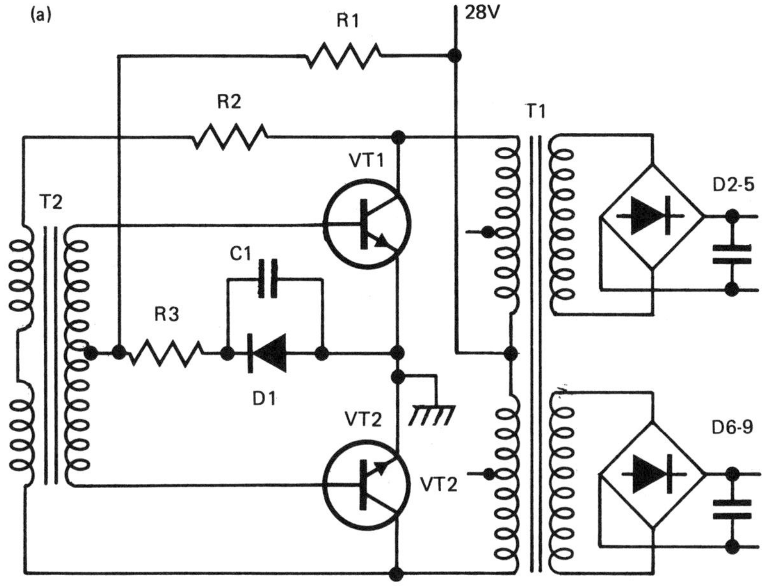

R1	–	1kΩ 1W
R2	–	390Ω 3W
R3	–	4Ω 1W
C1	–	1.5μF
D1	–	1N4001
D2-9	–	BYW10-600
VT1, VT2		TIP31B
T2	–	IDT61*

*Gardners Transformers Ltd.

Power lost per transistor = 6W

T1 35mm Pot Core FX2243
Winding Order
First Secondary 124t @ 32 S.W.G. Tapped at 62t
Two Primaries Bifilar 12t @ 20 S.W.G. Tapped at 6t each
Second Secondary 124t @ 32 S.W.G. Tapped at 62t

(b) (c)

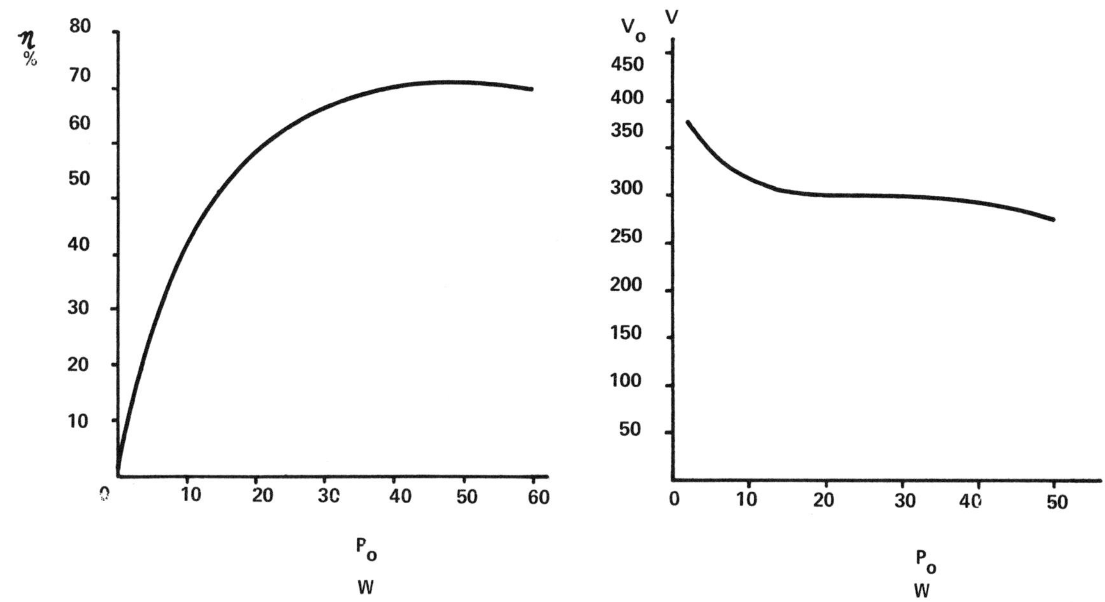

FIGURE 73 50W 20kHz 28V Two Transformer Converter

(a)

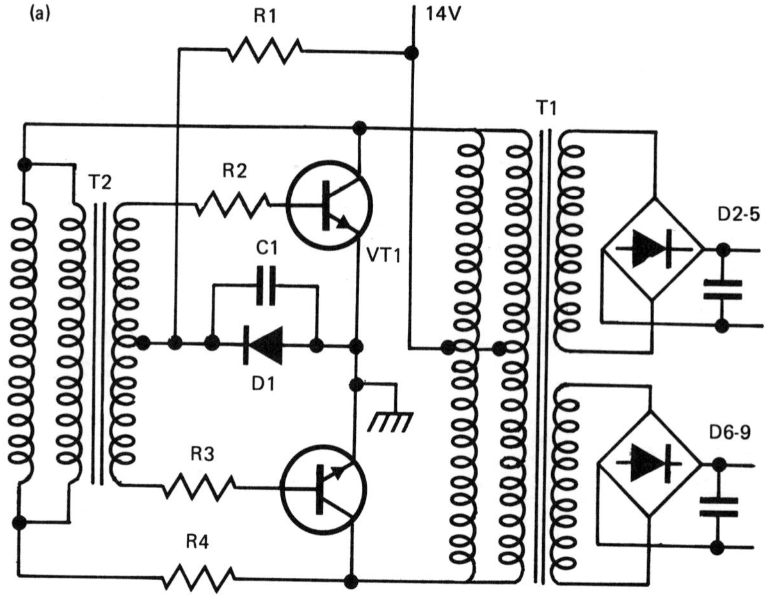

R1	–	470Ω	0.5W
R2, R3	–	2Ω	2W
R4	–	56Ω	5W
C1	–	1.5μF	
D1	–	1S020	
D2-9	–	BYW10-600	
VT1, VT2		BUY69C	
T2	–	IDT61*	
T1	–	As Figure 73	

*Gardners Transformers Ltd.

Power loss per transistor = 6W

(b) (c)

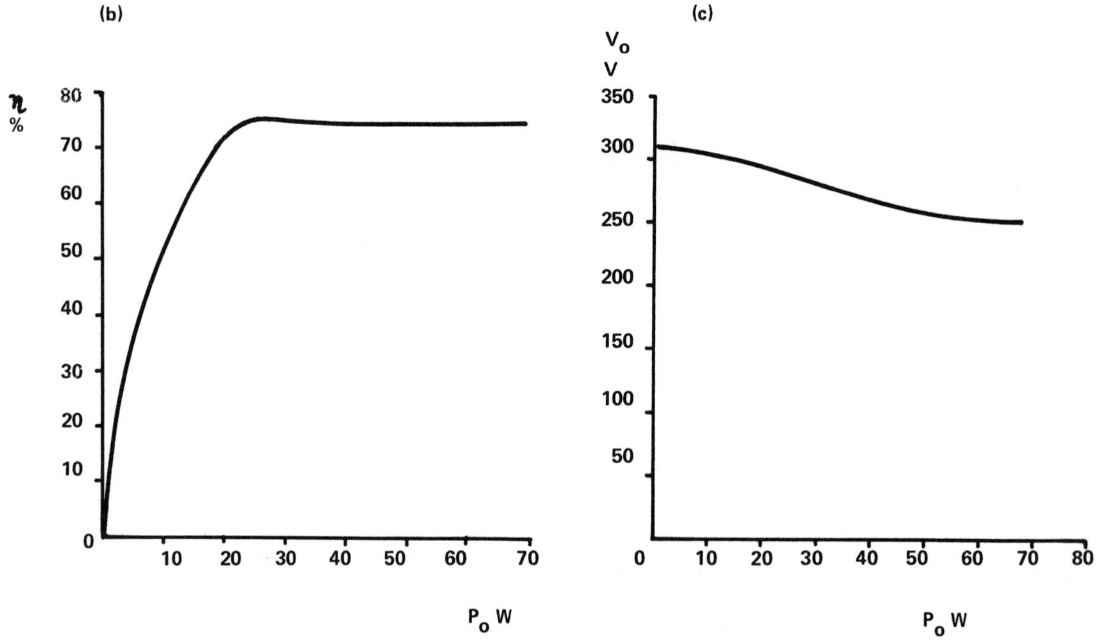

FIGURE 74 50W 20kHz 14V Two Transformer Converter

(a)

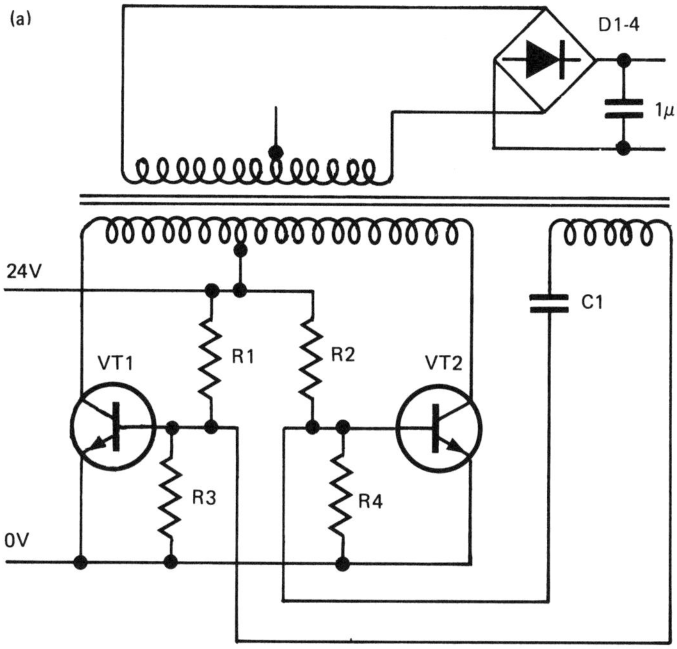

R1, R2	–	3k9Ω
R3, R4	–	100Ω
C1	–	47nF
D1-4	–	BYW10-800
VT1, VT2	–	TIP31B
T1	–	NH15*

*Gardners Transformers Ltd.

Power lost per transistor = 1.7W

(b) (c)

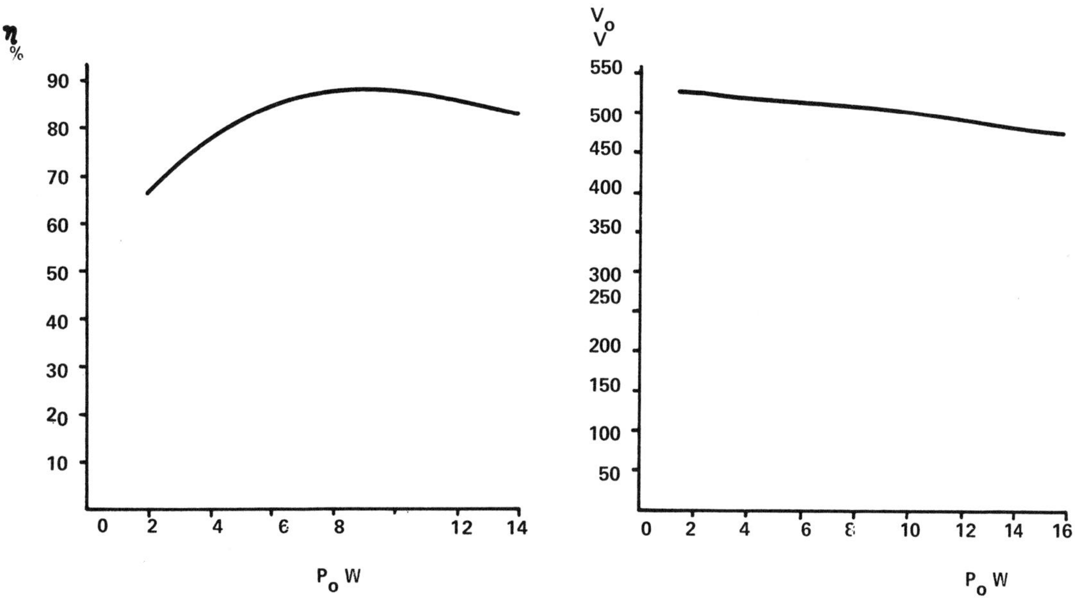

FIGURE 75 15W 20kHz 24V CR Timed Converter

(a)

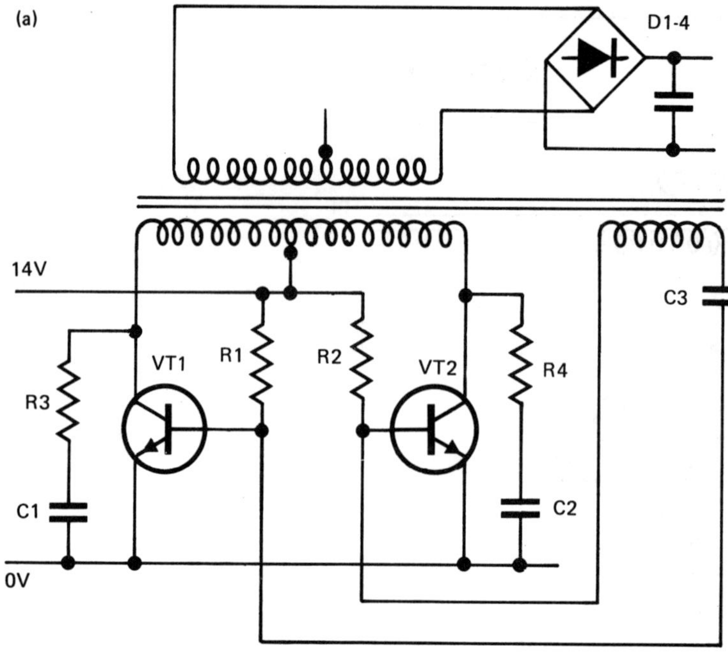

R1, R2	–	330Ω	1W
R3, R4	–	15Ω	
C1, C2	–	10nF	
C3	–	220nF	
D1-4	–	BYW10-800	
VT1, VT2	–	TIP31B	
T1	–	NH12*	

*Gardners Transformers Ltd.

Power lost per transistor = 3.6W

(b) (c)

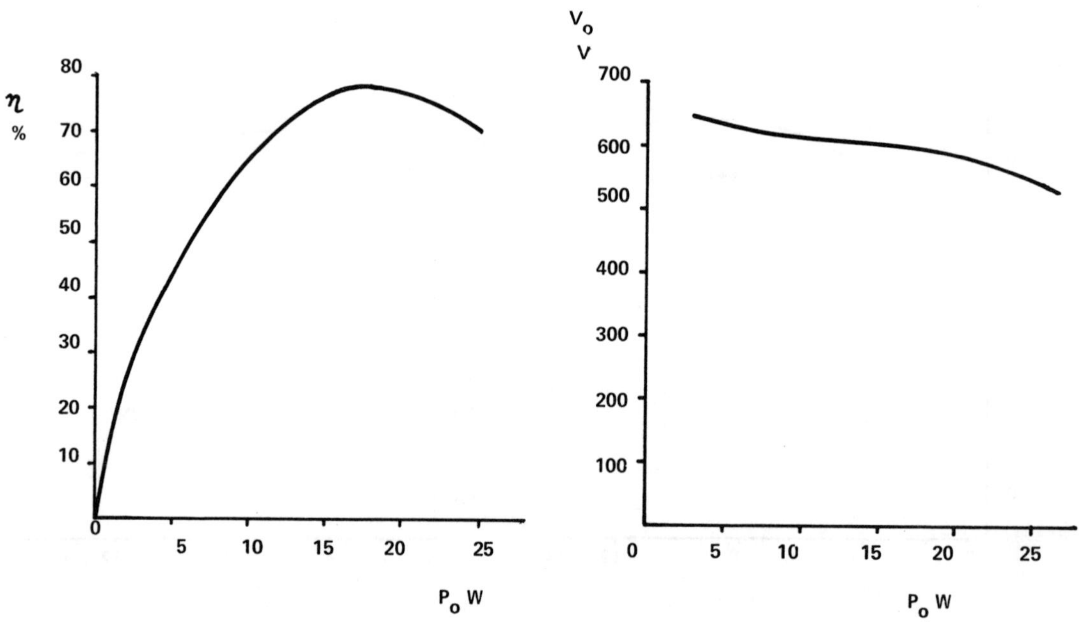

FIGURE 76 15W 20kHz 14V CR Timed Converter

REFERENCES

1. Thomas Roddam, *Transistor Inverters and Converters,* Iliffe Books Ltd.

2. *Manual of Inverter Transformers and Modules,* GT 21, Gardners Transformers Ltd.

3. E.W. Golding and F.C. Widdis, *Electrical Measurements and Measuring Instruments,* Pitman Press.

4. I. T. T., *Reference Data for Radio Engineers.*

5. Second Breakdown and Power Transistor Area of Operation, Application Note 167, Texas Instruments Ltd.

6. T. M. Montoux, W. Steffe, 'New Developments in Power Transistor Design', *Solid State Technology,* p 36-37, Sept 1967.

7. K. Rupra, Switching Mode Power Supplies, Chapter VII.

8. O. A. Kossov, 'Comparative Analysis of Chopper Voltage Regulators with L. C. Filter', *I.E.E.E. Trans on Magnetics* Vol MAG-4, No. 4, Dec 1968.

9. T. D. Towers, 'Switched Mode D. C. Regulators', *Wireless World,* pp 370-374, Vol. 74. No. 1396, 1968.

10. G. G. Avis, 'Application of transistors to D.C./D.C. Converters', *I.E.E. (International Convention on Transistors and Associated Semiconductor Devices),* Paper No. 2924E.

11. J. R. Nowicki, 'A D.C. Inverter and C.R. Timing', *Electronic Engineering,* Vol. 34, No. 413, pp 464-468.

12. J. R. Nowicki, 'Smaller D.C. Converters and Inverters', *Wireless World,* p. 42, March 1968.

13. P. J. Baxandall, 'Transistor Sinewave L.C. Oscillators', *Proc. I.E.E.* 106B Suppl. No. 16, pp 748-758, 1959.

14. C.J. Yarrow, 'Transistor Converters for the Generation of High-Voltage Low-Current D.C. Supplies', *Proc. I.E.E.* 106B, pp 1320-1324, May 1959.

15. C. Ridges, 'Voltage Stabilised Sinusoidal Inverters Using Transistors', *Proc. I.E.R.E.,* pp 109-127, August 1967.

16. V. Calchera, 'Sine v Square Waveform', *Wireless World,* Vol. 68, No. 10, pp 464-465. October.

17. G. Salters, 'A High Power D.C.–A.C. Inverter with Sinusoidal Output', *Electronic Engineering,* pp 586-591, September 1961.

18. F. Butler, 'Transistor Inverter (Frequency Stabilised Circuit for Running A Tape Recorder)', *Wireless World,* Vol. 67, No. 1, pp 17, January 1961.

19. D. R. Driver, 'Choice of Magnetic Cores for D.C. Converters', *Control,* Feb. 1967.

20. E. C. Snelling, *Soft Ferrites, Properties and Applications,* Iliffe Books Ltd.

21. K. A. Macfadyan, *Small Transformers and Inductors,* Chapman and Hall Ltd.

22. C. J. Dakin, C. E. G. Cooke, *Circuits for Digital Equipment,* Iliffe Books Ltd, pp 11-19.

23. M. C. Jackson, 'A Stable 10kV Reversible Power Supply, *Electronic Engineering,* pp 4523, July 1966.

24. D. L. Waidelich, C. L. Shackford, 'Characteristics of Voltage Multiplying Rectifiers', *Proc. I.R.E.,* pp 470-476, August 1940.

25. G. Ehni, 'Low Cost 400W Converter', Application Report, CA123, Texas Instruments Inc.

SECTION 3

AUDIO CIRCUITS

XI A HIGH FIDELITY PREAMPLIFIER

by Ian Hardcastle

This section describes a preamplifier which has been designed to interface between various signal sources and the power amplifiers described in the following chapter (XII). Principal features of this preamplifier are given in the specification.

The modular approach to the design enables simplified versions of the full preamplifier to be made. Particular attention has been paid to ensuring low harmonic distortion, even at high output levels, high overload capability and low noise.

SPECIFICATION

Inputs for:

(a) Magnetic pickup cartridges with R.I.A.A. microgroove characteristics.

(b) Tape replay with D.I.N.3.75 in.s^{-1} characteristic.

(c) Flat input for microphone.

(d) Ceramic pick-up cartridge, flat response with 2MΩ input impedance.

(e) Radio and auxiliary, with flat response, input impedances up to 2MΩ.

Bass and treble boost and cut controls giving gains of ± > 10dB at 100Hz and 10kHz.

Fixed Slope, variable frequency low-pass filter with component values given for turn-over frequencies of 5kHz, 10kHz, 20kHz and 40kHz.

Integral stabilized power supply which enables it to share an unregulated power supply with the power amplifier.

Distortion less than 0.03% at nominal output level, less than 0.1% at 20dB overload and less than 0.3% at 34dB overload.

Overload capability > 34dB on all inputs.

Signal to noise ratio > 60dB on magnetic pickup input.

THE CIRCUIT

In figure 1 the circuit diagram for one complete stereo channel is shown, together with sections of the other channel where necessary to show the cross channel switching and the tape output arrangements clearly. A detailed parts list is given in table 1.

The amplifier consists of three cascaded stages, each stage being a feedback amplifier. For low level inputs (magnetic pick-up, tape replay and microphone) all three stages are used. For the higher level input, only the latter two stages are used. This arrangement has two advantages:

(a) It saves having to try to get a low gain from the first stage when the high level inputs are used. This can only be done by attenuating the input (which degrades the signal to noise ratio) or by using a very low feedback resistor (which would load up the first stage output to such an extent that its overload capability would be reduced to practically nothing.)

(b) It enables ceramic pick-ups to be fed into a high input impedance F.E.T. stage which gives minimum noise and allows the cartridges internal frequency response shaping arrangement to work properly.
Each individual stage of the amplifier will now be considered in detail.

The First Stage

The requirements of the first stage are:

1. High open loop gain so that the frequency response of the stage will be determined by the characteristics of the feedback network only.

2. Low Noise.

3. Low output impedance so that the output can swing up to 5V r.m.s. even when loaded with the magnetic feedback network, which has an input impedance of only 4k5Ω at 20kHz.

4. High rejection of power supply ripple and noise to ensure low noise and prevent instability due to feedback via the power supply.

Originally, the two configurations shown in Figure 2 were measured to see which would fulfil these requirements best. The results of these measurements showed that the two circuits were similar as far as open

AUDIO PREAMPLIFIER Mk II (Channel A)

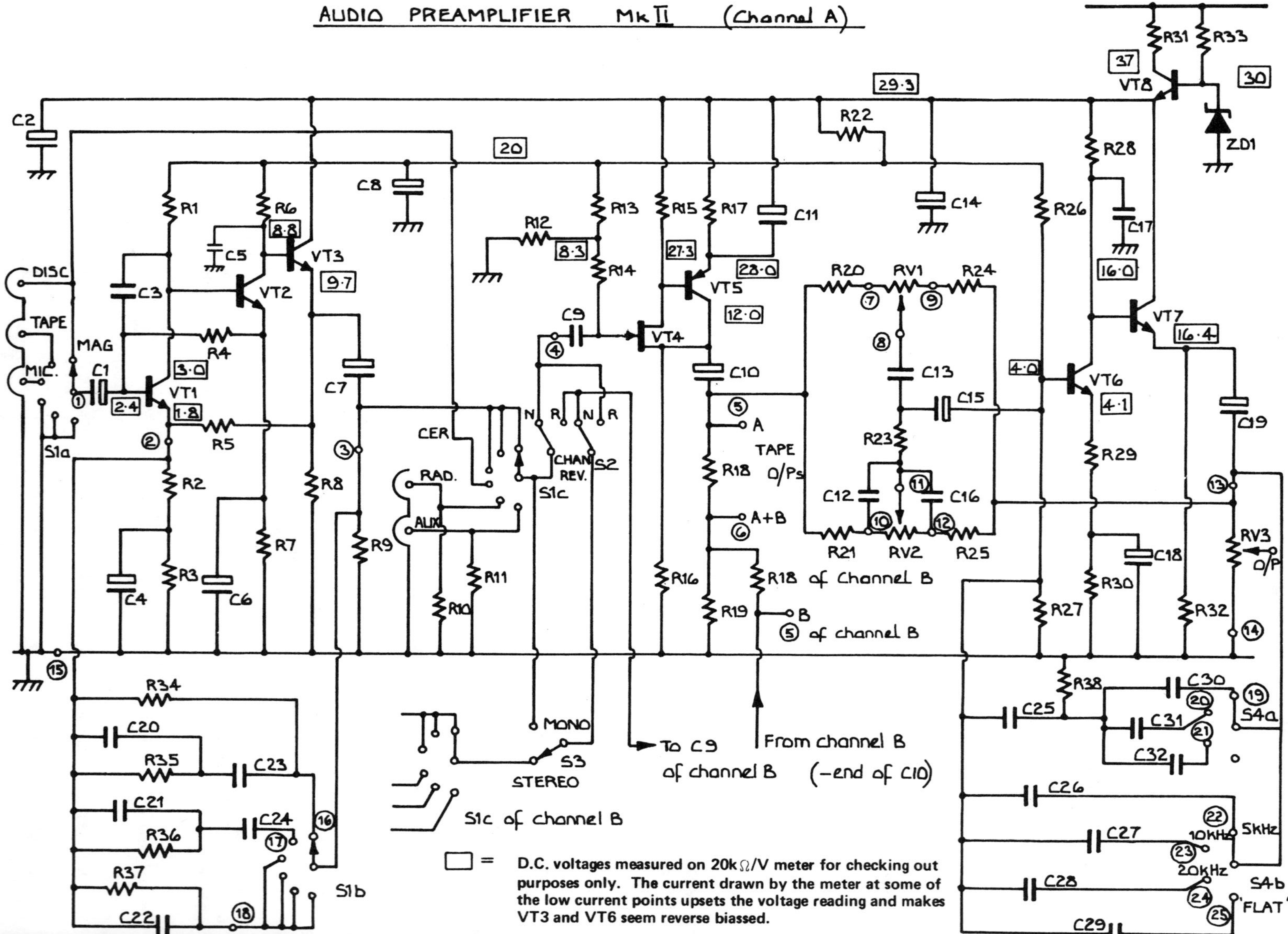

□ = D.C. voltages measured on 20kΩ/V meter for checking out purposes only. The current drawn by the meter at some of the low current points upsets the voltage reading and makes VT3 and VT6 seem reverse biassed.

Table 1

Capacitors

Number	Value F	Voltage V	Tolerance %	Type	Number	Value F	Voltage V	Tolerance %	Type
1	4μ	40	–	E	17	47p	125	–	P.S.
2	100μ	40	–	E	18	100μ	6V4	–	P.S.
3	470p	125	–	P.S	19	12.5μ	25	–	E
4	50μ	6V4	–	E	20	2n7	125	2	P.S.
5	100p	125	–	P.S	21	100p	125	5	P.S
6	50μ	6V4	–	E	22	100p	125	5	P.S
7	4μ	40	–	E	23	6n8	63	2	P.S
8	640μ	25	–	E	24	4n7	63	2	P.S.
9	10n	125	–	M	25	2n2	125	2	P.S.
10	4μ	40	–	E	26	330p	125	5	P.S.
11	100μ	6V4	–	E	27	220p	125	5	P.S.
12	33n	63	2	P.S	28	100p	125	5	P.S.
13	1n5	125	2	P.S	29	10p	125	10	P.S.
14	400μ	40	–	E	30	6n8	63	2	P.S
15	4 μ	40	–	E	31	2n2	125	2	P.S
16	33n	63	2	P.S	32	470p	125	2	P.S.

E = Electrolytic P.S. = Polystyrene M = Metallized Film

For accurate reproduction of correction curves, suggested tolerances should be used. One of each capacitor is required per mono system, two for stereo.

Switches

Number	Function	Type Required	
		Mono	Stereo
S1	Selector	3 pole 6 way	6 pole 6 way
S2	Channel reverse	not required	2 pole 2 way slide
S3	Mono/Stereo	not required	1 pole 2 way slide
S4	Filter Frequency	2 pole 4 way	4 pole 4 way

Resistors

Number	Value Ω	Quantity required for system		Number	Value Ω	Quantity required for system	
		Mono	Stereo			Mono	Stereo
1	150k	1	2	20	3k3	1	2
2	820*	1	2	21	10k	1	2
3	15k	1	2	22	33k (see text)	1	2
4	56k	1	2	23	15k	1	2
5	680k*	1	2	24	3k3	1	2
6	100k	1	2	25	10k	1	2
7	27k	1	2	26	330k	1	2
8	3k9	1	2	27	100k	1	2
9	100k	1	2	28	27k	1	2
10	100k	1	2	29	150	1	2
11	100k	1	2	30	8k2	1	2
12	220k	1	2	31	(see text)		
13	220k	1	2	32	2k7	1	2
14	2M2	1	2	33	(see text)		
15	4k7	1	2	34	1M5*	1	2
16	2k7	1	2	35	33k*	1	2
17	560	1	2	36	18k*	1	2
18	470k	0	2	37	39k*	1	2
19	100k	0	1	38	2k2	1	2

All 1/4W 5%, 1/2W may be used
For accurate reproduction of correction curves, resistors marked * should be 1% or 2%.

Potentiometers

Number	Function	Type Required	
		Mono	Stereo
RV1	treble	100kΩ linear	dual ganged 100kΩ linear
RV2	bass	100kΩ linear	dual ganged 100kΩ linear
RV3	volume	10kΩ log.	dual concentric 10kΩ log to provide volume and balance function.

Semiconductors

Number	Type
VT1	BC184L
VT2	BC184L
VT3	BC184L
VT4	BF244B
VT5	BC214L
VT6	BC184L
VT7	BC184L
VT8	BC182L
ZD1	1S3030A

One of each device is required per mono system, two for stereo.

Alternative types

BC182L	BC182
BC184L	BC109
	BC184
	2N3707
BC214L	BC214
	2N2605
	2N4058
BF244B	BF245B
	TIS59 yellow

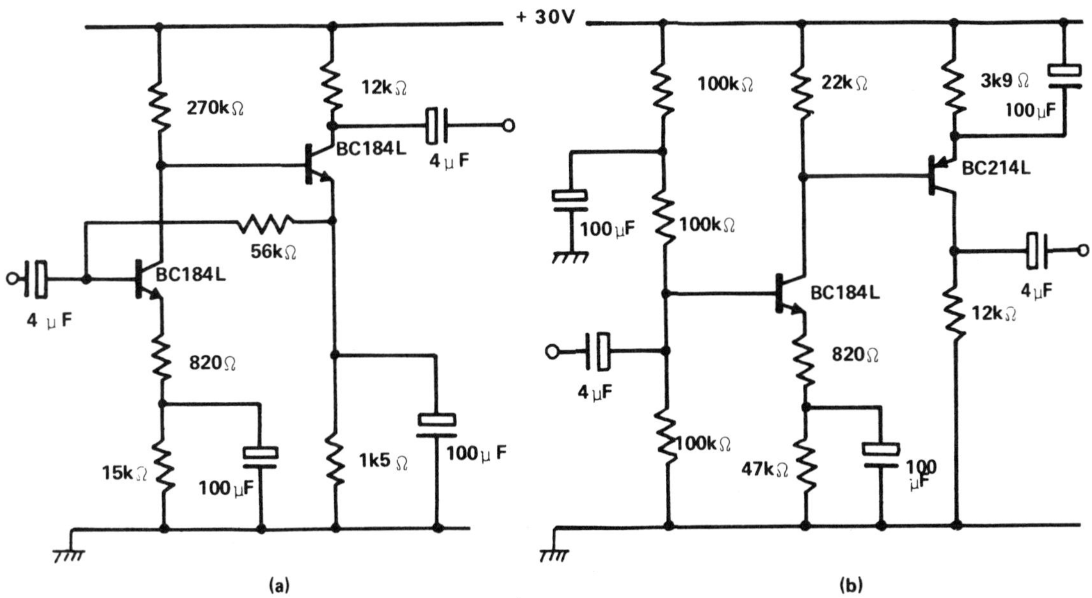

FIGURE 2

loop gain, output swing and noise was concerned. The power supply ripple rejection of the complementary circuit (Figure 2(b)) was superior to that of the dual npn circuit. This situation was reversed by splitting the collector load of transistor VT1 and decoupling the midpoint to ground by means of a large capacitor. This effectively prevents the injection of ripple into the base of transistor VT2. The output swing capability of neither circuit was good enough, however, the output waveform being considerably distorted at less than 2V r.m.s. This situation was remedied by adding to the stage a further transistor VT3 as an emitter follower. The low output impedance of this configuration enabled the output swing requirements of the stage to be met. Its high input impedance allows the collector resistor of VT2 to be increased from 12kΩ to 100kΩ, which improves the open-loop gain of the stage. The decrease in the collector current of VT2 from 1mA in the simple circuit of figure 2(a) to 100μA gives a reduction in noise because the source impedance R1 seen by VT2 is more optimum for low noise when its collector current is 100μA than it is at 1mA.

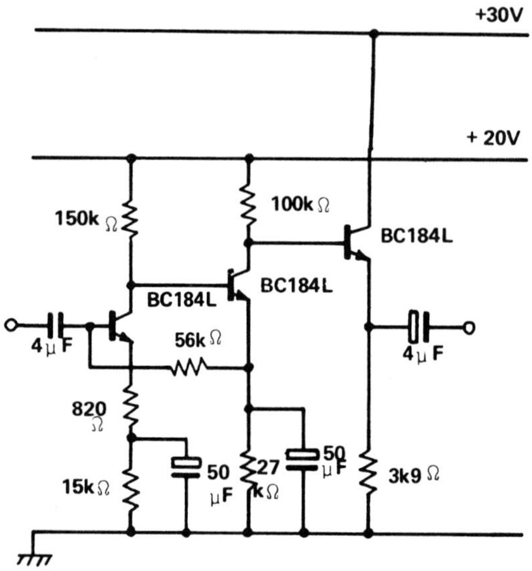

FIGURE 3

The resistor causing the levelling off of gain at 50Hz, R5, is directly connected between the emitters of VT3 and VT1, instead of via capacitor C6 like the rest of the feedback components, to ensure that the gain does not rise again at very low frequencies. Because different values of R5 are required for the tape and magnetic characteristics, the value of R5 required for tape is used. This is shunted by R34 to obtain the correct value for the magnetic characteristic. In the microphone position, a flat response from very low frequencies up to 40kHz is obtained.

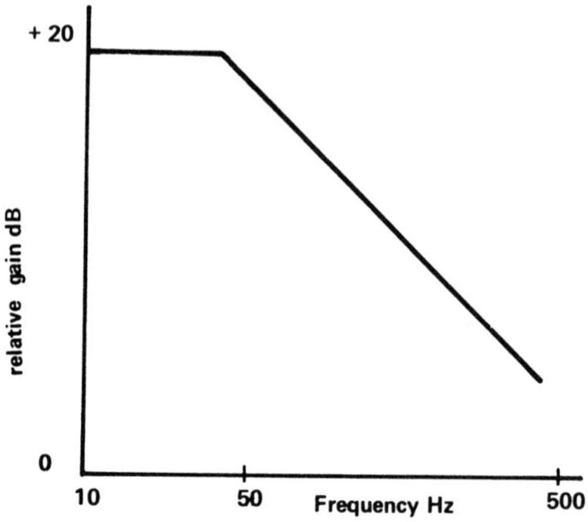

FIGURE 4

The low collector currents of VT1 and VT2 enable the collector resistors to be connected to the highly de-coupled 20V supply rail. With this arrangement no ripple voltage could be detected on the emitter of VT3 when a 1V r.m.s. 50Hz ripple was applied to the main 30V supply rail.

The d.c. voltage on the emitter of transistor VT3 is stabilised by d.c. feedback through resistor R4 from the emitter of VT2 to the base of transistor VT1. The advantage of this method is that as there is no connection between the base of VT1 and the supply rails, supply noise is prevented from appearing on the amplifier input. The complete circuit of the first stage amplifier is shown in figure 3.

The closed loop a.c. gain is defined by resistor R5 and the frequency selective networks consisting of resistors R34 - R37 and capacitors C20–24 connected between the emitter of VT1 and the emitter of VT3 via switch S1b.. These networks provide the frequency response shaping necessary to provide the magnetic pick-up and tape replay characteristics. At the low frequency end, both these frequency characteristics are of the form shown in figure 4.

All these inputs have a nominal sensitivity at 1kHz of 2mV, which may be considered too high for some magnetic cartridges. Alternative values of the feedback components C20, C23 R34 and R35 are given in table 2 for a sensitivity of 5mV. Also given in table 2 are alternative values of C24 and R36 to give tape replay time constants of 70μs, (for 7.5in.s^{-1}) and 120μs (for some 3.75 in.s^{-1} and 1.875 in.s^{-1}).

It should be noted that the tape replay curve does not have any high frequency boost to correct for tape head losses at high frequencies. This is because the amount of correction required depends on the type of head used. This facility may be added (at the expense of further complicating switch S1) by shunting resistor R2 with a series CR network. Network CR2 should have a time constant of about 40μs (say) and CR about 5μs.

The sensitivity of the microphone input may be adjusted by varying the value of R37.

Deviation of the amplifier's disc correction response from the standard R.I.A.A. Curve published in BS1928 (1961) is shown for the two sensitivities in figure 5. Tape and microphone replay responses are shown in figure 6. It can

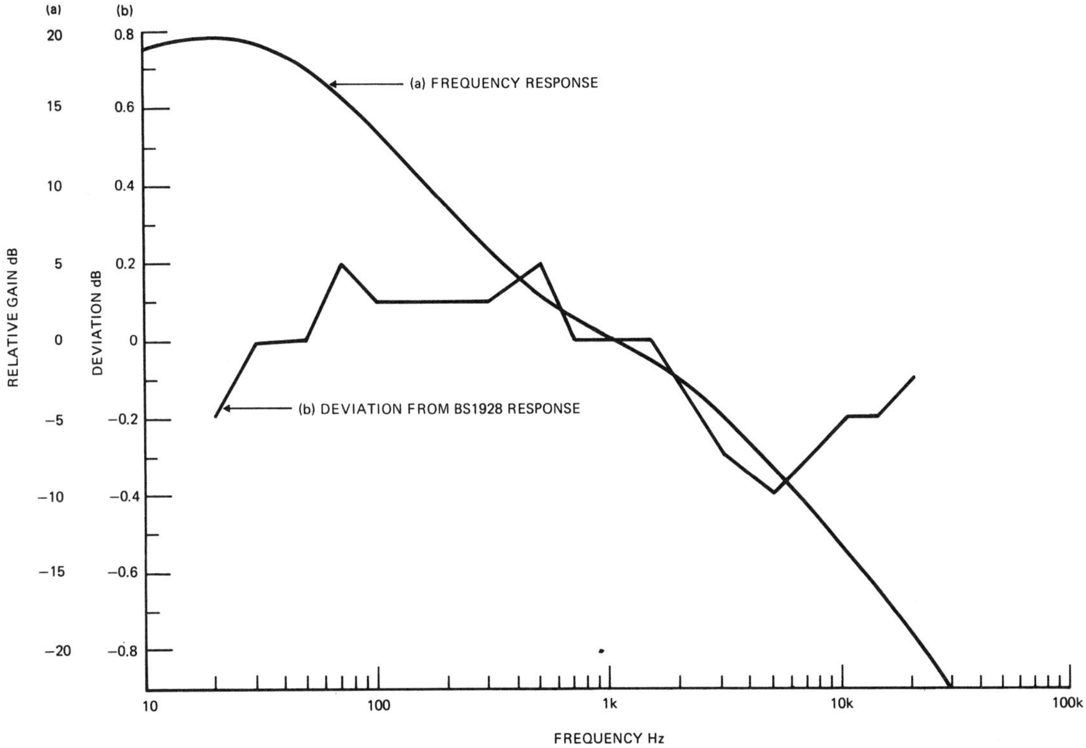

FIGURE 5(a) Magnetic Replay (a) Frequency Response (b) Deviation Curve Input = 2mV for 100mV out

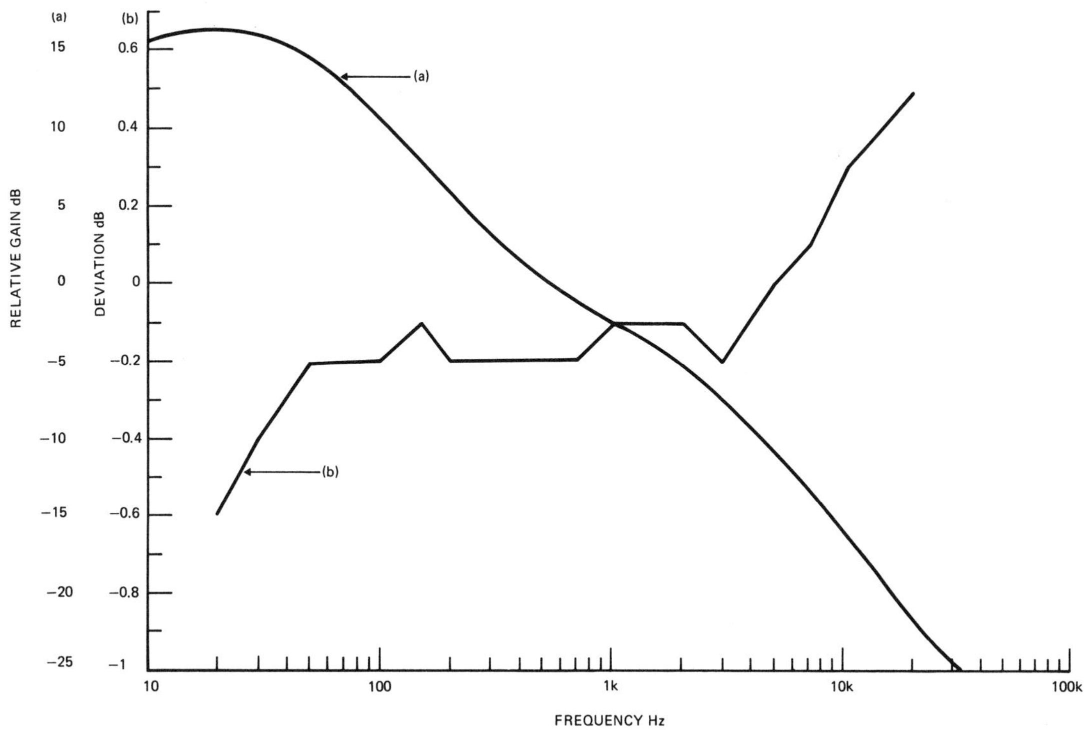

FIGURE 5(b) Magnetic Replay (a) Frequency Response (b) Deviation Curve Input = 4.3mV for 100mV out

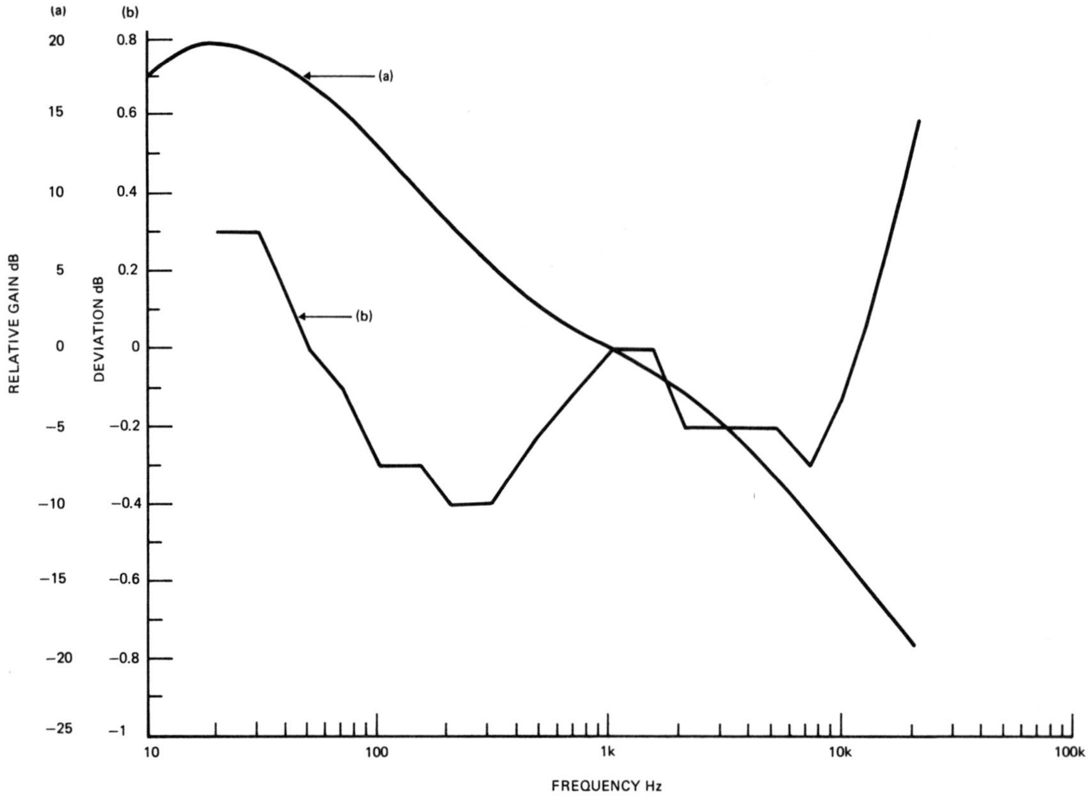

FIGURE 5(c) Magnetic Replay (a) Frequency Response (b) Deviation Curve   Input = 5.8mV for 100mV out

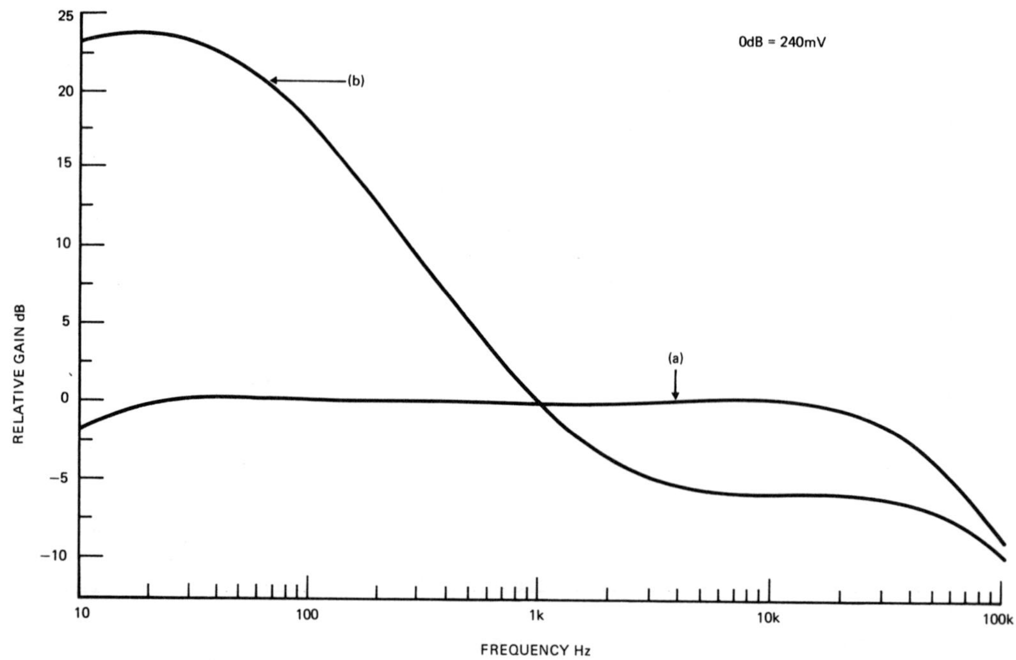

FIGURE 6 Frequency Response (a) Microphone Input (b) Tape Replay

Table 2.

Sensitivity mV	R34 kΩ	R35 kΩ	C20 nF	C23 nF
2	1M5	33k	2n7	6n8
5 { 4.3	270k	15k	5n6	15n
5.8	220k	13k	8n2	22n

(a) **Magnetic Replay**

Replay Time Constant µs	R36 kΩ	C24 nF
70	15k	4n7
90	18k	4n7
120	27k	4n7

(b) **Tape Replay**

be seen that these have a high frequency roll-off at about 85kHz and 40kHz respectively. This is due to the 100pF capacitors C21 and 22 which have been added to ensure stability of the amplifier when the input is open circuited.

The Second Stage

The output of the first stage together with the three high level inputs are fed into the second stage via the selector switch S1c. Between the selector switch and the stage input are switches S2 and S3. Switch S2 enables the input to channel A to be transferred to the second stage of channel B and vice versa, thus providing a channel reverse facility. Switch S3 enables the input to channel A to be fed into the second stage of both channel A and channel B.

The switches should be omitted from a mono only version of the preamplifier and may be omitted from the stereo version. In both these cases, the wiper of switch S1c should be connected to capacitor C9.

The main requirements of the second stage are that it should have an input impedance in excess of 2MΩ to ensure a flat frequency response from a ceramic cartridge without having to use a frequency selective feedback loop. The stage must also have a low output impedance so that full overload output voltage may be delivered into the same control stage whose input impedance can be as low as 3kΩ under some conditions.

To provide the necessary high input impedance and a good signal to noise ratio with a high impedance source, at F.E.T. is used in the second stage.

However, it is not sufficient to use this device on its own as a source follower as excessive distortion results from the large variation of source current which occurs during the ± 7.5V excursion of source voltage required for full overload output.

This difficulty is remedied by adding an extra pnp transistor VT5 which is driven from the drain of the F.E.T. VT4. The emitter of VT5 is decoupled to the supply rail by capacitor C10. As the output voltage of VT4 is derived with respect to the supply rail, this arrangement enables the stage to have high open loop gain and prevents any differential rail noise voltage appearing between base and emitter of VT5. This stops supply rail noise entering the system at this point. Feedback is applied from the collector of VT5 to the source of VT4 by directly connecting them together. Hence the two devices have a common source resistor/collector load R16. In this configuration, the large a.c. variation in transistor VT4's drain current which would cause distortion in a straight source follower configuration, is reduced by a factor of typically 200 by the current gain of transistor VT5. The distortion in this stage is, therefore, less than 0.03%.

The large output voltage requirement of this stage necessitates a close definition of the d.c. voltage level on the collecter of transistor VT5. As the running current of transistor VT4 is about 67µA, the output voltage is roughly $V_G - V_P$ where V_G and V_P are the gate voltage and pinch off voltage respectively of transistor VT4. To enable the output voltage to be closely defined a device with a closely defined V_P is required for VT4. The BF244 family provides an ideal device for this as A, B and C selections with a 2.5 : 1 spread of I_{DSS} (which is closely related to V_P) are available. The B version of the device is specified for use in this preamplifier giving an output voltage defined within ±1V.

The d.c. level of the gate of transistor VT4 is defined by the potential divider formed by resistors R12 and R13 connected between the 20V rail and ground. Connecting these resistors to the stabilised rail prevents main rail noise being injected into the amplifier via the gate of VT4. Alternative values of R12 and R13 to accommodate the A and C versions of the BF244 are given in Table 3.

Table 3

V_{GS}-V	BF244 type	R12 kΩ	R13 kΩ
0.4–2.2	A	270	220
1.6–3.8	B	220	220
3.2–7.5	C	180	270

The voltage at the junction of R12 and R13 is fed on to the gate of transistor VT4 through resistor R14, which defines the input impedance of the amplifier. The value of this resistor is 2.2MΩ, which gives an input impedance too high for normal radio and auxiliary inputs. Resistors R10 and R11 are placed between these inputs and ground to give a nominal 100Ω input impedance. Of course, the value of these resistors may be charged to give any input impedance desired, up to a maximum of 2.3MΩ. (No locations for resistors R12 and R13 are provided on the printed circuit board; it is suggested that they be mounted on the input sockets themselves).

The signal is fed onto the gate of VT4 via capacitor C9. As there is 2.2MΩ (R14) between this capacitor and the gate potential divider chain, it is essential that a low leakage capacitor is used for C10, as any leakage current will cause a voltage drop across R14 and upset the d.c. level of the gate of transistor VT4 and hence the stage output voltage. The output of the second stage is fed into the tone control stage and also into the tape output socket.

The purpose of this output is to provide an equalised signal from any of the sources which may be plugged into the preamplifier, amplified to the 100mV level but unaffected by the tone controls, suitable for feeding the high level input of a tape recorder. An additional facility offered is the A + B output which allows signals from stereophonic sources to be added together and fed into a monophonic tape recorder. Because of the attenuation introduced by the summing network (R18, R19 and R18) the output level is nominally 20mV, when fed into a 100kΩ input. The summing network does introduce some interchannel cross-talk, which is about –60dB with respect to the signal level in each channel and is thus well below that of most stereophonic sources. Fanatics may omit this feature or introduce a switch between C10 and R18 on one of the channels.

If the preamplifier is required for record playing with a magnetic cartridge only the second stage may be omitted completely. The output of the first stage should be taken from the junction of capacitor C7 and R9, via switches S2 and S3 if desired, to the junction of resistors R18, R20 and R21. Due to the increased loading on the output of the first stage a slight reduction in the overload capability will occur.

The Third Stage

The third stage consists of tone controls and filters placed round a high gain inverting amplifier formed by transistors VT6 and VT7. The feedback tone controls used are based on the Baxendall configuration with values modified to suit the lower impedance levels encountered in these transistor circuits. This type of control was chosen in preference to a passive interstage type due to its better signal to noise ratio and the lower demands the former type makes on the last stage of the amplifier (mid range gain of 1 required for Baxendall controls instead of about 15 for passive type).

The feedback tone controls work in the virtual earth mode so that attenuation may be obtained when the controls are in the 'cut' position and thus require an inverting amplifier. This amplifier consists of transistors VT6 and VT7, VT6 provides the gain and inversion, VT7 is an emitter follower which gives the low output impedance necessary to drive the feedback network with full overload output voltage. The open loop gain of the amplifier is limited to 50dB by the 150Ω resistor R29. This resistor defines the input impedance of the stage and also simplifies the high frequency stabilization of the stage, which is accomplished by means of capacitor C17.

The d.c. base voltage of transistor VT6 is defined by resistors R26 and R27 which are connected between the 20V stabilized rail and earth. Connecting these resistors to the 20V rail prevents the injection of power supply noise into the base of VT6. Power supply noise is injected into the base of VT7 but as there is only unity gain between this point and the output the contribution of this to the overall signal–noise ratio is small.

Also, placed in the feedback path of this stage is a bridge T high pass network, giving the amplifier a low pass filter characteristic. The values of one of the arms and the bridging capacitor are varied by switch S4, giving roll-offs at 5kHz 10kHz, 20kHz and 'flat'. Some or all of these positions may be omitted if desired. In the latter case, the capacitor C29 should be retained to ensure stability of the stage. The bridging capacitor is necessary to produce a sharper initial slope in the roll-off. The final slope of the roll-off is about 8dB/octave which is not as steep as some L–C filters achieve but is sufficient and does not ring excessively on transients. Curves showing the variation of output with various settings of the tone and filter controls are shown in figures 7, 8, 9 and 10.

The output of the third stage is fed via the volume control into the power amplifier. The reasons for siting the volume control of the amplifier are a combination of desirability and necessity:

1. As the noise level of the power amplifier is small compared with that of the preamplifier, both signal and noise coming from the preamplifier are attenuated equally by the volume control, maintaining a constant signal to noise at all settings of the volume control.

2. The only alternative siting of the volume control in this preamplifier (without adding further stages) is between the second and third stages. This position has the advantage of eliminating the necessity for the third stage to swing to full overload output voltage, but has the disadvantage that the tone control characteristics are altered by the setting of the volume control.

It will be noticed that there is no balance control shown in the circuit diagram. The balance control is omitted to increase the overload capability of the amplifier, since in driving a volume control from a balance control of similar impedance set mid-way, an attenuation of approximately 0.5 is obtained. This effectively halves the overload capability of the amplifier. In place of the balance control, it is recommended that a dual concentric volume control be used instead to obtain individual control over the level in each of the channels.

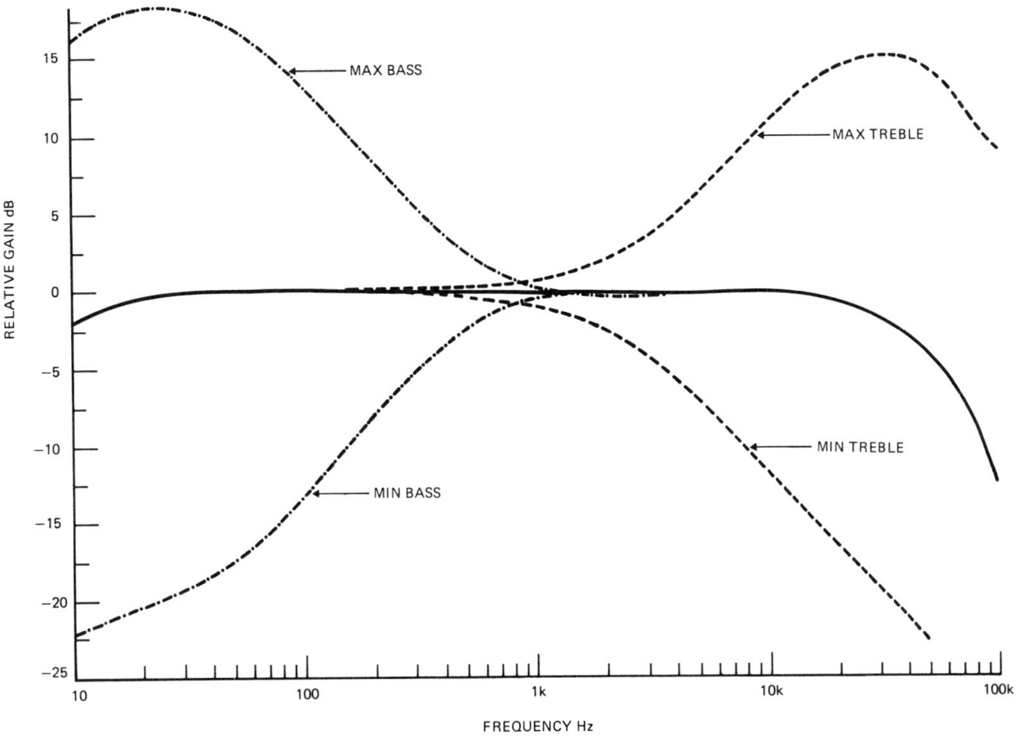

FIGURE 7 (a) Tone Controls – Microphone I/P Filters Flat

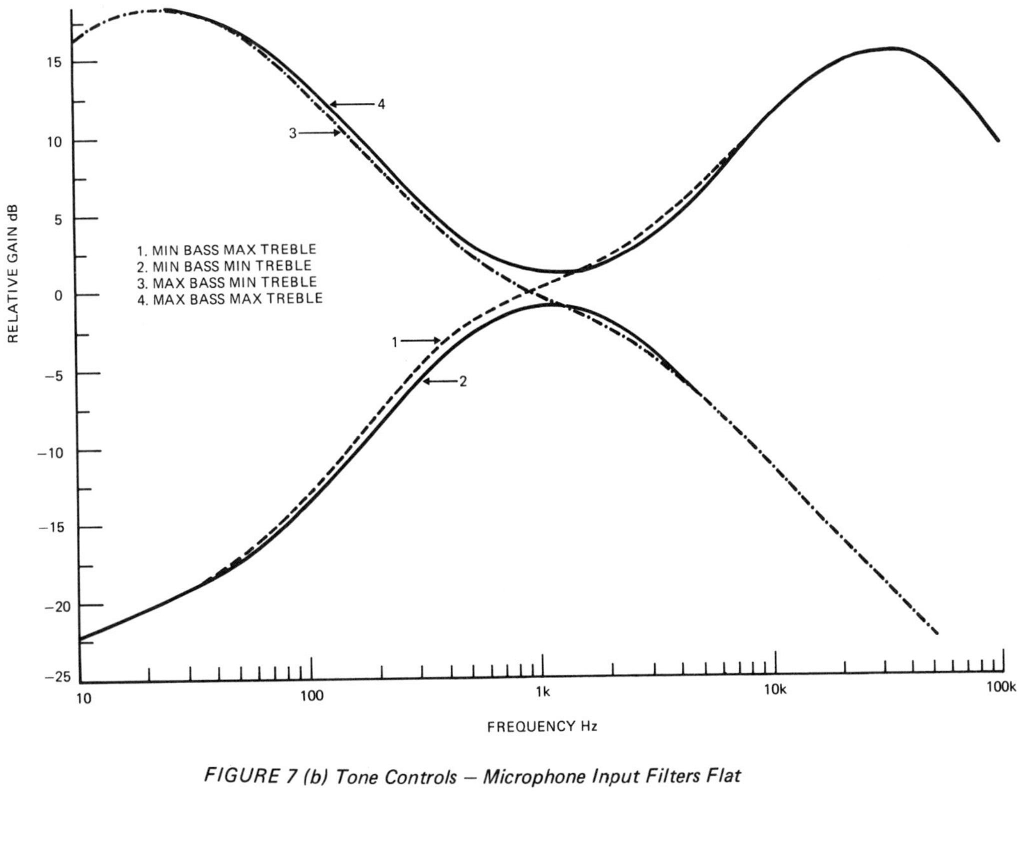

FIGURE 7 (b) Tone Controls – Microphone Input Filters Flat

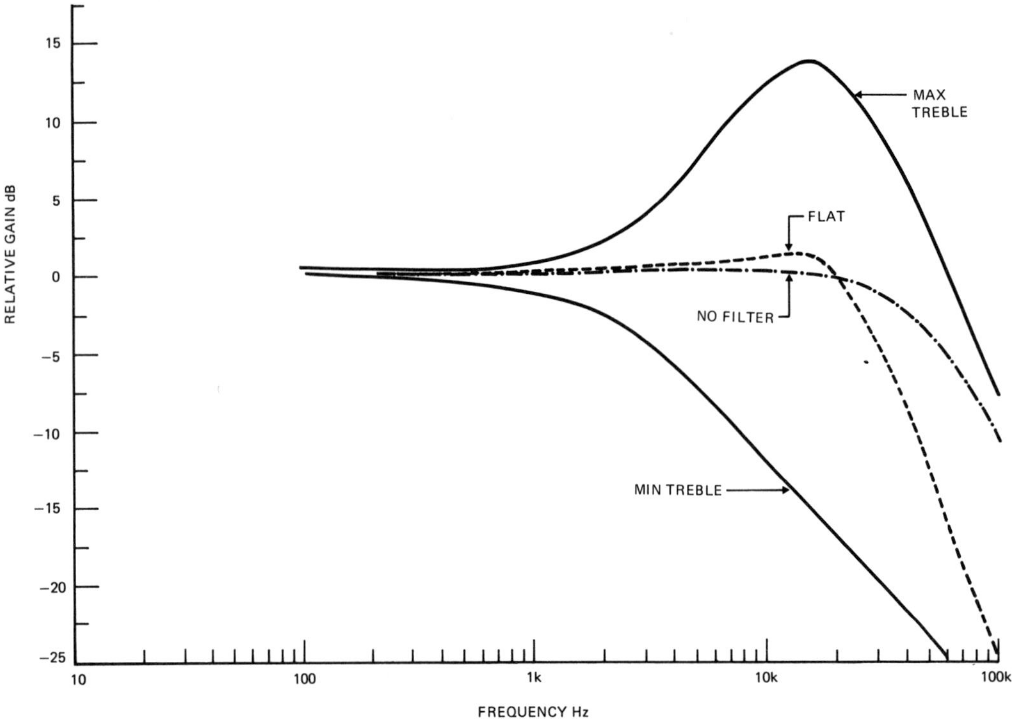

FIGURE 8 20kHz Filter Frequency Response

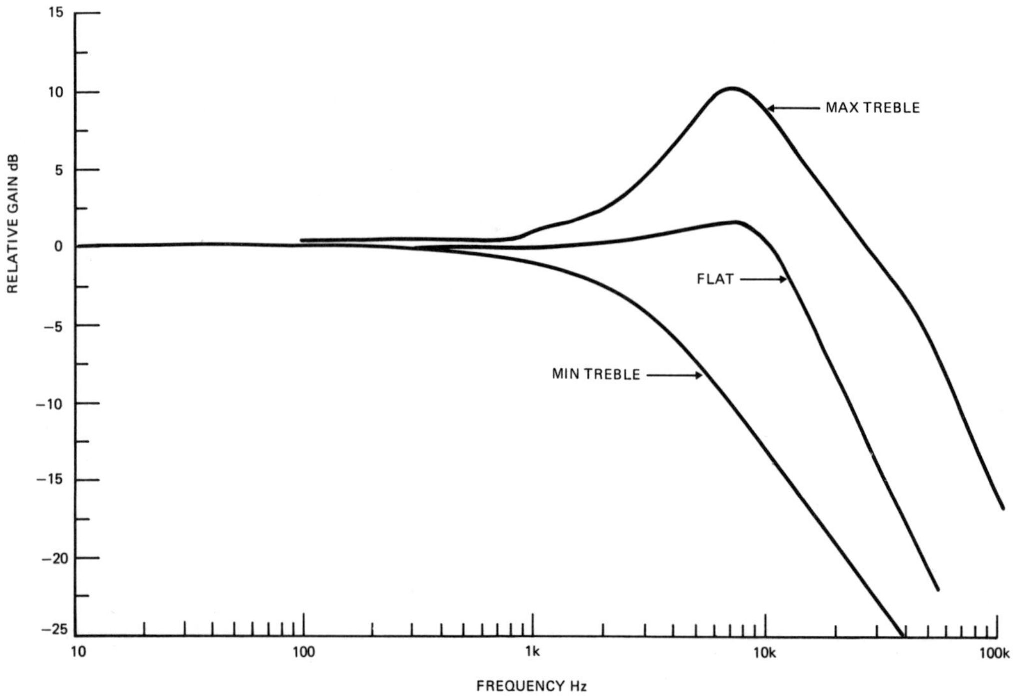

FIGURE 9 10kHz Filter frequency response

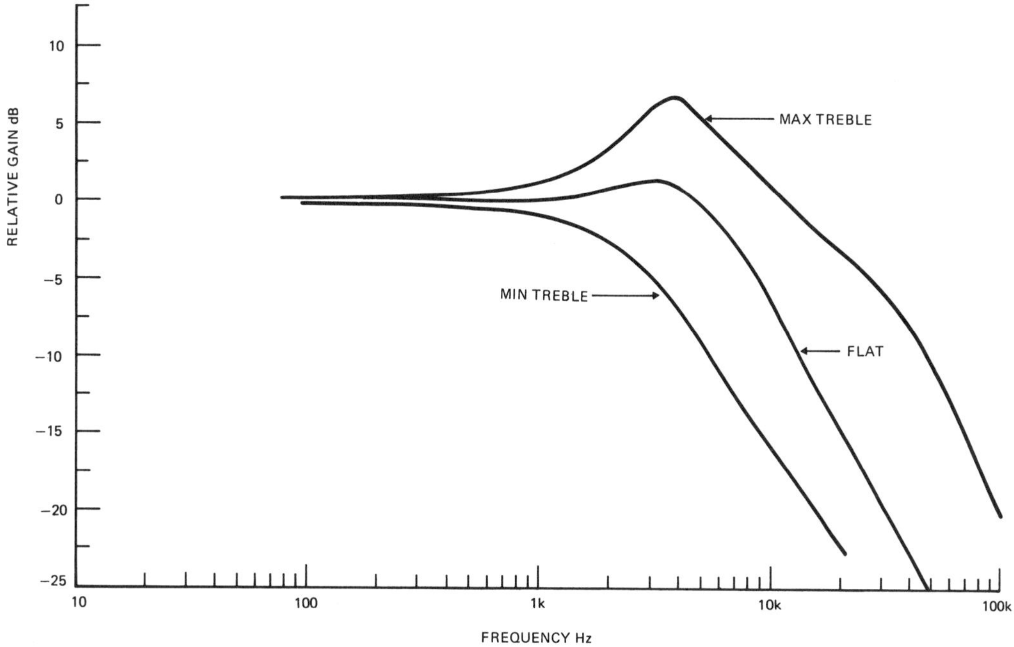

FIGURE 10 5kHz filter frequency response

The Power Supply

This pre-amplifier is designed to be used with any of the power amplifiers in chapter XII of this book. Because of the wide range of supply voltage that these amplifiers use, two different arrangements for the pre-amplifier power supply have to be made.

The preamplifier can share the power supply of the 10W 15Ω, 15W 15Ω, and 30W 8Ω amplifiers using the zener diode and transistor voltage dropper arrangement shown in figure 1. This rather complex configuration was chosen to allow the preamplifier to share the same power supply as the power amplifier. The power amplifier power supply has two main forms of variation: (a) a 100Hz sawtooth variation whose amplitude increases with increasing output amplitude, (b) a large-scale voltage droop which occurs with prolonged running at high power. Measurements showed that a simple resistor – capacitor voltage dropper network will adequately cope with the sawtooth variation but l.f. instability resulting from main amplifier power supply voltage droop occurred. This instability could only be cured by using the zener diode transistor arrangement shown in figure 11(a), although the other two configurations shown in figure 11 were also tried unsuccessfully.

A resistor R31 is used in the collector of transistor VT8 to reduce dissipation in the transistor, especially during switch-on when there is maximum base drive and a large capacitor to charge.

Values of this resistor and that supplying current to the zener diode (R33) for different power amplifiers are given in table 4.

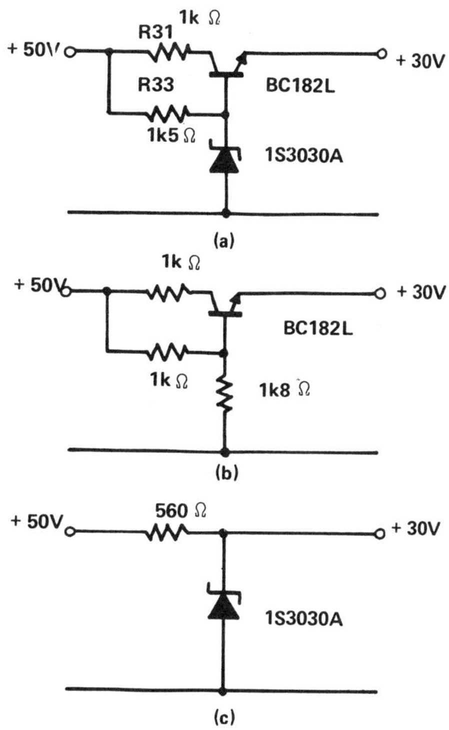

FIGURE 11

Table 4

Amplifier		R31 Ω	R33 Ω
P_{out}	R_L		
10W	15Ω	**470**	**680**
15W	15Ω	**1k**	**1k5**
30W	8Ω	**1k**	**1k5**

A small economy may be made by using a common voltage dropper transistor, zener and resistor for both preamplifiers of a stereophonic system. In this case, the value of resistor R31 should be halved. The 10W 8Ω and 15W 8Ω power amplifiers have power supply voltages too low for dropping down to the required preamplifier voltage rail. For these amplifiers, a separate low current 30V winding on the mains transformer or a separate small mains transformer should be used with the rectifier-capacitor circuit shown in Figure 12.

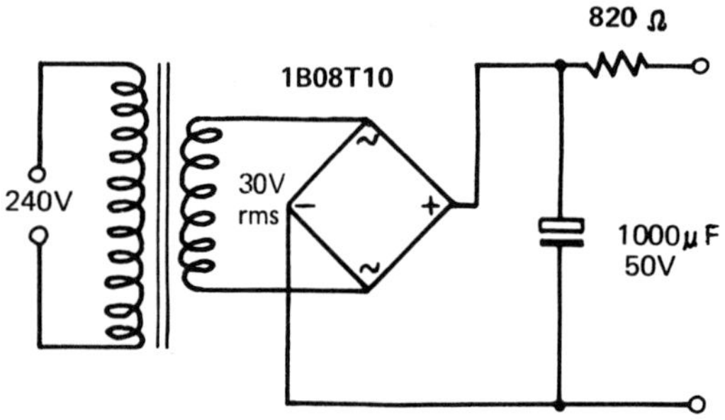

FIGURE 12

In this case resistors R31, R33, zener diode ZD1 and transistor VT8 should be omitted.

In this circuit several points, such as the bases of transistors VT2, VT3, VT4 and VT6 are especially vulnerable to noise pick-up from the power supply. In the past it has been customary to connect these points individually to the rail through two resistors in series, the centre point of which is decoupled to earth by a large electrolytic capacitor. As the current level and current swing at these points is relatively low, economies may be made by connecting them through resistors to a second supply rail, derived from the main rail but very well decoupled by a single capacitor. In this amplifier a 20V secondary rail, dropped down from the main rail by resistor R22 and decoupled by capacitor C8, is used. To maintain the versatility of the amplifier values of resistor R22 for various combinations of amplifier stages are given in table 5.

Table 5

Amplifier			R22
1st stage	2nd stage	3rd stage	kΩ
X			47k
	X		220k
		X	220k
X	X		39k
	X	X	110k
X		X	39k
X	X	X	33k

CONSTRUCTION

A suggested printed circuit board is shown in figure 13 with a component layout in figure 14. This layout is for a stereo preamplifier with two power amplifiers of the type described in the next chapter. The preamplifier may be split off from the power amplifier along the dotted line shown. A single preamplifier may be made by splitting the preamplifier section along the common earth rail. The printed circuit is designed round the following component types:

Resistors - preamp.	Iskra ¼W carbon film, except those marked * which should be Electrosil TR5.
Power amp	Iskra ½W carbon film, except R15 and R16 which should be Radiospares 1W W/W
Presets (RV1 & RV2)	Radiospares min. preset
Capacitors electrolytic	All except C2, C8, C14 (preamp) and C3 & C7 (power amp.) Mullard C426 series. C2, C8, C14 (preamp) C3 (power amp) Mullard C437 series
Polystyrene	C12, C13, C16, C20, C23, C24, C25, C30, C31 Mullard C295 series All others - Iskra, Sufflex or Radiospares polystyrenes
Metallized Film	Erie M310 or Radiospares Poly C
Switches — S1 and S4	Make up from Radiospares "Min Maka Switch"
S2 and S3	Radiospares slide switch
Potentiometers RV1 & RV2	Radiospares tandem ganged controls
RV3	dual concentric Plessey moulded track Plessey moulded track controls may also be used for RV1 and RV2
Input Sockets	DIN 5 pin (preferably)
Output Sockets	DIN 2 pin speaker sockets

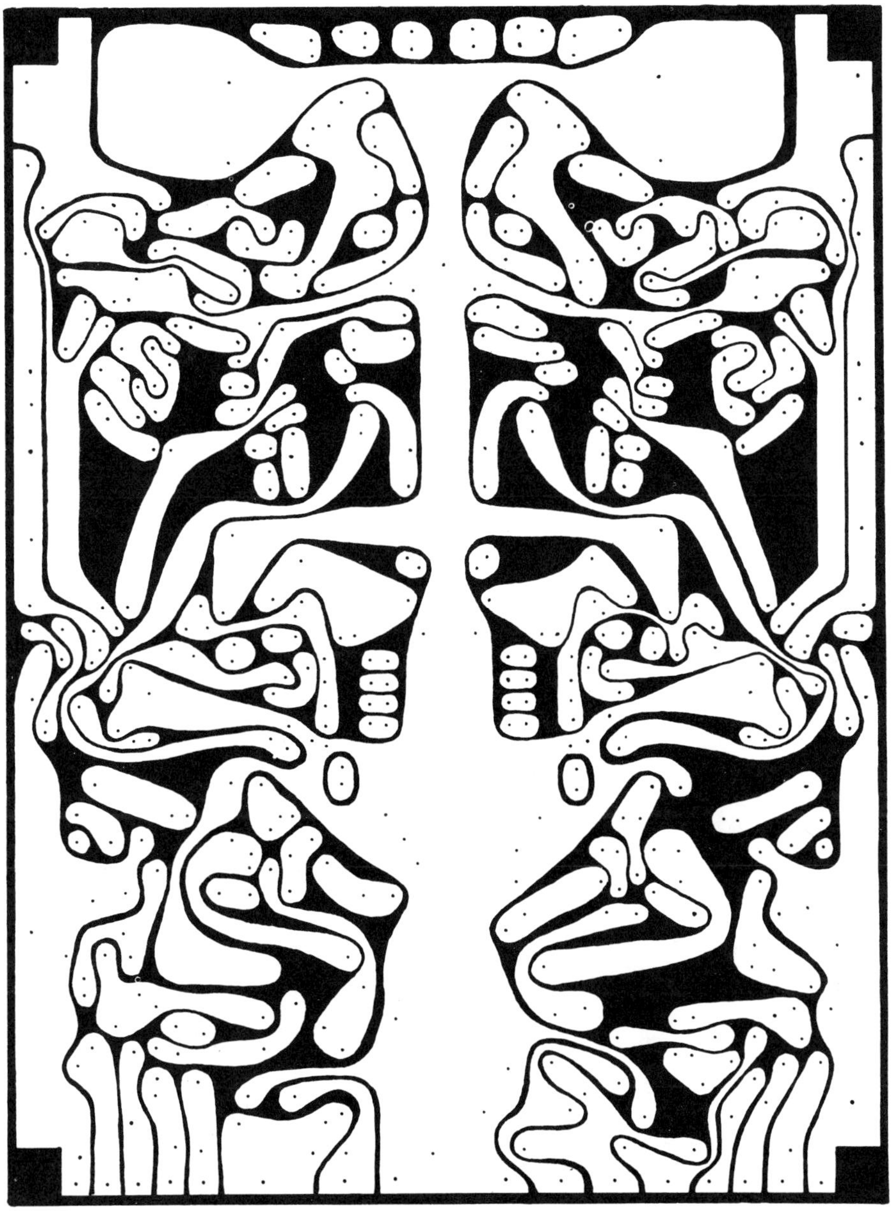

1. Etch away black areas
2. This is a view of etched and pattern as seen looking from component side of board, i.e. etched side is actually a mirror of this.

FIGURE 13 MK II Preamplifier with MkV Power Amplifier

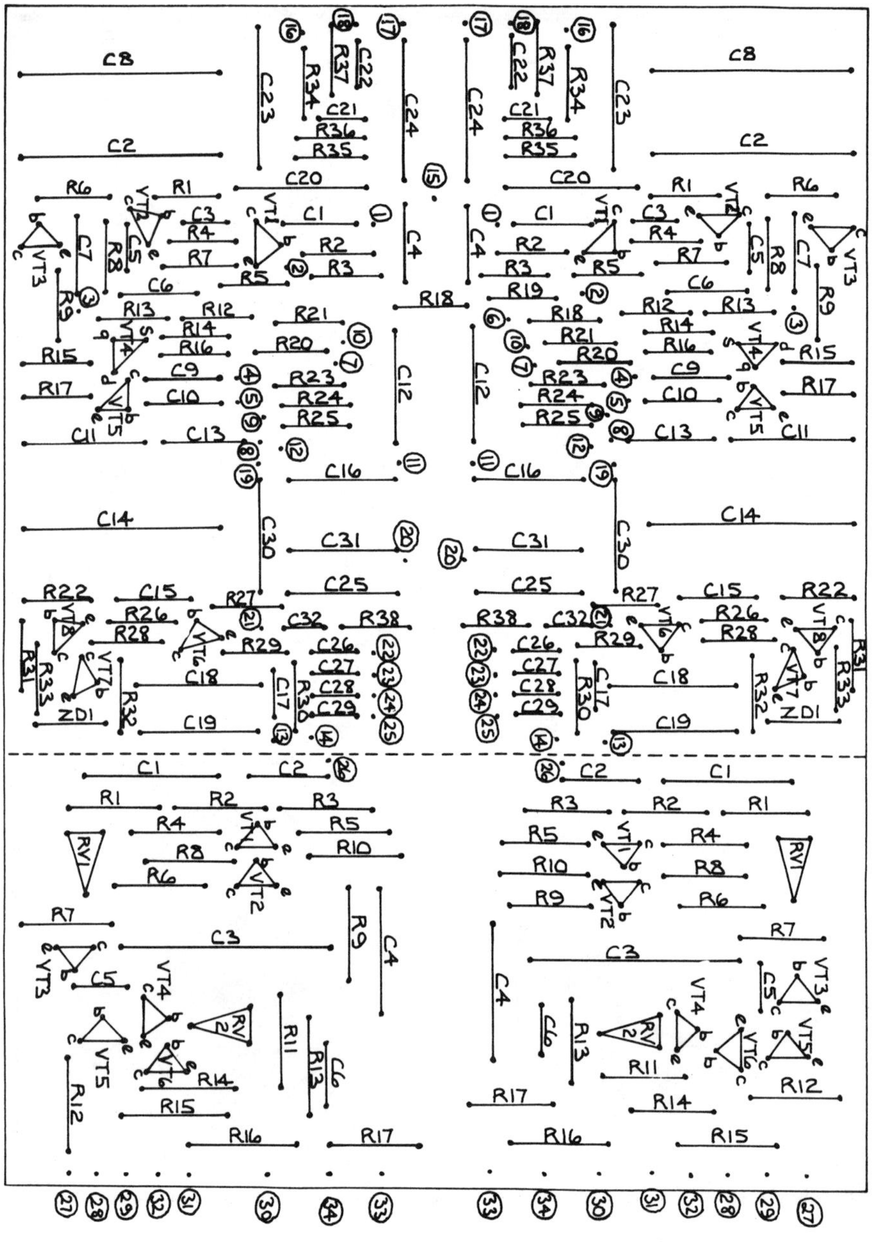

FIGURE 14 Mk II Preamplifier Component Layout

If printed circuit facilities are not available, the printed circuit layout may be used as a basis for a wired up version constructed on perforated board. Leads of considerable thickness > 20 s.w.g. should be used for the power and earth rails.

After the circuit board is constructed, make the connections between the board and the controls and switches, inputs, outputs and power supplies. The controls should be mounted on a front panel and it is preferable to lay them out so that leads between the board and controls are kept fairly short. It is essential to keep output and mains leads away from input leads. If very low noise is desired, screened wire should be used for the inputs and connections between the board and front panel, at least for connections associated with the first stage of the amplifier. In most cases, however, it is possible to get away with unscreened wire. Connect the earth rail of the preamplfier to the metalwork somewhere near the input socket and connect

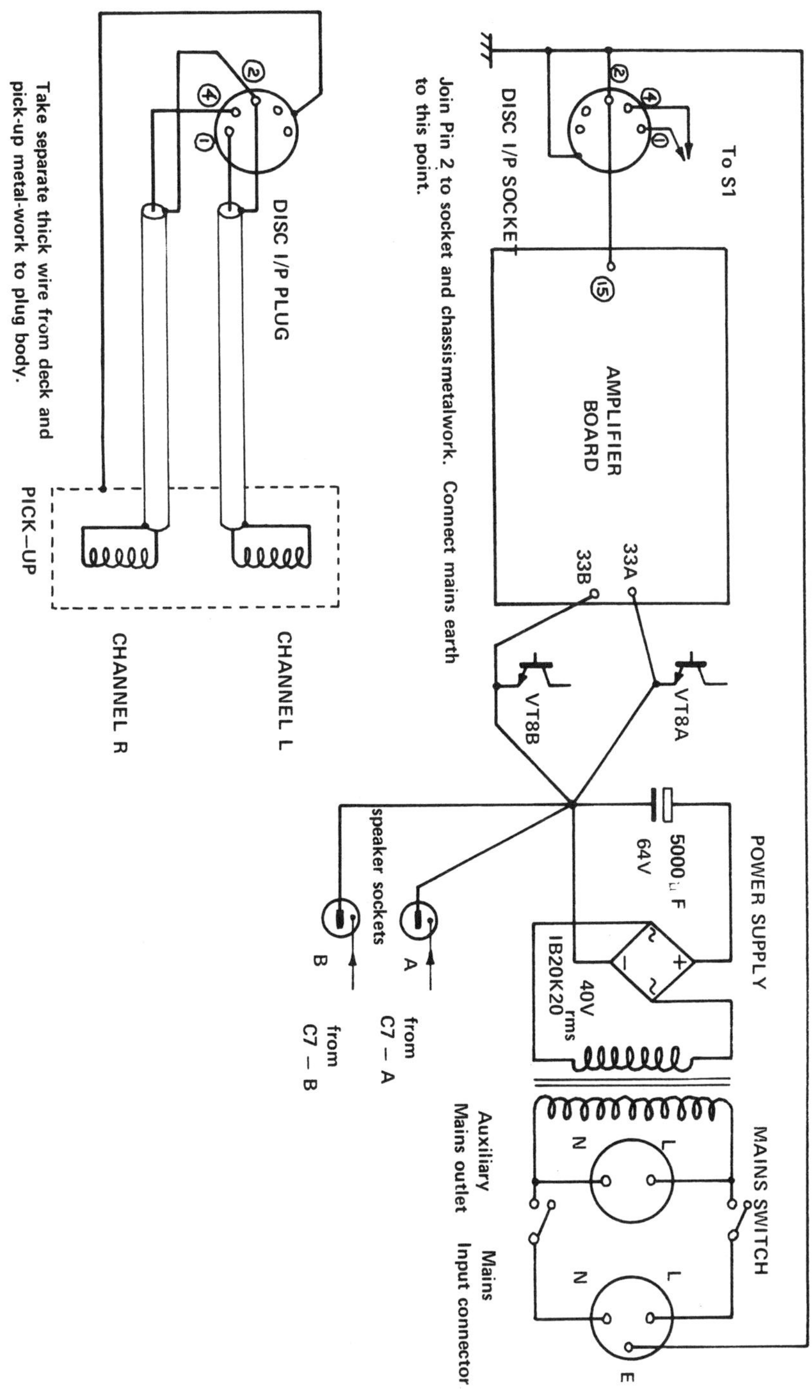

FIGURE 15 Earthing Arrangement

mains earth to this point, also. Connect all the high current negative connections (power transistor emitters and speaker returns) to one point the negative terminal of the power supply reservoir capacitor.

A point on the printed circuit board earth rail near the power amplifier outputs should be connected to this point also.

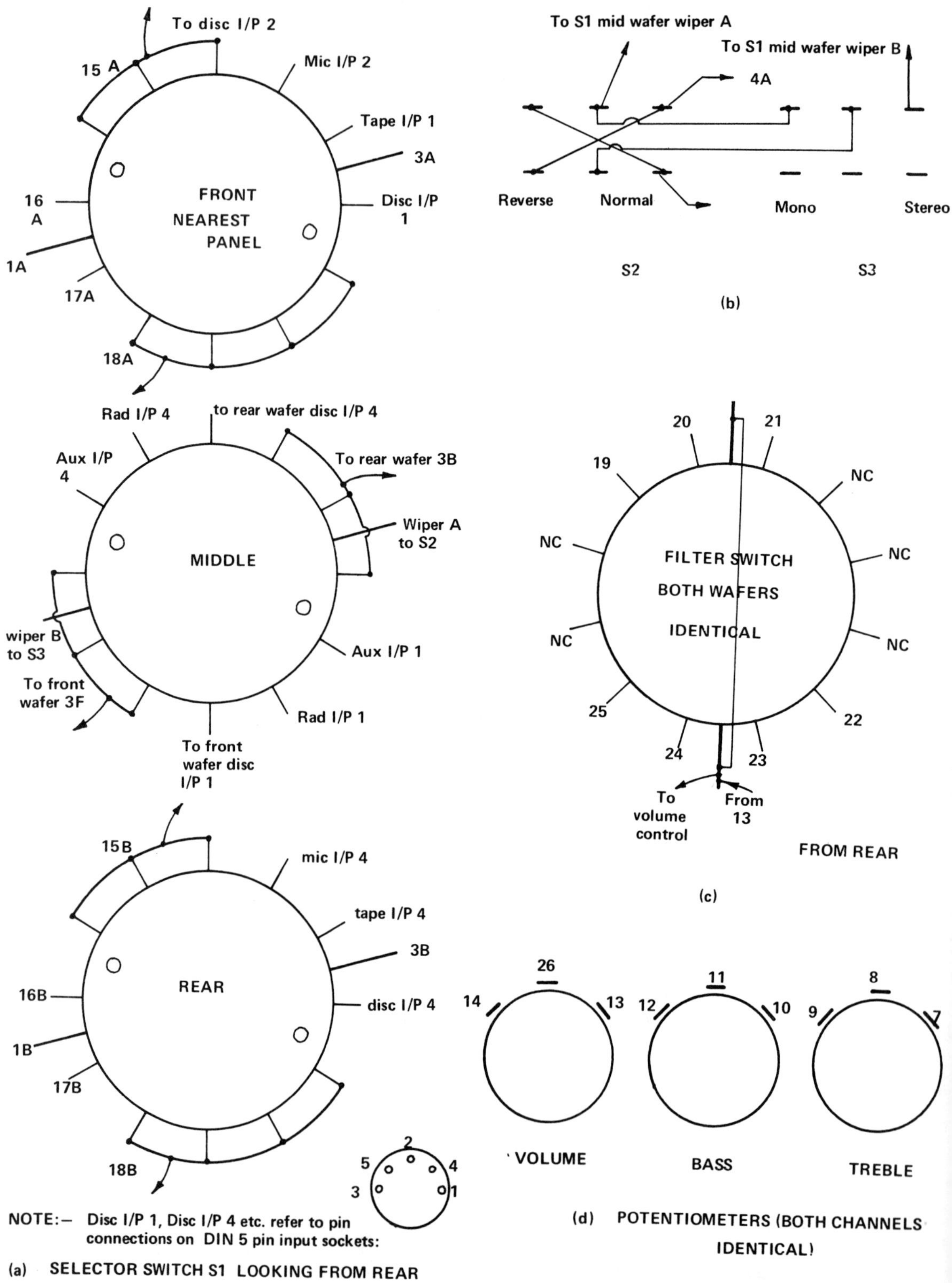

FIGURE 16 Switch Wiring Data

CHECKING OUT

To avoid expensive mistakes checking out and testing the system should be carried out carefully and systematically.

The Preamplifier

(a) Disconnect the preamplifier and power amplifier from the power supply and check that the no-load voltage is less than 60V.

(b) Reconnect the power supply to the preamplifier. Check the voltages on the 30V and 20V rails with a d.c. voltmeter. These should be correct to within ±2V.

(c) If a signal generator and oscilloscope or a.c. millivoltmeter are available insert a signal into one of the inputs and check the outputs of each of the three stages. Vary the frequency input selected and tone controls and plot out response curves, etc.

If such equipment is not available, the approximate d.c. voltages shown in figure 1 should be checked with a voltmeter.

Performance

Input sensitivities - for 100mV output @ 1kHz

1. Disc replay –
(a) magnetic (i) 2mV
(ii) 4.3mV
(iii) 5.8mV

R.I.A.A. standard frequency response. For deviation from standard see figure 5.

Input impedance > 50kΩ

(b) ceramic 100mV

Flat response

Input impedance: 2.2MΩ

2. Tape Replay: 2mV

DIN 90μs, 3180μs response curve, see figure 6.

Input Impedance : > 50kΩ

3. Microphone: 2mV

Flat response

Input Impedance > 50kΩ

4. Auxiliary and Radio: 100mV

Flat response

Input Impedance: Nominally 100kΩ may be increased up to 2M2

5. Tone and filter controls – see figures 7 – 10

Harmonic Distortion %

(a) Microphone Input

Frequency / O/P Voltage	100Hz	1kHz	10kHz
300mV	0.02	0.02	0.02
3V	0.02	0.013	0.11

(b) Magnetic input

Frequency / O/P Voltage	100Hz	1kHz	10kHz
300mV	0.02	0.02	0.02
3V	0.026	0.036	0.11

Overload factor dB, T.H.D. = 0.3%

	100Hz	1kHz	10kHz
Magnetic I/P	34.5	35.1	37.4
Microphone I/P	35.6	35.6	33.8

Signal/Noise ratio:

Magnetic	64dB
Tape	58dB
Microphone	61dB
Others >	70dB

Power Supply Requirements:

30V @ 15mA smoothed and stabilised, per channel or

50V @ 28mA rough per channel

XII POWER AMPLIFIERS (1 - 100 W)

by Ian Hardcastle

This chapter describes a number of audio power amplifiers which have been designed to fulfil a wide range of requirements. Special attention is paid to the design requirements of the output and driver stages and of the feedback arrangements. The three types of amplifiers covered are:

1. A number of variations on a 2 or 2.5W amplifier intended for TV sound output, record players, tape recorders, etc., together with a 1W version intended for domestic portable radio audio outputs.
2. Various hi-fi power amplifiers with powers > 10W based on the design published in a previous Application report. This redesigned version offers better performance and improved stability – especially with awkward loads.
3. Two versions of a 100W amplifier intended for hi-fi, guitar amplifier and public address applications.

These designs will be dealt with in turn.

PART 1

1,2 AND 2.5W AMPLIFIERS

In this section a number of variations on a basic 2.5W amplifier are described, together with de-rated versions giving 2W, but which require no setting up. All the amplifiers, except the 1W, run from a 35V rail, such as may be found in the present generation of colour television sets or be obtained from an overwinding on a record player or tape recorder motor. A 35Ω loudspeaker is used with these.

The 1W amplifier runs from a 9V battery into an 8Ω loudspeaker.

All the designs which are described in this chapter use what is basically a Class B push-pull output stage. This configuration is shown in Figure 1.

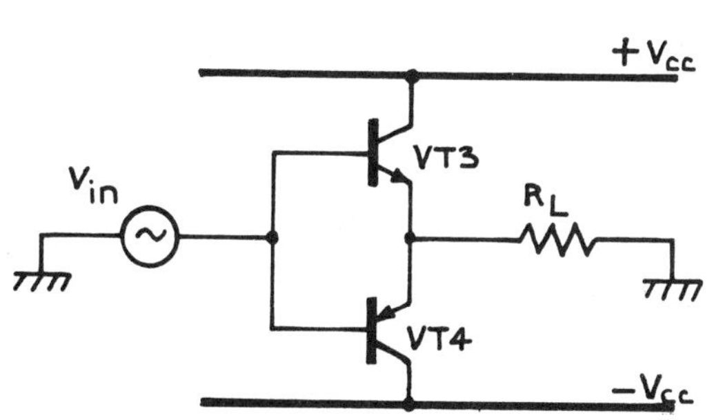

FIGURE 1

In this, when the input is zero, both transistors are turned off and no current flows. As the input voltage swings positive, transistor VT3 turns on (assuming zero V_{BE} for the moment) and current flows into the load. As the input voltage increases to its peak and then turns towards zero, the current into the load increases and decreases accordingly until, at zero input, the current returns to zero and VT3 turns off. As the input voltage swings negative, VT3 remains turned off and VT4 turns on, pulling current out of the load. VT4 remains on throughout the negative half cycle of the input voltage and turns off when the input returns to zero. This arrangement has relatively high efficiency as the current drawn from power supplies is proportional to the current driven into the load, with zero current being drawn when there is no output.

As an illustration of the procedure required for the design of a class B output stage, a simple amplifier delivering 2.5W into a load will be used. Often, the load impedance will be specified but in this case where the amplifier and speaker are most likely to be built as a unit and a non hi-fi speaker is required, the choice of speaker impedance can be left to the designer. The first thing to do is to draw up a table showing peak output current and peak to peak output voltage for 2.5W into various load impedances.

From this information the rail voltage and output transistors can be chosen and the most economical line up selected. The choice table is shown in Table 1.

Table 1

Speaker Imp.Ω	4	8	16	35	40
I_O (peak) (A)	1.12	0.79	0.559	0.378	0.353
V_O (pk – pk) (V)	8.94	12.64	17.89	26.45	28.28
V_{CC} (V)	15	20	25	32	35
NPN output	TIP31	TIP29	BC142	TIS90	TIS90
PNP output	TIP32	TIP30	BC143	TIS91	TIS91

From the table, it is obvious that on cost consideration, the choice lies between the 35 and 40Ω version, as the output devices are cheapest. The 35Ω version is the final choice as it allows a greater voltage margin between the breakdown voltage if the transistors and the minimum power supply voltage required. Positive and negative power supplies are required to ensure that no current flows through the load under quiescent conditions. A single power supply may be used by coupling the load to the output mid-point through a capacitor and attaching the other end of the load to the positive power supply or to earth. The capacitor is charged to half the supply voltage and causes the voltage across the load to swing about the supply or earth by an amount equal to the swing of the output mid-point. The value of this capacitor must be such that its impedance is low compared with the impedance of the load at the lowest frequency at which it is desired to have full power output; its current rating must be such that it can handle the full output current of the amplifier.

The mean dissipation, P, in each output transistor is given by the expression:

$$P = \frac{V_{CC} \cdot V_o}{2\pi\sqrt{2} \cdot R_L} - \frac{V_o^2}{8 R_L} \cdot \left[1 + \frac{R_E + R_S}{R_L}\right]$$

Where R_S is the source resistance of the power supply and V_o the output voltage (r.m.s.)

Differentiating this in V_o gives a maximum for P when

$$V_o = \frac{V_{CC} \cdot R_L}{\sqrt{2} \cdot \pi (R_E + R_S + R_L)}$$

which corresponds to an output power of 1.48W and a maximum device dissipation of 0.88W, assuming a zero power supply source resistance.

The basic driver and output stage is shown in Figure 2.

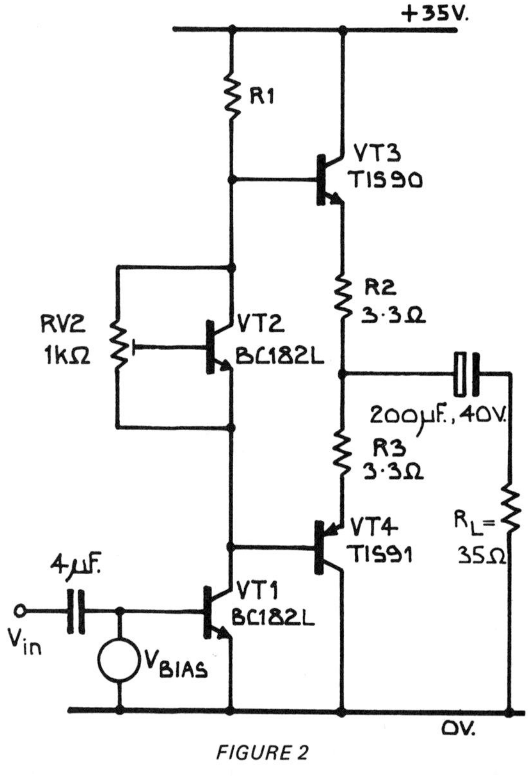

FIGURE 2

The TIS90 and TIS91 output transistors are special high dissipation silect transistors having a copper collector lead which, when it is soldered down to a large pad on a printed circuit board, enables each device to dissipate the necessary 900mW at up to 80°C heat sink and lead temperature. These transistors have a base emitter voltage of about 0.7V each. This means that, if they are used in the circuit shown in Figure 1 (and this argument also applies to the circuit shown in Figure 2), the input voltage has to rise to about 0.7V before VT3 turns on and the output starts to follow the input. When the input voltage has passed its maximum and falls back towards zero, the output returns to zero when the input is still at +0.7V. The input has to fall to about −0.7V before VT4 turns on and the output voltage can follow the input again. The output stops following the input when the input voltage rises beyond −0.7V. This effect produces an output waveform exhibiting cross-over distortion as shown in Figure 3.

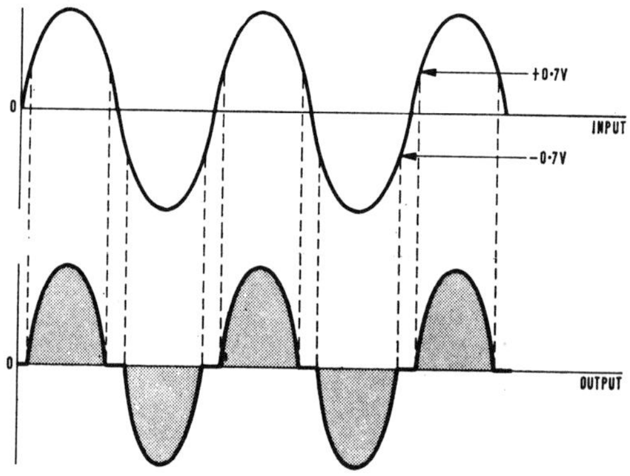

FIGURE 3

Cross-over distortion is most noticeable for low level inputs where the 1.4V gap is most significant compared with the output voltage. In practical amplifiers, the amount of cross-over distortion found is somewhat less than would be expected from the rather over-simplified argument above. The main reason for this is that the output stage is current driven rather than voltage driven and this gives a smoother transition through the cross-over region. The amount of cross-over distortion can be reduced by applying large amounts of negative feedback, but this tends to be less effective at high frequencies where effects due for the output transistors requiring a finite time to turn on and off become noticeable. Also the driver stage may not be fast enough to cope with the sudden change in its operating conditions which occurs at the cross-over point. A more effective way of reducing cross-over distortion is to run the output stage in class AB in which, at zero input, the output transistors are not turned fully off, but run with a small DC quiescent current of about 2mA. To set up this quiescent current the bases of the output transistors must have a voltage placed across them such that the required quiescent current flows. This quiescent current is stabilised by voltage feedback due to the current flowing through the resistors R2 and R3 in the emitters of the output transistors. The full output current at the peaks of the output is also dropped across these resistors and thus, to enable the output voltage swing requirements to be met from the available power supply voltage rail, the value of the feedback resistors must be kept small. The relatively high thermal resistance of the output transistors (compared with larger power transistors) means that, during operation, the junction temperature of the output devices changes considerably, with a consequent large change in the base emitter voltage.

The decrease in base emitter voltage with increase in temperature increases the quiescent current and thus the dissipation in the devices. This situation can lead to a thermal runaway condition and consequent device failures. This situation may be avoided by using, to set the interbase voltage, an element whose voltage accurately tracks the base emitter voltage of the output transistors with temperature. A fairly obvious choice for this would be two forward biassed diodes which could be placed in thermal contact with the output transistors. The main objection here, however, comes from manufacturing tolerances.

Therefore, if diodes with low V_F were used with high V_{BE} transistors, the quiescent current set would be practically zero; if high V_F diodes and low V_{BE} transistors were used, then the current would be high enough to cause overdissipation in the transistors. To make allowance for these differences, the value of the emitter resistors R2 and R3 would have to be so high that a serious loss of output swing capability would result. Replacing one of the diodes with a variable resistor would allow the quiescent current to be set up to the required value, but, the current once set, would vary widely as the temperature of the output transistors changes. The 'ideal' element required is a diode whose forward voltage may be varied. This can be achieved by using a transistor in the manner shown in Figure 4.

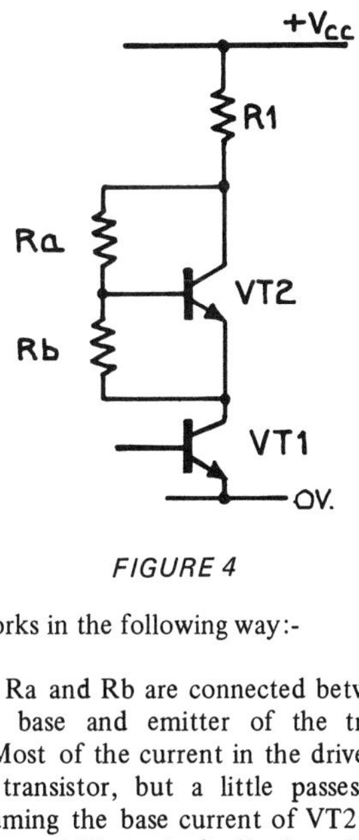

FIGURE 4

This device works in the following way:-

Two resistors Ra and Rb are connected between base and collector and base and emitter of the transistor VT2 respectively. Most of the current in the driver stage passes through the transistor, but a little passes through the resistors. Assuming the base current of VT2 is small, then the resistors Ra and Rb apply feedback to the transistor so that the ratio of its collector emitter voltage to its base emitter voltage is defined by the expression.

$$V_{CE}/V_{BE} = (Ra + Rb)/Rb$$

By making Ra approximately equal to Rb, the collector voltage of transistor VT2 can be set to about 2 V_{BE} which is the approximate interbase voltage required for this output stage. Thus transistor VT2 and the two resistors may be used to set the interbase voltage of the output pair, transistors VT3 and VT4, so that the quiescent current of the desired value flows. This arrangement also gives one further benefit.

If the biassing transistor VT2 is placed in thermal contact with one of the output transistors it may be used to stabilise the quiescent current of the output stage as the junction temperature of the output transistor changes. This effect is achieved because the temperature of the base emitter diode of the bias transistor follows that of the base emitter diode of the output transistor. Any change in the base emitter voltage of the bias transistor causes the collector emitter voltage to change by the factor $(R_a + R_b)/R_b$, which is in this case about 2. This enables the interbase voltage to change sufficiently to compensate for changes in the base emitter voltages of both output transistors, thus achieving the necessary quiescent current stabilisation. In practice, however, it is necessary to use a potentiometer instead of the fixed resistors Ra and Rb to allow the quiescent current to be adjusted to the required value, because of production tolerances in the base emitter voltages of the three transistors involved.

The Driver Stage

The purpose of the driver stage is to provide the necessary voltage and current drive for the output stage. It also has to produce a large linear voltage gain. When the output is at its most negative the peak collector current of VT4 is 360mA. This requires, for a minimum gain transistor, a base current of 9mA which must pass through the driver transistor VT1. This current transistor VT1 can handle easily. When the output is at the positive peak of its swing the collector current of VT3 is 360mA, which requires a maximum base current of 6mA. (The npn transistor has a slightly higher current gain than the pnp). This current flows through collector resistor R1 of the driver transistor. To ensure that the output voltage swing requirements of the circuit are met the voltage drop across resistor R1 must be less than 2V, which means that if VT1 turns 'off' completely at this point, the value of R1 must be less than 330Ω. This low value of R1 means that under quiescent conditions, VT1 must dissipate nearly 1W and at the negative peak of the output swing VT1 must not only sink the base current of VT4, but also a further 100mA from R1. This is not only undesirable from the point of view of dissipation in both the transistor and resistor but also leads to distortion due to the large changes in VT1's collector and base current. These disadvantages may be overcome by connecting the upper end of R1 to the power supply via the load as shown in Figure 5.

Under quiescent conditions the base voltage at VT3 is approximately $V_{CC}/2 + V_{BE}$ and hence the voltage across R1 is about $V_{CC}/2 - V_{BE}$. The voltage across the load swings about the positive rail and the base voltage of VT3 swings an equal amount about $V_{CC}/2$. Thus the voltage across, and hence the current through R1, remains constant as the output level changes. The value of R1 may now be designed so that the required base current for VT3 is provided when the voltage across R1 is $V_{CC}/2$. In fact the value of R1 chosen is slightly less than this to reduce the swing in VT1's collector current. The large increase in the value of R1 that this configuration makes possible reduces the quiescent current of VT1 to 8.2mA and its peak collector current to 17.2mA.

Simple Amplifier Stage

A very simple amplifier may be made by applying the D.C. and a.c. negative feedback components R4, R5 and RV1 to the basic driver and output stage already discussed. Resistors R4 and RV1 form a potential divider from the output mid-point to the base of VT1. Any tendency for the mid-point voltage to rise causes the base voltage of VT1 to rise: this causes the collector current of VT1 to rise and hence its collector voltage and the mid-point voltage to fall. The tolerance of the base emitter voltage of VT1 and the small range of quiescent mid-point voltage allowable for the necessary voltage swing to be achieved means that RV1 should be variable so that mid-point voltage may be set to the required value. Because the quiescent mid-point voltage is so dependent on the V_{BE} of VT1 the change in this voltage of 40mV/°C increase in VT1's temperature may be expected.

set so that symmetical clipping occurs in the output stage. Asymmetrical clipping, giving a considerable loss in output power, would otherwise occur.

Table 1.2

Power / Frequency	Distortion %			
	10mW	100mW	1W	2.5W
100Hz	0.27	0.42	0.98	3.70
1kHz	0.33	0.46	0.96	3.60
10kHz	0.45	0.35	0.76	2.60

Sensitivity: 0.47V for 2.5W
Input Impedance: 4k7

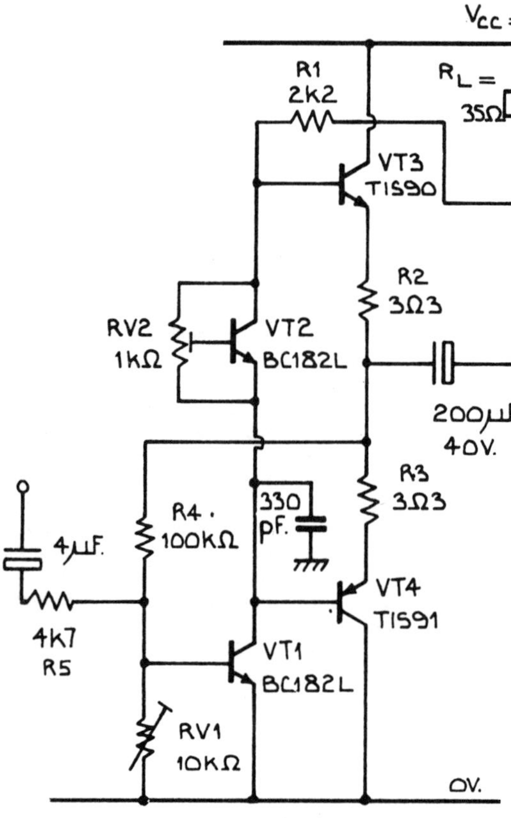

FIGURE 5

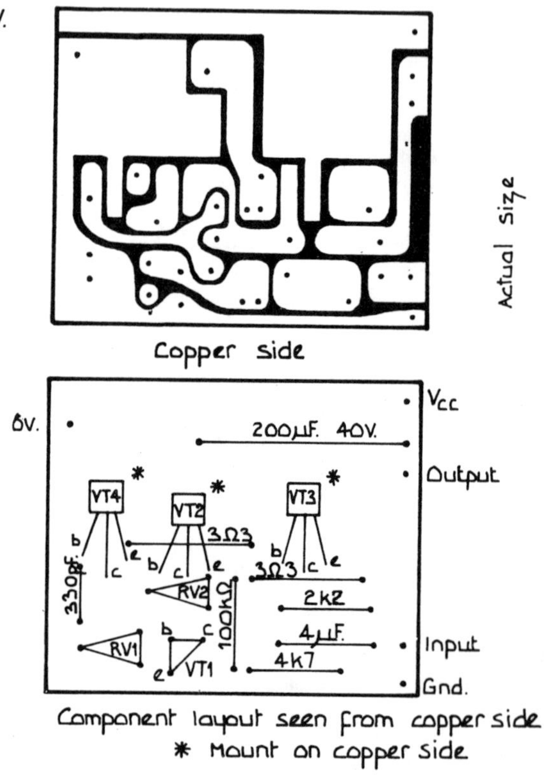

FIGURE 6

The quiescent mid point voltage is then defined by:-

$$V_Q = V_{BE1} \cdot \left[\frac{R4 + RV1}{RV1}\right]$$ which shows that

any tolerance in V_{BE1} is multiplied up by a factor of about 20. According to the data sheet for the BC182L, its V_{BE} can vary by as much as ±0.075V which would give a tolerance of ±1.5V on the setting of the mid point voltage. RV1 is made variable so that the mid point voltage may be

Performance figures for the circuit shown in Figure 5 are given in table 1.2. A suggested circuit layout showing the large copper areas required for heat-sinking the copper collector leads of the output transistors and the arrangements for ensuring the thermal contact of the biassing transistor VT2 with one of the output devices is given in Figure 6.

If a lower power output is tolerable, some of the output swing capability may be thrown away and the biassing arrangement shown in Figure 7 may be used.

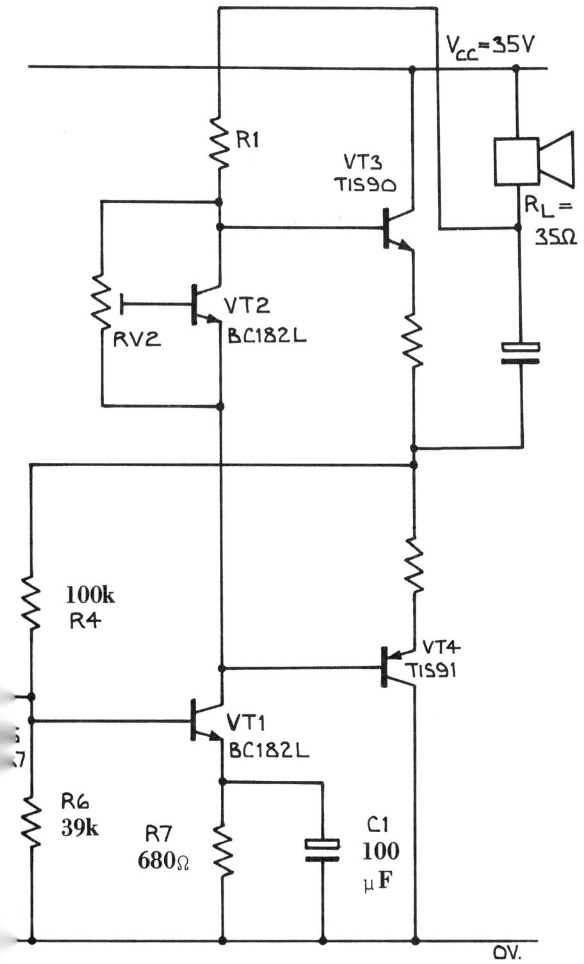

FIGURE 7

Calculator analysis shows the total worst case variation on the mid-point voltage on this 2W design is ±3.8V. This calculation assumes the worst case data sheet variations of V_{BE}s and 10% resistor tolerances. The resulting voltage variation is greater than that tolerable. However, it is unlikely that such a worst case combination will occur and therefore this amplifier design will give its rated 2W in most cases. The temperature coefficient of this mid-point voltage is about 3.1mV/°C. The emitter resistor R7 is decoupled by capacitor C1 to preserve the a.c. open loop gain of the amplifier.

In this design and the design shown in Figure 5, the base of transistor VT1 is a virtual earth point, hence the a.c. gain of the amplifier is defined by the ratio of resistors R1 and R5 and approximately equal to R4/R5. In the design shown in Figure 5, the gain is set to 20 and thus it requires 0.47V for 2.5W (9.35 V r.m.s.) output. The relatively large base current required by VT1 means that the value of R4 must be kept low (100kΩ). The value of R5 is defined by the gain required, giving a fairly low (4k7Ω) input impedance.

High Input Impedance Version of Simple Amplifier Stage

The input impedance may be increased by connecting another transistor, VT5 to VT1 in the Darlington configuration as shown in Figure 8. This configuration has the advantage of increasing the input impedance and gain of the stage with the addition of only two further components, the extra resistor being used to limit VT5's collector voltage so that a low voltage high gain device may be used in this position.

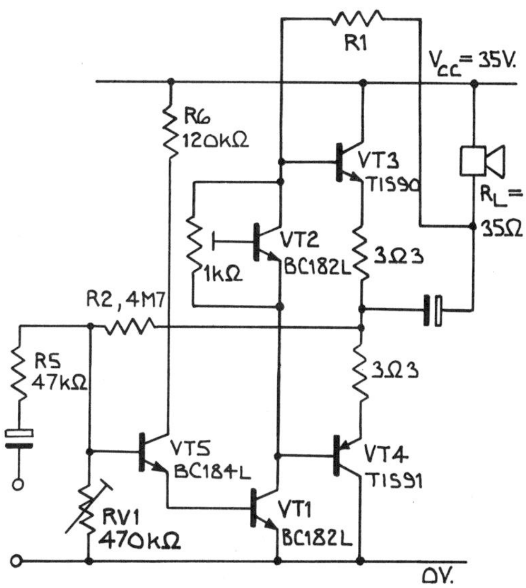

FIGURE 8

This design has an a.c. gain of 100 giving an input requirement of 93mV for 2.5W output, with input resistance of 47kΩ, which is a worthwhile improvement on the basic circuit. The main disadvantage of this configuration is that RV1 must still be variable to set the mid-point quiescent voltage. The temperature coefficient of the mid-point voltage is twice that of a simple circuit, as two V_{BE}s are now involved in definition of this voltage.

These disadvantages may be overcome at the expense of the output power by the adoption of similar modifications to those shown in Figure 7 (with an appropriate change in the value of R4 to compensate for the extra V_{BE} across it.)

Simple Amplifier with Preamplifier

A version of the amplifier having a preamplifier stage which gives a considerable increase in sensitivity and a slight increase of input impedance over the basic circuit shown in Figure 5 is shown in Figure 9. This circuit uses considerably more components that the circuit shown in Figure 8, but has the advantage of having a smaller temperature coefficient of quiescent midpoint voltage. Performance figures for this circuit are shown in Table 1.3 and a printed circuit layout for this circuit is given in Figure 10.

Table 1.3

Freq. \ Power	Distortion % 10mW	100mW	1W	2.5W
100Hz	0.42	0.80	2.4	4.1
1kHz	0.40	0.78	2.3	3.2
10kHz	0.62	0.64	1.8	2.7

Sensitivity: 31.4mV for 2.5W
Input Impedance: 22kΩ

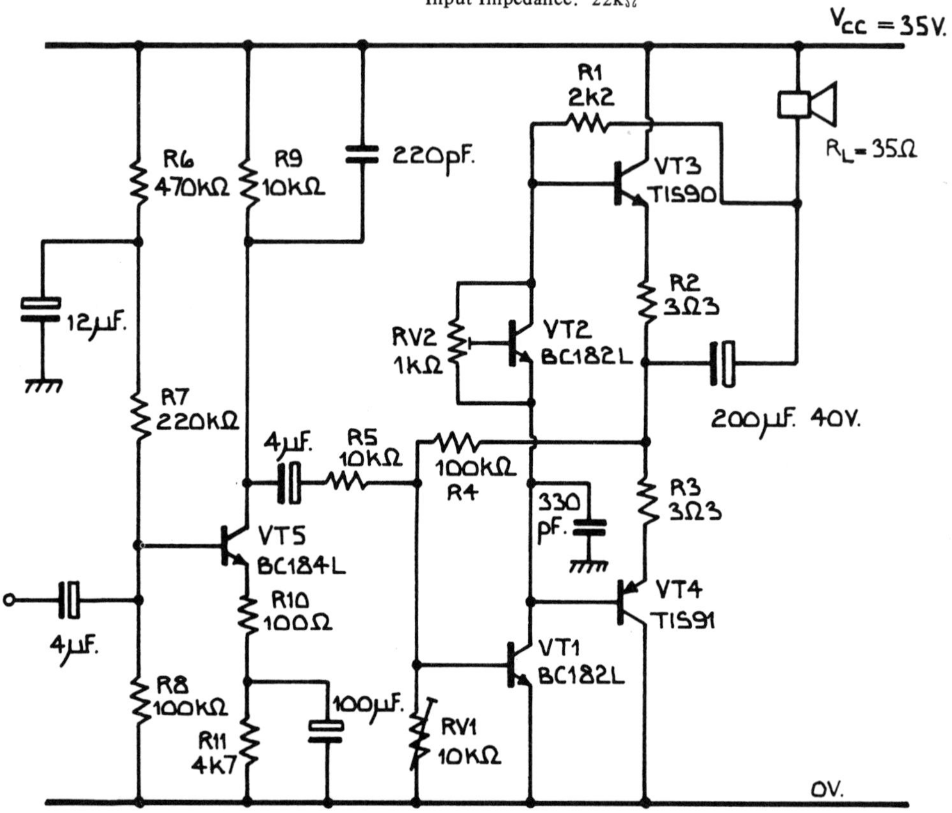

FIGURE 9

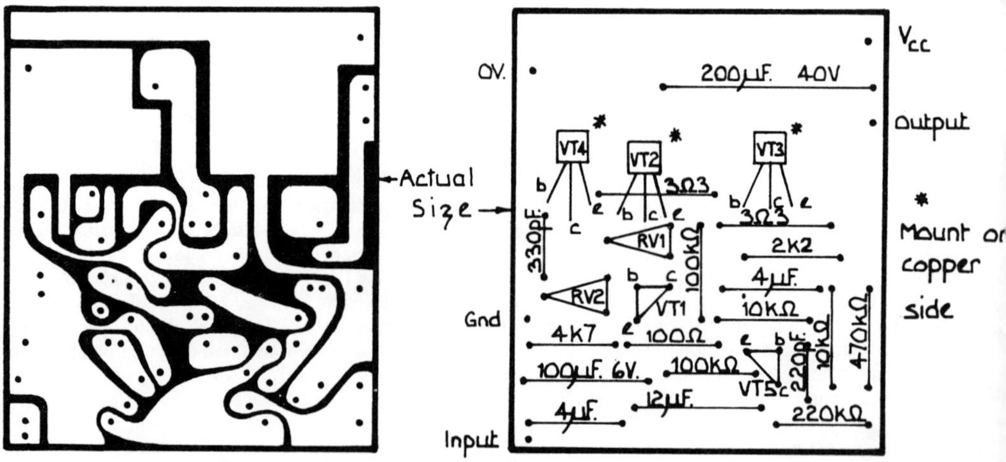

FIGURE 10 Copper side — Component layout seen from copper side.

Improved Amplifier Configurations

If the expense of using an extra transistor can be justified, a better way of using it than in the Darlington configuration and a more economical way than in a simple a.c. pre-amplifier is the circuit shown in Figure 11. This circuit has a higher closed loop gain and input impedance that the basic circuit shown in Figure 5. Because the extra transistor is D.C. coupled and used to give current gain, the open loop gain of the circuit is higher and hence its distortion is lower.

The open loop gain of the amplifier is given by:

$$A_V = \frac{\beta_1 . \beta_3 . R_L}{r_{e5}}$$

where β_1 and β_3 are the current gains of VT1 and VT3 (or 4) respectively and r_{e5} is the intrinsic emitter resistance of VT5) and is typically about 1000Ω.

Also the D.C. level of the output mid-point is closely defined without the need for adjustable components. Unfortunately, in this design a number of extra passive components (including two electrolytic capacitors) as well as one more transistor are needed, and hence, in some cases, the use of the Darlington circuit of Figure 8 may be justified on economic grounds. In the circuit shown in Figure 11 the driver circuit is turned upside-down and pnp transistor used. The load is connected to earth instead of to the positive supply so the arguments which were applied to the npn driver may be applied similarly to this pnp driver stage. The base of the driver stage is directly coupled to the collector of the extra input transistor VT5. The large spread of current gain of transistor VT1 means that its base current will also have a large spread. To reduce the variation in the collector current of VT5 that this would cause, the base emitter diode of transistor VT1 is shunted by 2k7Ω resistor R7 which reduces the possible range of VT5's collector current from 4.8 : 1 down to 1.35 : 1.

The base voltage of transistor VT5 is defined by the potential divider chain formed by resistors R4, R5 and R6. The junction of resistors R4 and R5 is decoupled to earth by capacitor C2 to prevent hum and noise from the power supply being fed into the base of transistor VT5 and thus from appearing on the output.

The emitter of transistor VT5 is connected to the output mid-point via a 10kΩ resistor R8 and to earth via the 100Ω resistor R9 and capacitor C3. This arrangement applies to both D.C. and a.c. negative feedback to the circuit. For D.C. C3 may be considered as an open circuit, thus the D.C. level of the output mid-point is defined by the base voltage of VT5 (set by the potential divider R4, R5 and R6), its V_{BE} and the voltage drop due to its collector current flowing through resistor R8. The first two of these quantities are virtually static and as has been stated above, the addition of R7 minimises the variation of VT5's

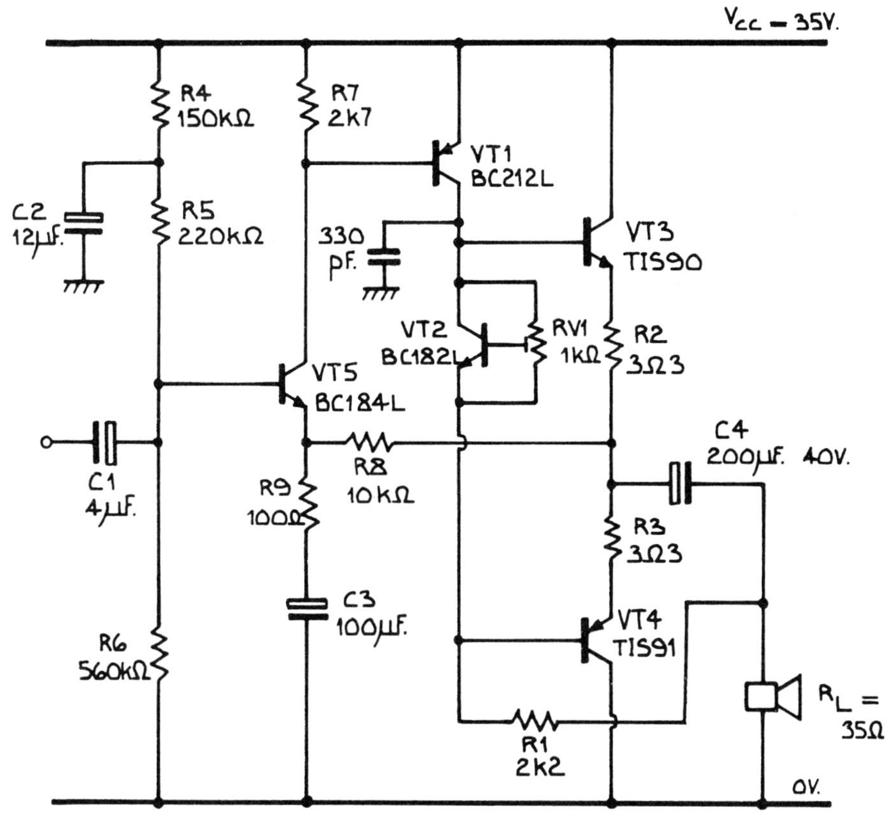

FIGURE 11

collector current. The D.C. level of the output midpoint is, then, well defined, and the predicted temperature coefficient is +6mV/°C. Any tendency for the output mid-point voltage to rise reduces the collector current of VT5 and hence the base current to transistor VT1, causing VT1 to turn off, and its collector voltage to fall which corrects the tendency for the mid-point voltage to rise.

For a.c. the emitter of transistor VT5 may be considered to be grounded via the 100Ω resistor R9, so that the signal from the output appearing there is attenuated in the ratio of 100/10100. This sets the a.c. closed loop gain to 101. As the input frequency is reduced, the reactance of capacitor C3 increases until it is comparable with R9, when the closed loop gain of the amplifier starts to fall. The low frequency −3dB point is defined by the time constant R9. C3.

Performance figures for this circuit are given in table 1.4 and a suggested printed circuit layout is shown in Figure 12.

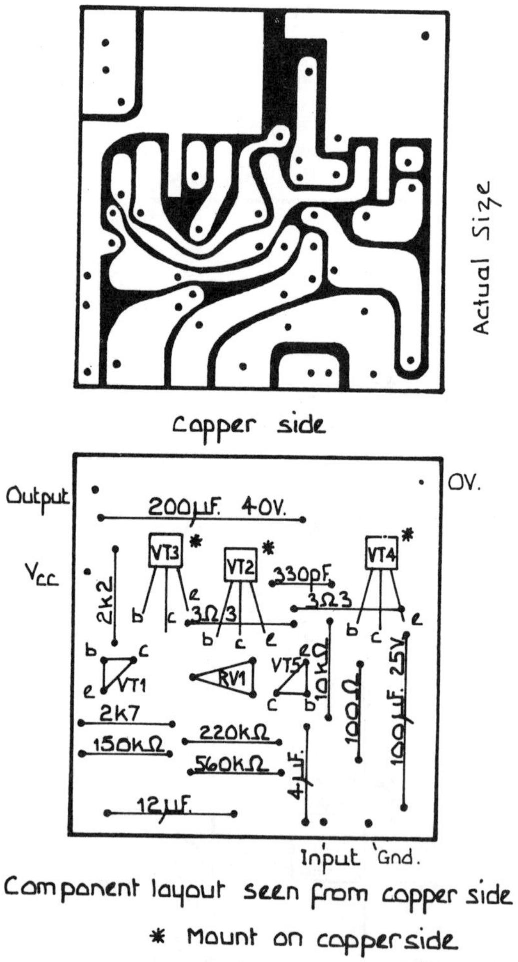

FIGURE 12

Table 1.4

Freq. \ Power	Distortion %			
	10mW	100mW	1W	2.5W
100Hz	0.6	0.13	0.16	0.84
1kHz	0.06	0.07	0.15	0.78
10kHz	0.12	0.09	0.12	0.62

Input Impedance: 180kΩ Normal version
1M1Ω High Input Impedance version of Figure 1.13
Sensitivity: 100mV for 2.5W

High Input Impedance Version

A further modification to this circuit may be made to increase its input impedance. The input impedance seen at the base of transistor VT5 is high due to the high ratio of open loop gain to closed loop gain of the amplifier, but this impedance is shunted by the impedance of the potential divider formed by resistors R5 and R6. For a.c. these resistors are decoupled to earth by capacitor C3. This impedance is low compared with the impedance looking into VT5's base because of the necessity to keep the current through R4, R5 and R6 high compared with transistor VT5's base current. This is to avoid changes in VT5's base voltage due to its large gain spread.

The input impedance of the amplifier may be increased by making a modification to the input stage shown in Figure 13.

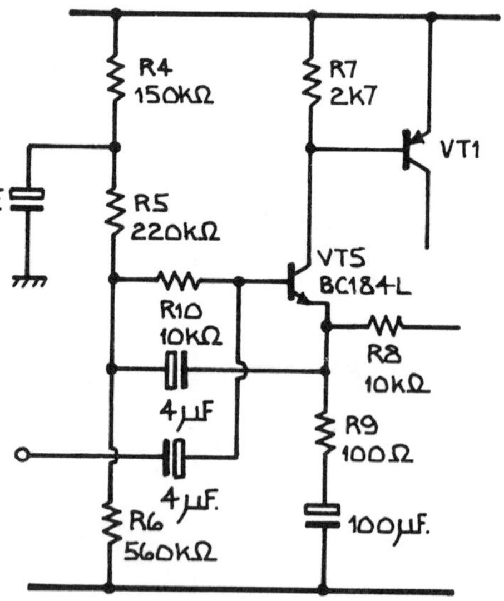

FIGURE 13

In this circuit, the a.c. voltage occurring on the emitter of transistor VT5 is fed onto the junction of resistors R5 and R6. This voltage is in phase with the input voltage, but slightly less than it in amplitude by the ratio $(A - A^1)/A$ where A is the open loop and A^1 the closed loop gain of the amplifier. Thus the input impedance looking back into the 10kΩ resistor R10 is multiplied up by the ratio A/A^1. For the input impedance of the stage as a whole, this resistor is shunted by the impedance seen looking in the base of transistor VT5, which is high.

With this bootstrap feedback arrangement, the input impedance is greater than 1MΩ. This makes the amplifier suitable for use in low cost record players as it has a high enough input impedance and sensitivity to be driven directly (via a volume and tone control) from a medium output crystal or high output ceramic cartridge. Performance figures for this amplifier are given in table 1.4 and a suggested printed circuit board layout in Figure 14.

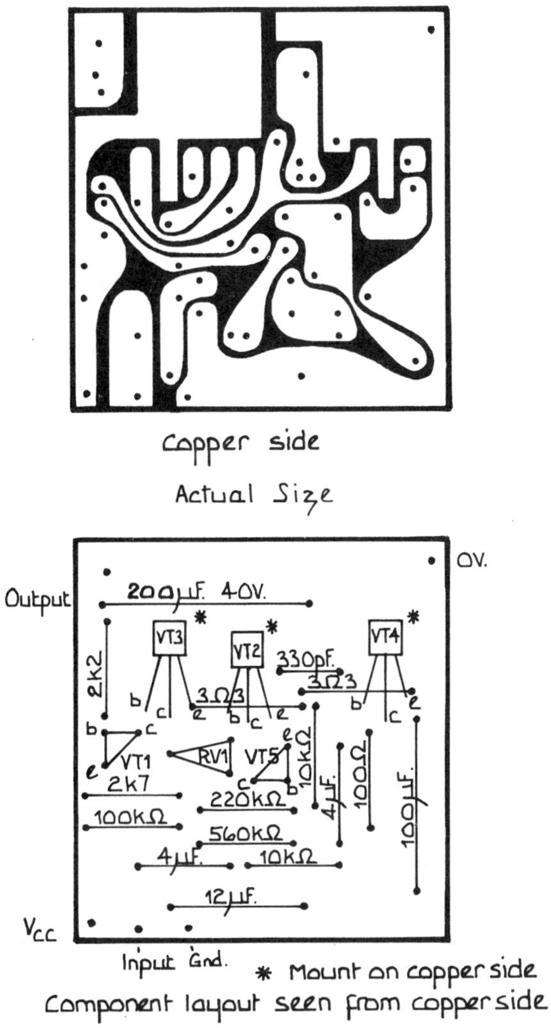

FIGURE 14

Version requiring no setting up of quiescent current

From a mass production point of view, all the amplifiers so far described suffer from an important disadvantage. The quiescent current in the output stage néeds to be set to the correct value by potentiometer RV2 after assembly. This is probably a more difficult adjustment to have to do than the mid-point voltage setting which has already been eliminated, as it is easier to measure a voltage on test than it is to interrupt the collector circuit of VT3 in order to measure its collector current. By limiting the power output of the amplifier, the emitter resistors of VT3 and VT4 may be increased to a point where in spite of the tolerances in the V_{BE}s of transistors VT2, VT3 and VT4, the quiescent current in VT3 and VT4 is large enough to eliminate cross-over distortion, but not great enough to cause over-dissipation.

If a tolerance of ±50mV is allowed on each of the V_{BE}s (which is less than the data sheet allows) and a maximum quiescent current of 10mA is allowed, emitter resistors of 27Ω are required, which limits the available output power to 1.25W. If such a reduction in output power can be tolerated, satisfactory operation of any of the circuits so far described may be obtained by making R2 = 27Ω, R3 = 27Ω, removing potentiometer RV2 and making Ra = 620Ω and Rb = 470Ω (see Figure 4).

A more expensive way of eliminating RV2 is to use the circuit shown in Figure 15.

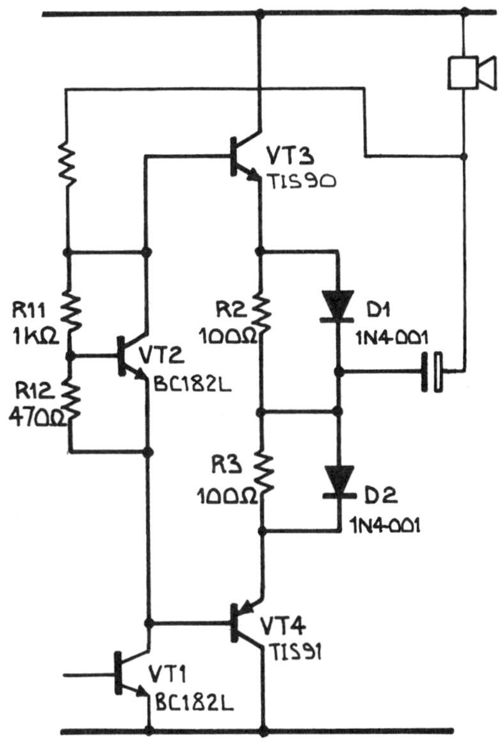

FIGURE 15

In this circuit the resistors R11 and R12 set the collector emitter voltage of transistor VT2 to about 2.1V. This means that the voltage across the 100Ω resistors R2 and R3 is about 0.4V each and the quiescent current in the output stage is thus about 4mA. If a signal is applied to the driver transistor the voltage on bases of transistors VT3 and VT4 start to change and current is fed into the load. This current passes through resistor R2 or R3 and causes the voltage across the resistor to increase. When the voltage across the resistor R2 or R3 increases sufficiently to forward bias diode D1 or D2, the diode starts to conduct. Any further increase in the current fed into the load is supplied through the appropriate diode. The voltage drop across the diode and resistor is limited to the diode forward voltage, which is unlikely to exceed 1.0V at full output current. Shunting the emitter resistors with diodes allows one to benefit from having high value emitter resistors at the quiescent current level, without the penalty of the voltage drop at full output that having high value resistors would otherwise incur. A slight disadvantage of this method is that, because the collector emitter voltage of transistor VT2 is about three times its base emitter voltage, over compensation for the thermal change of V_{BE} in the output transistors occurs and the quiescent current is slightly reduced as the temperature of the output devices increases. This does increase the thermal stability of the output stage however.

The 2W circuit shown in Figure 7 was modified to eliminate the quiescent current setting potentiometer. A sample of ten different devices (taken from different families to eliminate optimistic results arising from the similar V_{BE}s that are likely to be found in a batch of devices from the same family) was tried in this circuit. The maximum variation of quiescent current was ±0.5mA.

Distortion figures for the modified circuit are shown in Table 1.5(a).

These figures may be compared with those taken on the manually set quiescent current version of Figure 7, i.e. Table 1.5(b).

They are worse than those of the unmodified circuit due to residual cross-over distortion arising from the non-linear nature of the diode-resistor combination in the emitters of VT3 and VT4, but about three times better than if the stage was run in pure class B.

High Frequency Stability

All the transistors used in these circuits have transition frequencies up in the hundreds of megahertz region. The large amounts of feedback applied to these circuits means that they tend to be unstable at high frequencies. This instability may be eliminated by adding the small capacitors indicated to roll off the high frequency gain of the circuit. The actual value of capacitor required may depend to some extent on the layout used and so some modification to the values indicated may be necessary.

Power Supply Requirements

All the circuits described so far in this report require a 35V power supply. Supply current requirements for the various power levels are shown in the curve in Figure 16. These sort of current and voltage levels can be provided by simple over windings on record player and tape recorder motors.

The circuits shown in Figures 11 and 13 are fully hum proofed and so produce low hum output levels on power supplies with relatively small reservoir capacitors. In the rest of the circuits a full ripple voltage appears across the speaker, because the D.C. level of the output is referenced to earth, but the speaker is connected from this point to the positive rail. With these circuits a higher hum level must be tolerated or more smoothing used on the power supply.

Possible variations

The majority of designs in this part of the chapter have been 2.5W designs. This power level is about the maximum power which can be extracted from the TIS90/91 complementary pair without exceeding their ratings. By changing the circuit values, the circuits described in this section may be derated to lower power levels.

Table 1.5(a)

Total Harmonic Distortion

Power W / Freq. kHz	2.0	1.0	0.3	0.1	0.03	0.01
0.1	2.7	0.64	0.90	1.60	2.20	2.70
0.3	2.65	0.63	1.00	1.53	2.14	2.60
1	3.75	0.62	0.85	1.52	2.10	2.60
3	2.95	0.62	0.83	1.50	2.70	2.85
10	2.75	0.63	0.78	1.40	2.05	2.50

Table 1.5(b)

Total Harmonic Distortion

Power W / Freq. kHz	2.0	1.0	0.3	0.1	0.03	0.01
0.1	2.5	0.66	0.35	0.35	0.46	0.42
0.3	2.0	0.62	0.36	0.33	0.37	0.40
1	2.43	0.84	0.37	0.38	0.40	0.39
3	1.80	0.64	0.38	0.43	0.35	0.37
10	1.70	0.63	0.36	0.33	0.33	0.33

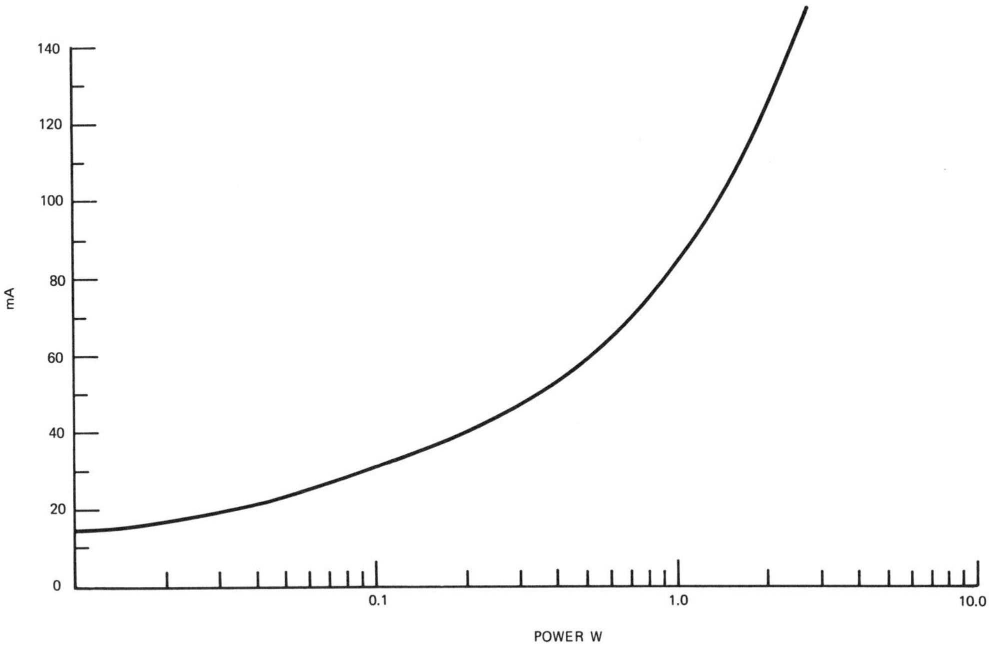

FIGURE 16 Current Drawn from 35V Supply R_L=35Ω

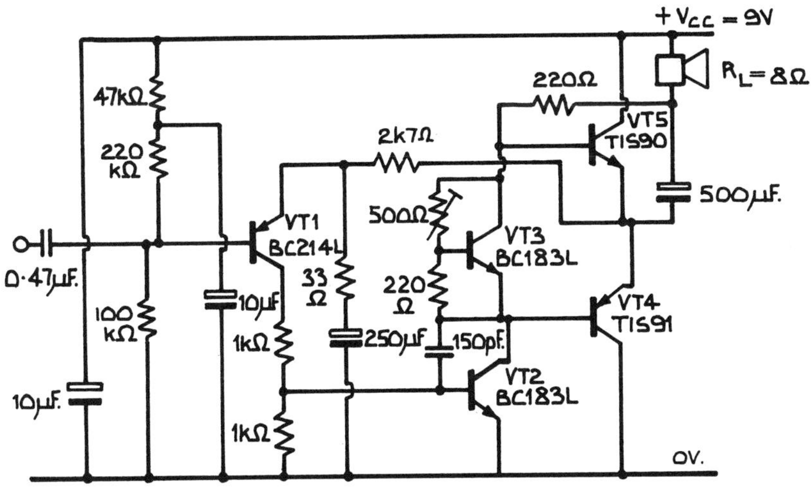

FIGURE 17

At these levels, lower power supply voltages and speaker impedances may be used. The 1W design shown in Figure 17 uses a 9V power supply and a 8Ω load resistor. It is based on an amplifier described in a report published by the Texas Instruments Applications Laboratory in Germany.

As can be seen, the basic configuration used is similar to that shown in Figure 11, except an npn driver transistor and pnp input transistor are used. Performance figures for this amplifier are given in table 1.6 and figure 18 and suggested printed circuit layout is shown in Figure 19.

Table 1.6

Supply Voltage	9V
Maximum Supply Current at 1W	155mA
Output Power at 10% T.H.D.	> 1W
Input Voltage for 1W	45mV
Frequency Response (–3dB)	80Hz – 100kHz
Input Impedance	50kΩ

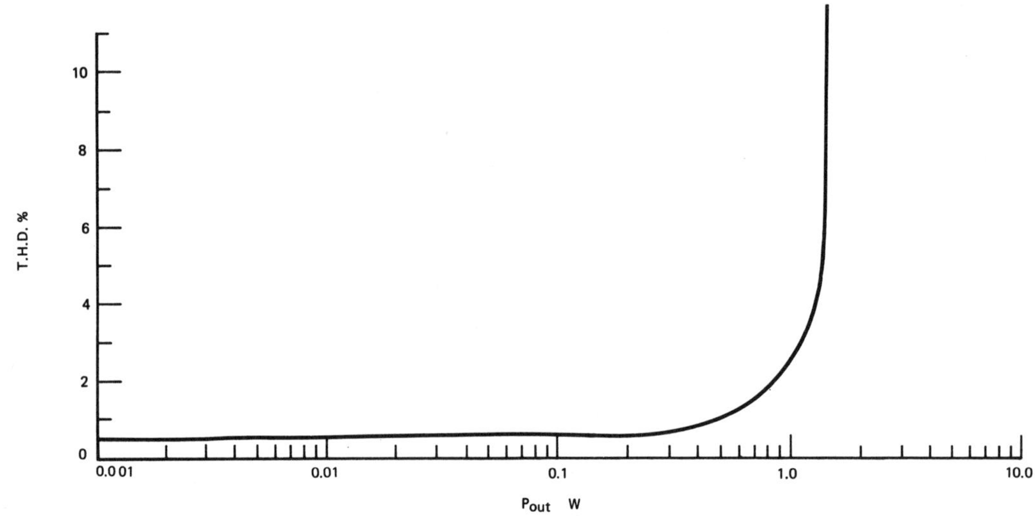

FIGURE 18 TOTAL HARMONIC DISTORTION VS OUTPUT POWER FOR 1W/8Ω AMPLIFIER

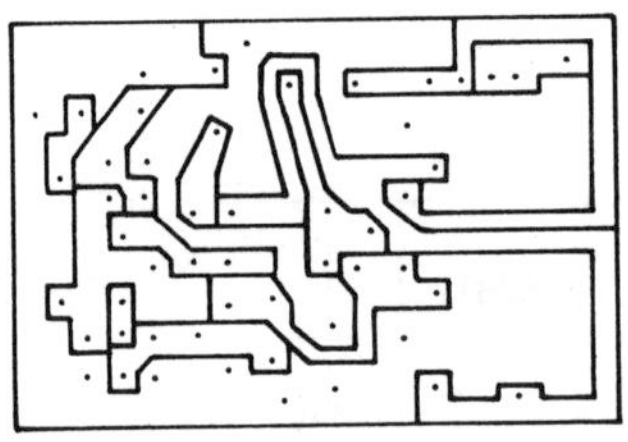

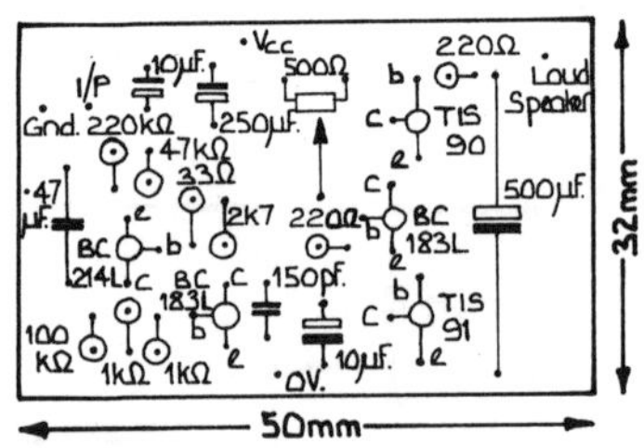

FIGURE 19

PART 2

AMPLIFIERS BETWEEN 10 – 100W

This section deals initially with designs for 10, 15, 20 and 30W amplifiers.

They are intended to work into 8 or 15Ω loads, to conform to a Hi-fi specification (distortion less than 0.1%, noise less than −80dB, stable into inductive, resistive and capacitive loads) and to work from simple unregulated power supplies. The basic voltage and current requirements of the output stages are summarised in table 2.1. A discussion on modifications to the basic design so that the output may be increased beyond 30W concludes the section.

The Output Stage

The basic circuit which is to be used for powers up to 30W is shown in Figure 20 component values for the individual amplifiers are given in table 2.2. All these designs feature transistors from the TIP31 to 34 and TIP41 and 42 range of plastic power transistors. These transistors are available in 3, 6 and 10A versions with 40, 60, 80 or 100V breakdown voltages. The low current gain of 3A transistors at high collector currents necessitates the use of the 6A transistors in the circuits with peak output currents greater than 1.5A. The 40V transistors are specified for use in the circuits with 32 and 35V power supplies, but the 60V versions are specified for the amplifiers requiring 40V (because an unregulated supply may well exceed 40V) and 50V supplies.

Table 2.1

POWER W	LOAD IMP Ω	RMS VOLTAGE V	PEAK VOLTAGE V	PEAK–PEAK VOLTAGE V	RMS CURRENT A	PEAK CURRENT A	POWER SUPPLY V
10	8	8.94	12.61	25.22	1.12	1.58	32
10	15	12.25	17.31	34.62	0.82	1.15	40
15	8	10.95	15.50	31.00	1.37	1.94	36
15	15	15.00	21.20	42.40	1.00	1.41	50
20	8	12.64	17.88	35.78	1.58	2.24	42
30	8	15.60	22.05	44.10	1.95	2.76	50

Table 2.2

P_{out} (W)	R_L Ω	R4 Ω	R5 Ω	R6 Ω	R8 Ω	R9 Ω	R11 Ω	VT7	VT8	C7 μF	V_{CC} V
10	8	12k	3k3	12k	56k	1k2	3k9	TIP32	TIP31	2000	32
10	15	15k	3k9	15k	120k	820	10k	TIP32A	TIP31A	1000	40
15	8	15k	3k9	15k	82k	1k	5k6	TIP42	TIP41	2000	36
15	15	15k	4k7	15k	82k	680	8k2	TIP32A	TIP31A	1000	50
20	8	15k	3k9	15k	82k	820	5k6	TIP42A	TIP41A	2000	42
30	8	15k	4k7	15k	82k	680	4k7	TIP34A	TIP33A	2000	50

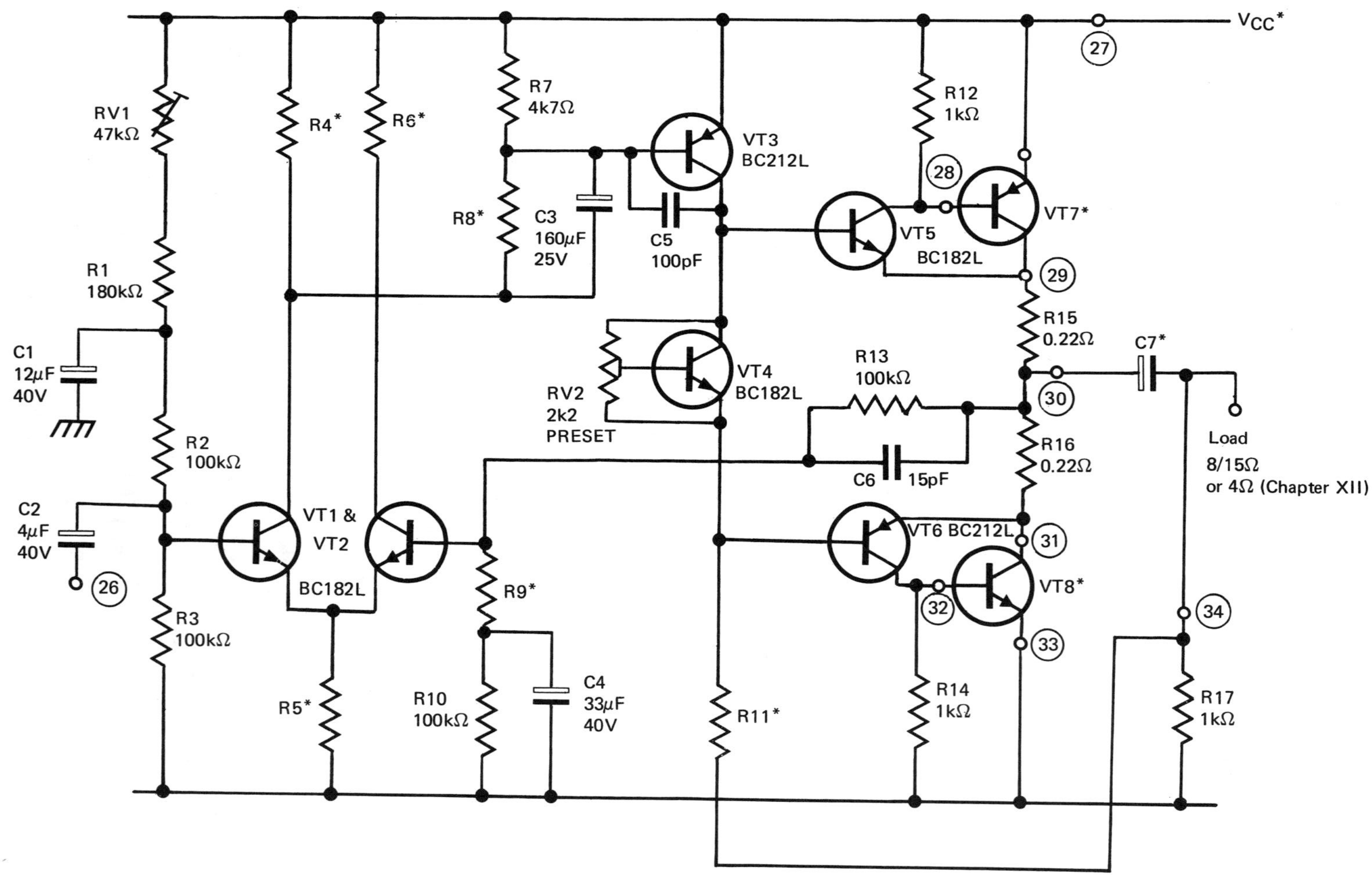

*See Table 2.2

FIGURE 20 Complementary 10/15/30 Watt Audio Power Amplifier Mk V

It would be possible to uprate the designs described in the first part of this chapter using these power transistors in the output stage to provide the level of power output required for the amplifier in this section, but most of the other requirements of the specification could not be met. Because of the relatively high current drain from the output stage, the driver stage would need to be run at much higher current levels, about 100mA for the 15Ω 15W design. This would mean that a small power device would have to be used for the driver transistor as its dissipation would be about 2.5W. Its static base current requirement would be about 1mA which would necessitate the use of low value feedback resistors, giving a low input impedance. Also these circuits would have insufficient open loop gain for enough feedback to be applied to reduce the distortion in the circuit to the required level. To overcome these difficulties both the current and voltage gain of the circuit are increased. The best place to increase the current gain of the amplifier is in the output stage as this also reduces the requirements of the driver stage. This can be achieved by increasing the current gain of the output transistors, i.e. specifying a higher gain transistor. This, however, is difficult because the gain of most silicon transistors operating at these current levels is fairly low. It is better to increase the effective current gain of each of the output transistors by operating an extra small transistor in combination with it in one of the configurations shown in Figure 21. In both these configurations the gain of each 'combination' output transistor is typically 5000 at 3A instead of 35 for the power transistor alone. This solution is fairly low in cost. Also as the driver devices operate in class B like the output devices, the power dissipation (which may be regarded as P_D/β_o where P_D and β_o are the power dissipation in and the current gain of the output transistors) in them is low, enabling small signal silect transistors to be used. Each of the configurations in Figure 21 has its own advantages and disadvantages, but in all the circuits described in this section of the chapter the configuration shown in Figure 21b is used as it has the following advantages over the other:–

(a) The output is driven through only two base emitter junctions instead of four. This means that the change of input voltage required to take the output voltage over the cross-over point is halved. The slope of the output current versus input voltage curve is steeper ensuring a faster turn on of the output stage at the cross-over point. Typical I_C versus V_{BE} curves for the two configurations are shown in Figure 2.3.

(b) Greater thermal stability of the DC quiescent current is possible. In the circuit in Figure 21 the bias transistor has to compensate for any temperature change in the base emitter voltages of the two power transistors and the two drivers. Because of the different dissipation in and thermal resistance of the power and driver devices, their junction temperatures will change differently and hence only approximate compensation is possible. It is also difficult to obtain thermal contact between the bias transistor and the output transistors as the latter are usually mounted on a heat sink, remote from the printed circuit board, necessitating the use of long leads to the bias transistor and the problem of attaching it to the heat sink. In the circuit in Figure 21b, the bias transistor has to compensate for temperature changes in the base emitter voltage of the two driver transistors only. It is easy to arrange for thermal contact of the bias transistor with one of the driver transistors, as both are mounted on the printed circuit board.

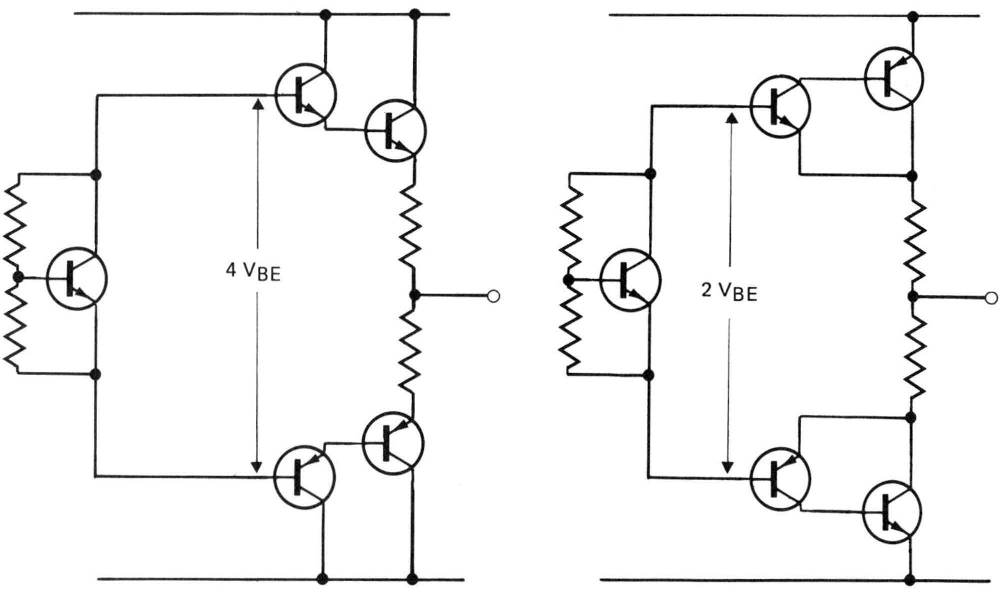

(a) DOUBLE DARLINGTON

(b) COMPLEMENTARY DARLINGTON

FIGURE 21

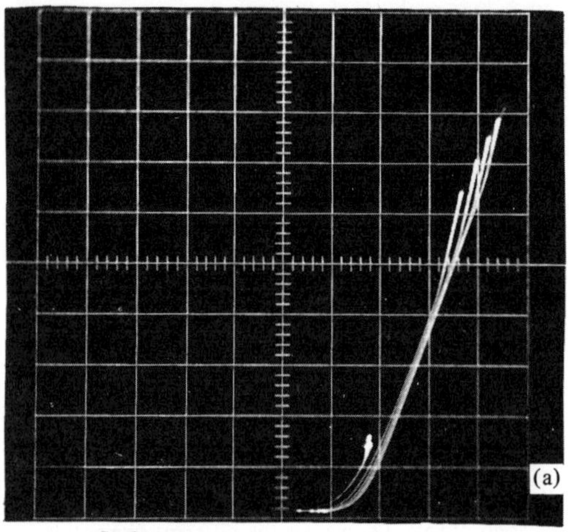

Scales:

Vertical↑ I_C 0.5A/cm
Horizontal →V_{BE} 0.2V/cm
I_B = 1mA/step

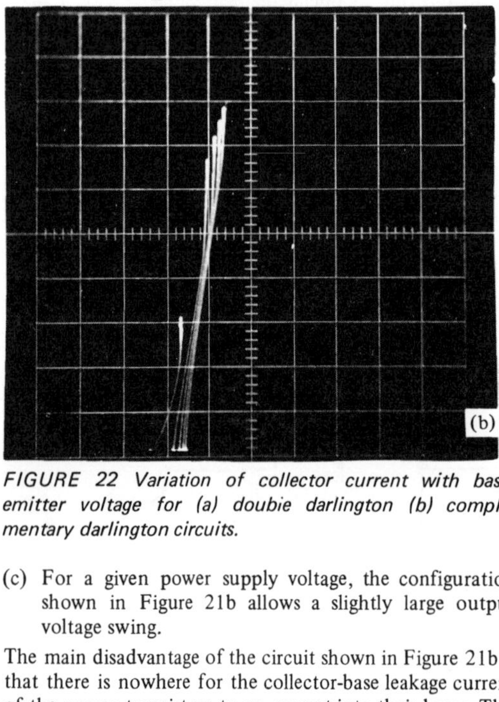

FIGURE 22 Variation of collector current with base-emitter voltage for (a) double darlington (b) complementary darlington circuits.

(c) For a given power supply voltage, the configuration shown in Figure 21b allows a slightly large output voltage swing.

The main disadvantage of the circuit shown in Figure 21b is that there is nowhere for the collector-base leakage current of the power transistors to go, except into their bases. This leakage current into the base increases the collector current and hence the dissipation, temperature and leakage current of the power transistors. It is possible, under extreme circumstances, for a thermal runaway condition to occur. This problem may be alleviated by shunting the base emitter junction of each of the output transistors with a 1kΩ resistor, which provides a path for the collector base leakage current and ensures thermal stability. These resistors also provide a path for the extraction of the stored base charge, thus helping to shorten the turn off time of the power transistors, improving the cross-over performance of the amplifier.

The Voltage Amplifier Stage

The voltage amplifier used in these circuits is similar to the driver stage used in the low power amplifiers. It has been renamed to avoid confusion with the small transistors which drive the output transistors used in the higher power circuits. The constant current sunk by resistor R11 is about twice the peak current required by transistor VT6. This extra current can be handled without over dissipating the driver transistor VT3, due to the smaller base current requirement of the output stage of this amplfier compared with that of the lower power amplifiers. It has the advantage of reducing the current swing in the pre-driver stage down to about 3 : 1 (half to one and a half times the stage quiescent current) enabling the stage to operate more linearly, which reduces distortion.

An alternative configuration for the voltage amplifier stage is shown in Figure 23.

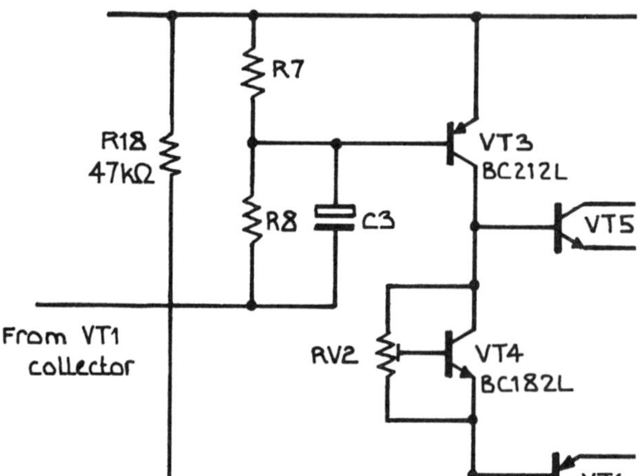

FIGURE 23

In this, the boot-strapped resistor R11, which acts as a constant current sink, is replaced by the transistor VT9. This transistor has a constant voltage placed on its base by the potential divider chain formed by resistors R18 and R19 and diode connected transistor D1, and thus acts as a constant current sink. The voltage across diode D1 roughly matches the base emitter voltage of transistor VT9 and ensures that the voltage across resistor R20, and thus the current sunk by transistor VT9, remains constant with changes in temperature, as any change in the base emitter voltage of transistor VT9 is cancelled by a similar change in the diode voltage. The two devices should be placed in thermal contact.

This arrangement has all the advantages that the boot-strapped resistor configuration has in reducing the current swing in the voltage amplifier transistor VT3. In addition to this, an improvement in cross-over distortion (which is present in all these amplifiers in spite of running the output stage in Class AB) is obtained. This improvement is possible because the transistor current sink maintains its high input impedance throughout the complete range of output swing, which ensures a fast turn on of the ongoing half of output stage at the cross-over point. This contrasts with the boot-strapped resistor arrangement whose input impedance tends to fall at the cross-over point because the voltage on the collector of the driver transistor VT3 changes more than the output voltage – thus the resistor cannot act as a constant current sink at the cross-over point.

A version of the 30W amplifier incorporating this modification has been built and tested. Its distortion figures are given in Figure 33 at the end of this section. It can be seen that the extra complexity and expense of this modification is justified by a worthwhile improvement in the distortion figures – especially at high frequencies.

The Input Stage

The input stage consists of transistors VT1 and VT2 in the long-tail pair configuration. This arrangement, although it uses one more transistor than the input stages discussed previously, offers the following advantages:–

(a) Excellent temperature stability on the DC level of the output mid-point, since any change in the base emitter voltage of transistor VT1 due to temperature changes is offset by a similar change in the base emitter voltage of transistor VT2. Similar value resistances are used in the input and feedback circuits connected to the bases of transistors VT1 and VT2 respectively, so that any changes in the base current requirements of transistors VT1 and VT2 due to temperature changes, produce nearly equal off-sets on the two sides of the circuit and thus prevent any drift of the output mid-point.

(b) A high input impedance to both the input and feedback circuits, allowing a low value capacitor to be used to decouple the feedback circuit.

The DC level of the base of transistor VT1 is set by a resistor chain consisting of resistors RV1, R1, R2 and R3. Potentiometer RV1 is used to adjust the voltage level on the base of transistor VT1. Because of the DC feedback on the amplifier, which will be discussed later, the voltage on the base of transistor VT1 sets the quiescent voltage of the output mid-point. Thus, by adjusting resistor RV1, the DC level of the output mid-point may be adjusted so that symmetrical clipping of the output waveform occurs when the amplifier is driven beyond its full rated output power. This adjustment allows up to about 35W to be squeezed out of the 30W amplifier. Alternatively, the full 30W may still be obtained by substituting a 200kΩ resistor RV1 and R1 and using close tolerance (2%) resistors for R1, R2, R3, R10 and R13.

The junction of resistors R1 and R2 is decoupled to earth by capacitor C1 to prevent hum and noise on the power supply from appearing on the base of transistor VT1 and hence on the output.

Distortion in the first stage is minimised by running it at a fairly high current. The DC voltage level on the base of transistor VT1 is about one quarter of the supply voltage which sets the tail current of the long tail pair to about 2.5mA. This means that the transistors VT1 and VT2 run with collector currents of about 1.25mA each. The base current requirement of transistor VT3 represents only a small percentage change in the collector current of transistor VT1, thus minimising distortion. This arrangement necessitates, however, the use of a level shift network to couple the signal from the collector of transistor VT1 to the base of transistor VT3.

A resistor is used to transfer the DC current required, but it is shunted by a large capacitor C3, ensuring that as far as a.c. is concerned, the base of transistor VT3 is directly connected to the collector of transistor VT1, giving a high open loop a.c. gain. The base-emitter diode of transistor VT3 is shunted by 4k7Ω resistor R7 to reduce the variation in current through R8 due to the gain spread of transistor VT3. This resistor also improves the transient response of the amplifier by providing a path which helps transistor VT3 to turn off quickly.

A zener diode level shift network was used in the earlier version of this amplifier, but it has been dropped in favour of the CR coupling network because the latter is:–

(1) Cheaper

(2) Quieter. Experience with running zener diodes at low zener currents has shown that, although they have low slope resistances, they are, nevertheless, noisy.

(3) When its power supply is switched on, the circuit will start to adjust the DC level of its output mid-point whilst the supply voltage is close to zero. With the zener diode, the power supply voltage has to rise to about +30V before the zener would pass current and the output voltage start to stabilise its level. This caused quite a disturbing thump in the loudspeaker, an effect which has been considerably reduced with the CR coupling network.

Feedback Arrangements

DC and a.c. negative feedback are applied by resistors R9, R10 and R13 and capacitor C4. For D.C. capacitor C4 can be ignored and the DC level of the output mid-point is divided by approximately 2 and applied to the base of transistor VT2. The action of the circuit is that the DC level of the output mid-point changes until the base voltages of the transistors VT1 and VT2 are equal. Any tendency for the mid-point voltage to rise (say) increases the base voltage of transistor VT2. This increases its collector current and hence decreases the collector current of transistor VT1, reducing the base drive to transistor VT3. This turns down the collector current of transistor VT3, reducing the voltage drop across resistor R11 and thus overcomes the tendency for the mid-point to rise.

For a.c. the lower end of R9 may be considered to be connected to earth by capacitor C4. The a.c. signal on the output is divided down in the ratio 0.68/100.7 by the potential divider formed by R13 and R9 and applied to the base of transistor VT2. The long tail pair is a subtractive arrangement and so this feedback signal is subtracted from the input signal.

Thus:–

$$V_{out} = AV$$

where A is the open loop gain of the amplifier and V is the differential input signal.

$$V_{out} = A(V_{in} - V_{out})R9/(R9 + R13)$$
$$V_{out}/V_{in} = A' = (R9 + R13)/R9 \quad \text{if A is large.}$$
$$= 100.7/0.68$$
$$= 148$$

This gives an input requirement of 105mV for 30W output.

As the input frequency is reduced, the reactance of capacitor C4 increases. When this impedance becomes comparable with the magnitude of R9, the closed loop gain starts to fall. The frequency at which the gain is –3dB with respect to the gain at 1kHz is defined by the time constant R9.C4. The values of R9 and C4 specified give a –3dB point of 7Hz, but the tolerance of C4 can increase the –3dB point to about 10Hz. This is not the –3dB point of the whole amplifier because there is a further time constant $C7.R_L$ which gives a –3dB point at about 12Hz. Two different values of C7 are specified, one for 8Ω and one for 15Ω loads. If low frequency response can be sacrificed for economy these values may be reduced accordingly.

High Frequency Stability

This circuit has gain up in the megahertz region and with the large amount of feedback which is applied, high frequency instability can occur. This is prevented by the capacitors C5 and C6, which roll off the open and closed loop gain respectively, ensuring stability with resistive, capacitative and inductive loads and with an open circuit input. The clean square wave response of the amplifier is illustrated in the photographs shown in Figure 24. (These photographs were taken from a Tektronix 547 oscilloscope with 1A1 plug in: bandwidth 50MHz.)

Construction and Heat-sinks

To ensure a stability and optimum performance of the amplifier, the following construction rules should be observed:-

a) Leads carrying large currents should be short and kept well away from the input of the amplifier.

b) Substantial leads or tracks should be used for the positive and negative rails. The positive power supply lead should be joined from the reservoir capacitor of the power supply to the emitter of transistor VT7 and then to the circuit board containing the rest of the components. The negative (earthy) power supply lead should be joined from the reservoir capacitor to the emitter of transistor VT8 and then to the board. The return lead from the load should be joined straight to the negative terminal of the power supply reservoir capacitor.

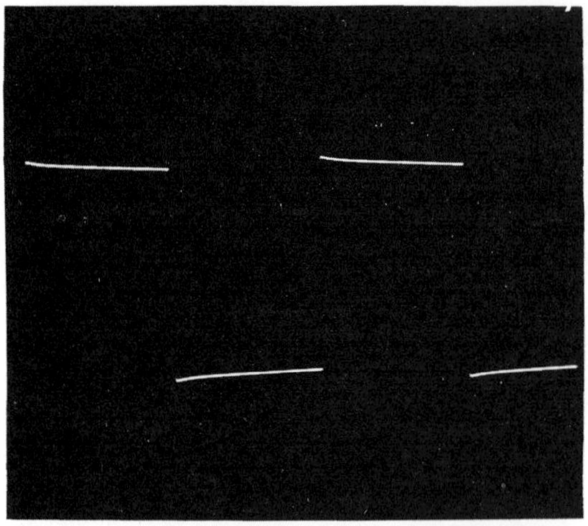

1kHz into 8Ω 40V pk-pk

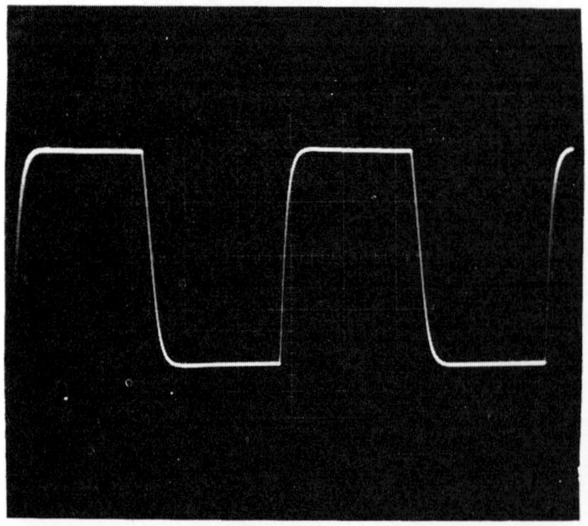

10kHz into 8Ω 40V pk-pk

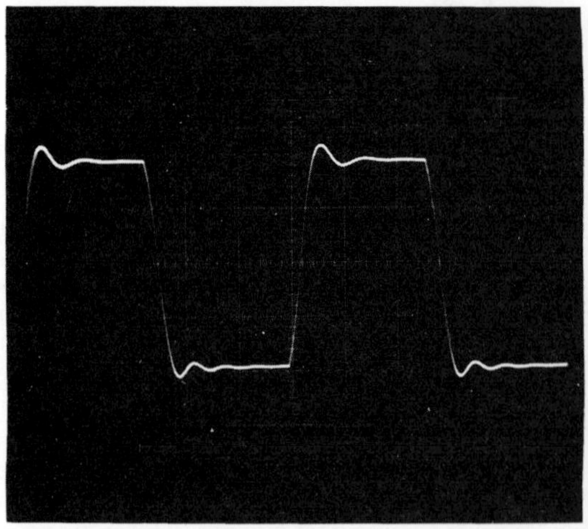

10kHz into 8Ω || 2μF 40V pk-pk

FIGURE 24

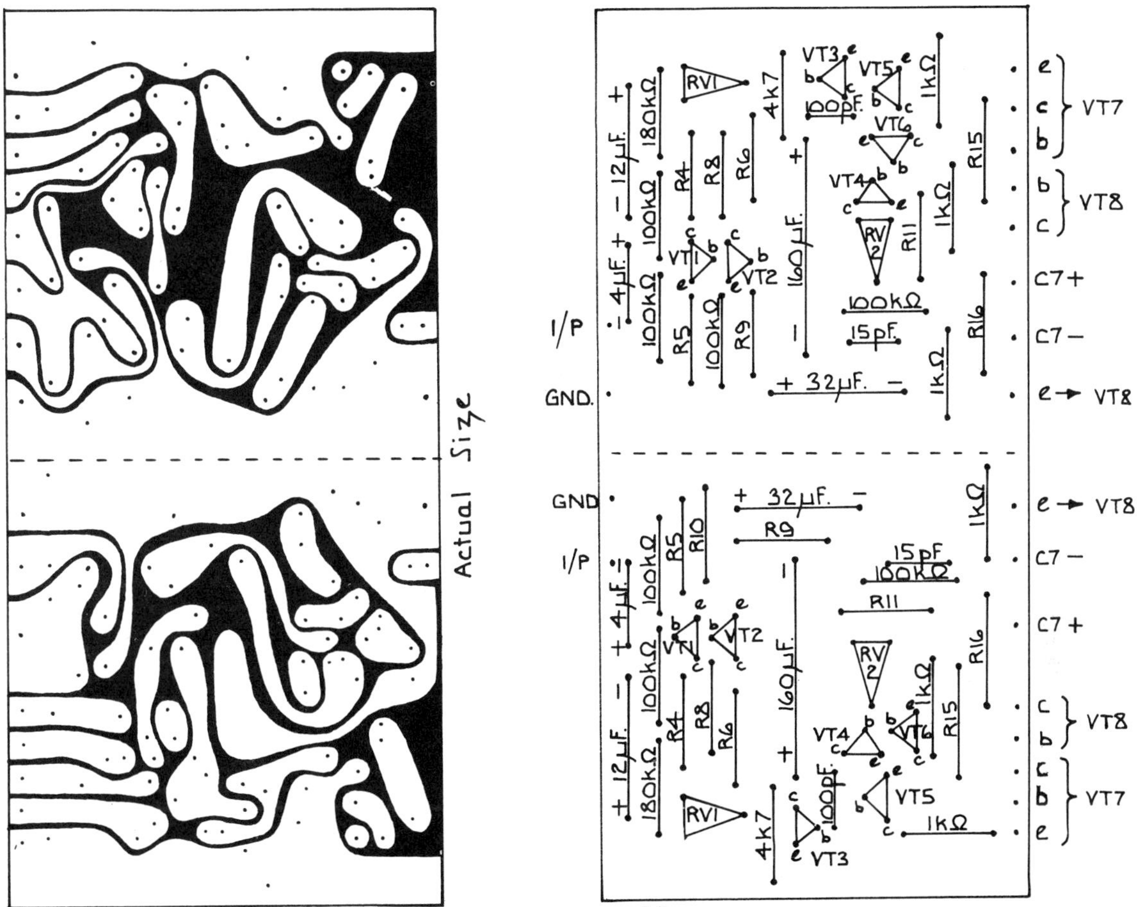

(a) COPPER SIDE (b) COMPONENT SIDE

FIGURE 25

A suggested printed circuit layout which will accommodate a stereo pair of any of the designs covered in this section is shown in Figure 25 (a) shows the board looking at the copper side, the black areas should be etched away. (b) shows the component layout on the upperside of the board.

In normal domestic use, the mean power dissipation in the output transistors is probably less than about 2W. It is sufficient to mount them on a small aluminium fin, about 4sq. ins. in area, to keep them cool. It should be remembered that if full power testing is done with the power transistors mounted thus, readings must be taken quickly with plenty of time allowed between readings to allow them to cool off.

If prolonged running at high power levels is contemplated, the power transistors must be adequately heat sinked. If it is assumed that the dissipation in each output transistor is one quarter of the total output power, the output pair in the 30W amplifier needs to dissipate a total of 15W. A 4in. length of extruded aluminium heat sink about 5ins. wide, (e.g. Radiospares heat sink), is recommended for this. With this heat sink, and allowing for the fact that mica washers have to be used under the power transistors, the junction temperature of the output transistors in this 30W into 8Ω version will be less than 55°C above ambient temperature at full output power, allowing the amplifier to be run at full power indefinitely. Adequate ventilation for the heat sink should be provided in the cabinet. The lower power amplifiers can make do with smaller areas of heat sink, e.g. a stereo pair of 15W amplifiers could use one Radiospares heatsink.

Setting-up Procedure

Before the amplifier is switched on for the first time its construction and wiring should be thoroughly checked out, preferably by someone else. The setting up procedure set out below should then be followed. Read through the instructions before beginning.

With the power supply disconnected from both pre- and power amplifier check that its no-load voltage does not exceed 60V.

(a) Connect only the power amplifier to the power supply through a 100mA meter and a 5kΩ wire wound potentiometer. Set the potentiometer to maximum resistance. Ensure that the bias current setting potentiometer RV2 is set at the end of its track connected to the collector of transistor VT4.

(b) Turn on the power supply and carefully decrease the resistance of the potentiometer. If the current shown by the meter increases to more than 50mA continuous switch off and check the amplifier out for faults.

(c) A healthy amplifier should draw a standing current of less than 15mA when the resistance of the potentiometer is reduced to zero.

(d) Having checked that the amplifier is working properly, the next task is to set the output stage quiescent current. This should not be done with the amplifier cold. A signal should run through the amplifier for a while to allow it to attain normal working temperature.

(e) Switch off the power supply, connect a suitable signal source across the preamplifier volume control using a screened lead. An audio oscillator is ideal for this, but if one is not available a suitable signal can be picked up from across the volume control of a transistor radio. Turn on the power supply, check that the standing current of the amplifier is less than 15mA, short out the current meter and turn up the volume control. Run the amplifier with a signal of a few volts across the load for about 10 mins. Switch off the signal source, unshort the current meter and adjust RV2 carefully until the current drawn is about 40 ±10mA. Care is needed to carry out this adjustment as the current increases quite sharply once the right part of the potentiometer track has been reached.

(f) With the input reduced to zero measure the supply voltage with a d.c. voltmeter. Transfer the meter to measure the d.c. level of the output mid-point and set this level to be half the supply voltage using the potentiometer RV1. Throughout this setting-up process an oscilloscope, if available, may be used to monitor waveforms and voltages. Test (f) may be performed at full power, giving a better setting of RV1 by driving the output to full power and monitoring the output voltage with an oscilloscope. RV1 should be adjusted for symmetrical clipping of the output waveform at just beyond the onset of clipping.

This setting up process is made considerably safer if a current limited power supply of suitable voltage and current rating for the amplifier in question is used. In this case, the potentiometer in series with the supply rail may be dispensed with. For the first switch on of the amplifier, the current limit should be set at about 100mA and the output voltage turned up slowly with an eye on the current meter. Be prepared to switch off if the current rises beyond 50mA, apart from a current surge which occurs at about 15V supply voltage. Setting up of the quiescent current and voltage should always be done on the amplifier's own power supply.

Finally, restore the power supply to the preamplifier, turn down the volume control, switch on and check that all the components work together as a system.

Power Supply Requirements

Power supply ripple is prevented from getting into the base of transistor VT1 by decoupling resistor R1 to earth by capacitor C3.

The output voltage on the collector of transistor VT1 is developed with respect to the positive power supply rail and fed into the base of transistor VT3 whose emitter is also connected to the positive rail. Therefore, no rail ripple voltage can be injected into VT3.
The collector voltage of transistor VT3 is developed with respect to earth. This voltage is passed by the output transistors to the load which is connected to earth. Thus the amplifier is completely hum-proof and can, therefore, be driven from a simple unregulated power supply of the type shown in Figure 26. The value of the smoothing capacitor should be as large as one can afford. It is recommended that it should be at least 1000μF/channel for the 15Ω verions and 2000μF/channel for the 8Ω versions. The transformer secondary voltage is somewhat more difficult to define as the regulation of the transformer secondary voltage is an important parameter here. Under quiescent conditions, the secondary voltage, when rectified and smoothed, should not exceed 60V across the smoothing capacitor when A-type output devices are used, or 40V with non A devices. At full output power the voltage on the smoothing capacitor should not fall below the appropriate values specified in table 2.1.
Typical secondary voltages for the different versions of the amplifier (measured at full output current) are given in table 2.3, together with the capacity and voltage of the smoothing capacitor used.

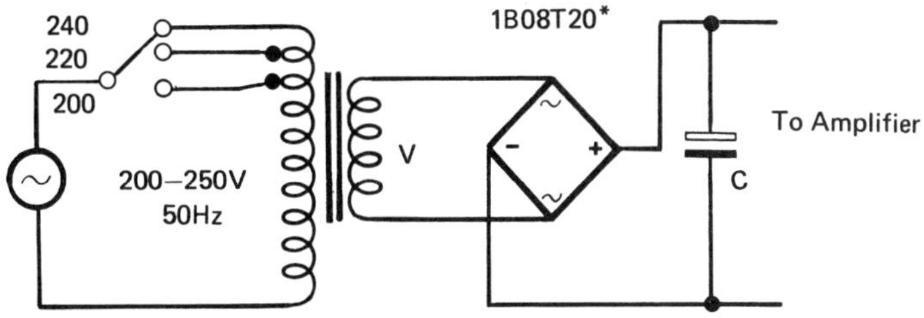

FIGURE 26

Table 2.3

AMPLIFIER		SECONDARY	CAPACITOR	
POWER (W)	LOAD Ω	VOLTS V	μF	Voltage V
10	8	24	1000	40
10	15	30	1000	64 (50)*
15	8	27	1500	40 (50)*
15	15	38	1000	64
20	8	29	2000	64 (50)*
30	8	38	2000	64

* If the transformer regulation is good, 50V working units may be used.

Performance

Brief performance figures showing distortion figures, signal/noise ratio and input sensitivity for all the amplifiers are given in figures 27 to 33 and table 2.4. A fuller characterisation of the 30W, 8Ω version of the amplifier has been done and the results of this are given in table 2.5.

Table 2.4

AMPLIFIER TYPE		HARMONIC DISTORTION See Fig:	SIGNAL / NOISE RATIO dB		FREQUENCY RESPONSE Hz				INPUT VOLTAGE FOR FULL P_{out} at 1kHz mV	INPUT IMPEDANCE Ω
P_{out} W	R_L Ω		R_s = 600Ω	R_s= 10kΩ	−3dB	−1dB	−1dB	−3dB		
10	8	2.8	93	80	15	50	35k	68k	110	46k
10	15	2.9	92	76	10	40	29k	56k	101	46k
15	8	2.10	93	78	15	50	32k	58k	110	48k
15	15	2.11	92	78	11	42	29k	54k	101	44k
20	8	2.12	93	78	15	51	32k	59k	104	47k
30	8	2.13	93	77	16	52	33k	62k	105	47k

Power Supply Requirements: See Figure 2.15.

Table 2.5

EXTRA DATA ON THE 30W/8Ω AMPLIFIER

1. OUTPUT IMPEDANCE AT 1kHz – 80 mΩ
2. INTERMODULATION DISTORTION

OUTPUT FREQUENCIES AND AMPLITUDES				INTERMODULATION DISTORTION %
Hz	V	Hz	V	
1100	7.6	300	7.6	0.020
11k	7.6	9k	7.6	0.080
10k	12	1k	3	0.028
10k	3	1k	12	0.022
10k	12	120	3	0.048
10k	3	120	12	0.018
1k	12	120	3	0.012
1k	3	120	12	0.012

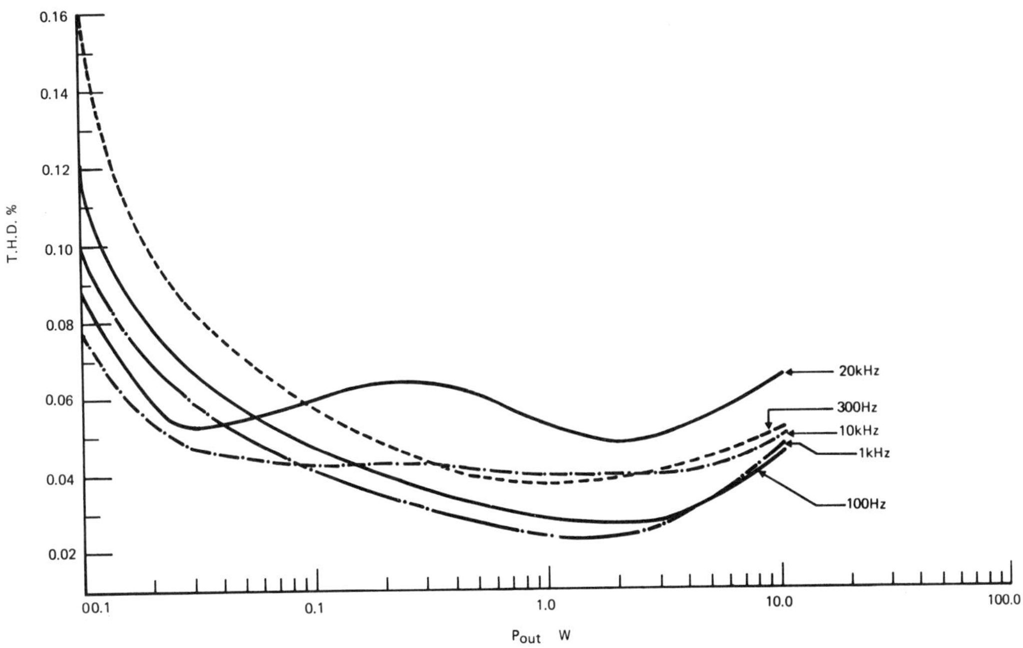

FIGURE 27 Total Harmonic Distortion VS Output Power for 10W 8Ω Amplifier

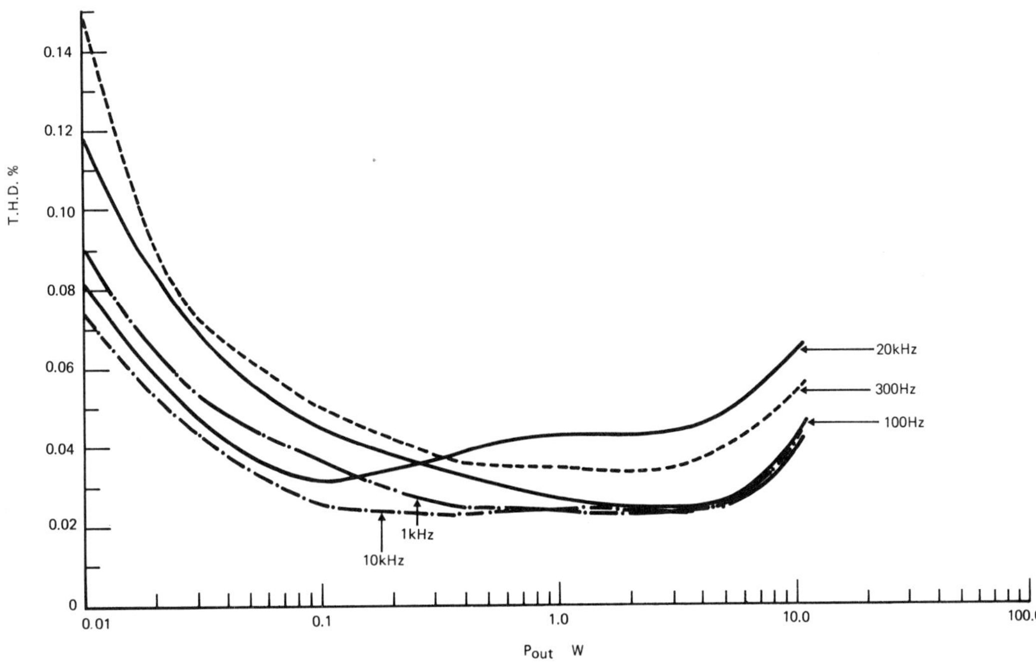

FIGURE 28 Total Harmonic Distortion VS Output Power for 10W 15Ω Amplifier

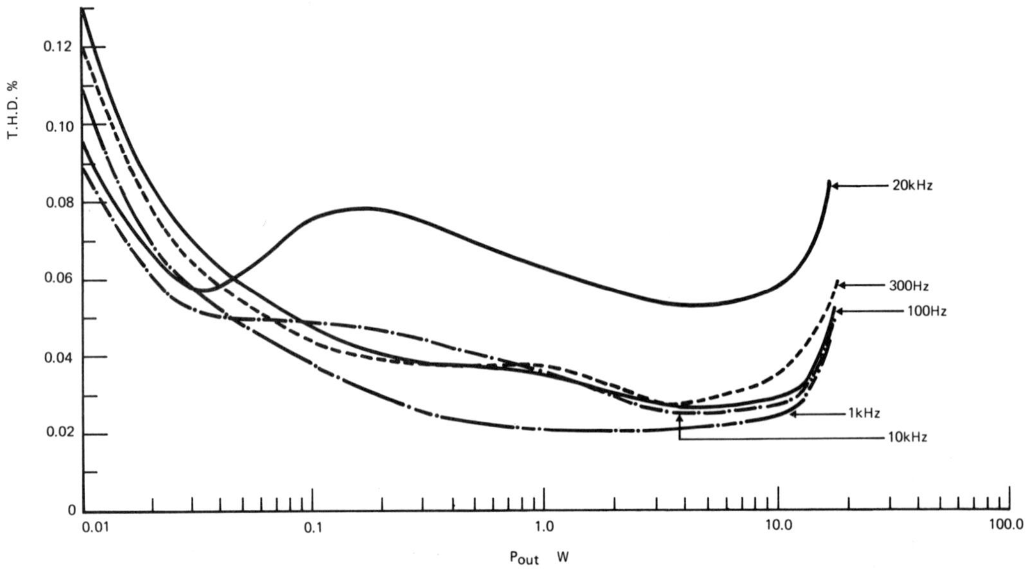

FIGURE 29 Total Harmonic Distortion VS Output Power for 15W 8Ω Amplifier

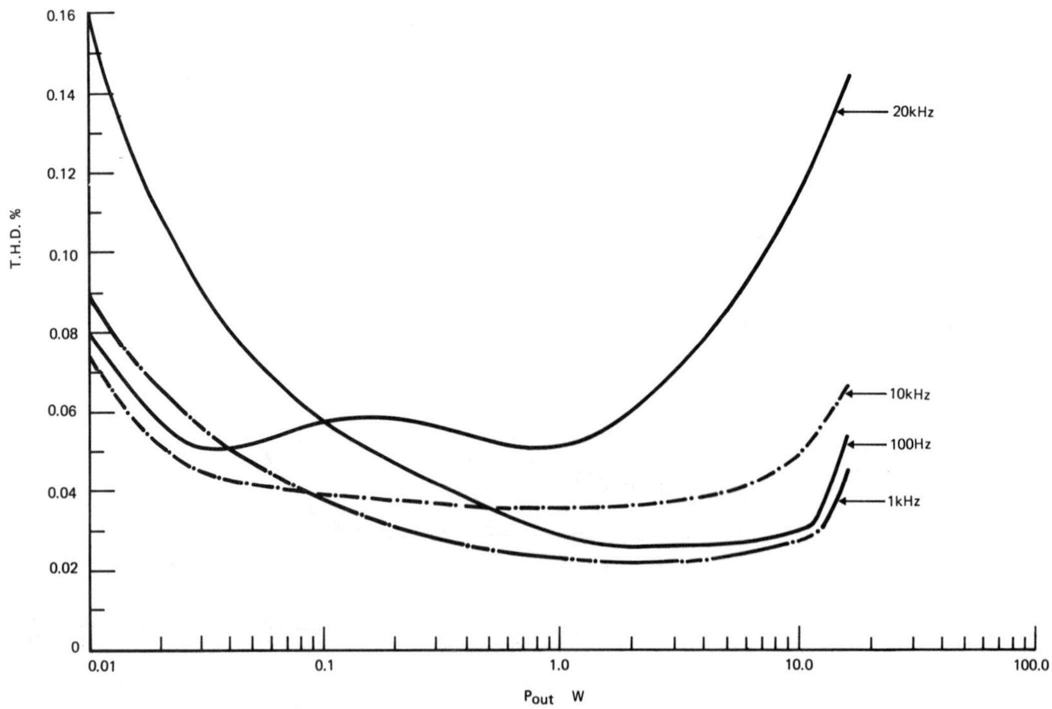

FIGURE 30 Total Harmonic Distortion VS Output Power for 15W 15Ω Amplifier

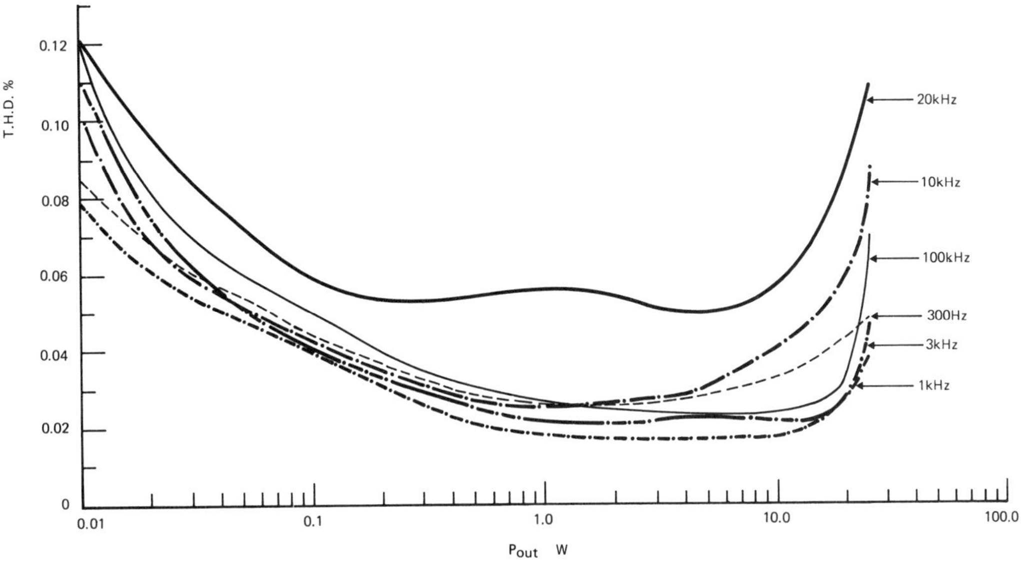

FIGURE 31 Total Harmonic Distortion VS Output Power for 20W 8Ω Amplifier

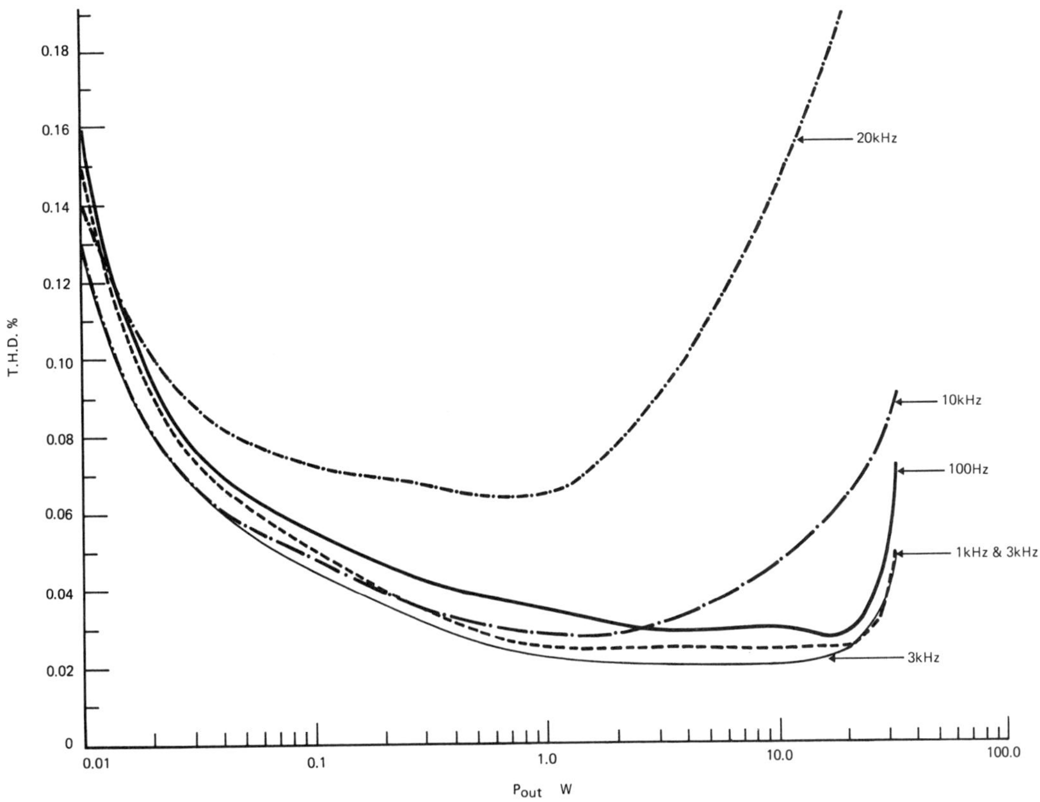

FIGURE 32 Total Harmonic Distortion VS Output Power for 30W 8Ω Amplifier

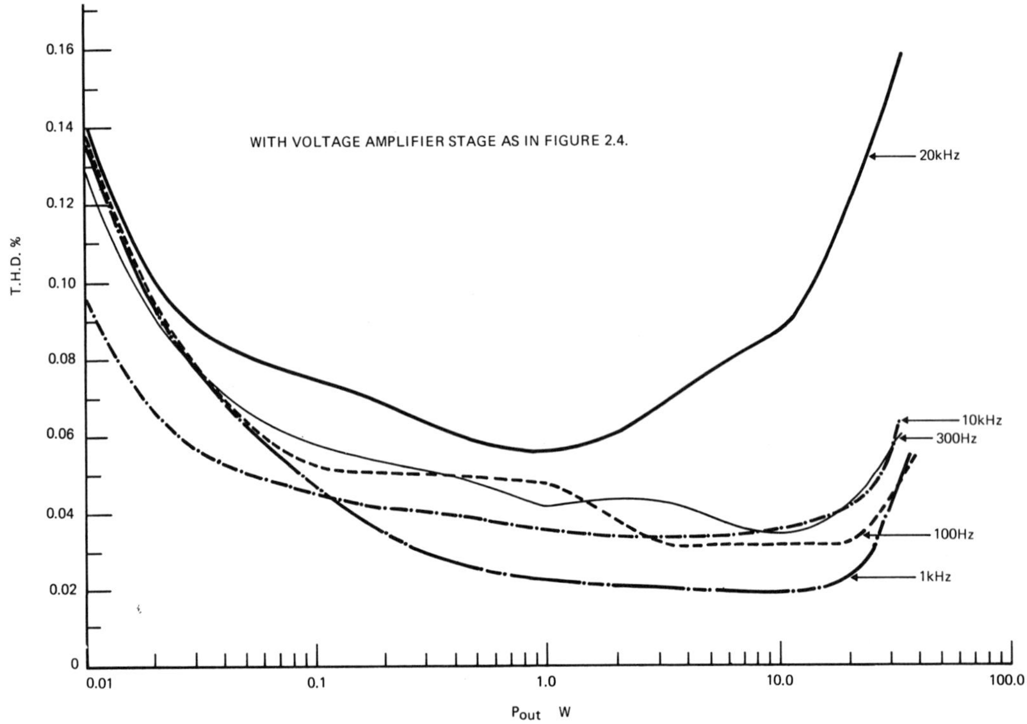

FIGURE 33 Total Harmonic Distortion VS Output Power for 30W 8Ω Amplifier

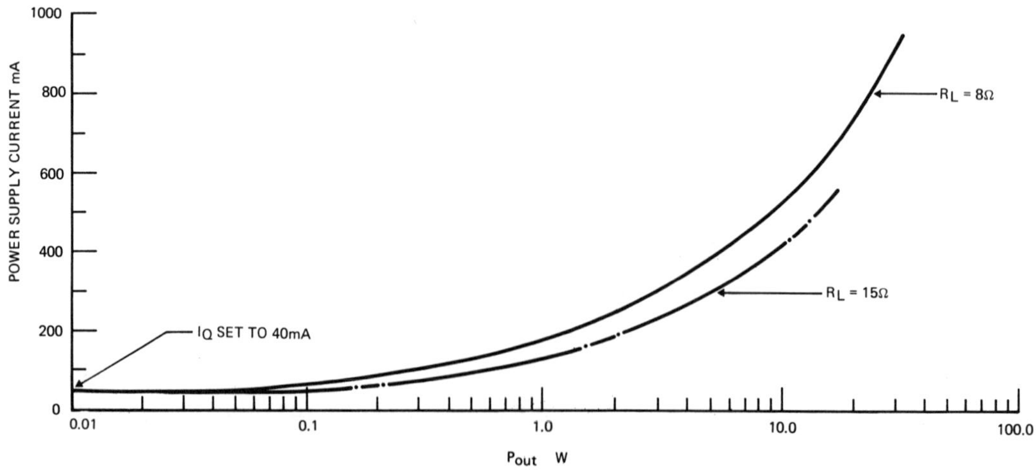

FIGURE 34 Power Supply Current VS Output Power

Overload Protection

For most domestic applications it is unlikely that overload protection is necessary. Under music power conditions, the output stages of the amplifiers will cope with 4Ω loads without failure, although the transistors in output and driver stages will be operating beyond their data sheet maxima. If a fuse of the value specified in Table 2.6 is included in the power supply line for each amplifier, it will protect the amplifier against a short circuit load connected before switch on.

A form of overload protection is desirable if the amplifier is to be used outside the domestic situation, where connection to short circuit loads, prolonged playing into low impedance loads, or the replacement of blown fuses by nails is most likely to happen.

The simplest form of overload protection is shown in Figure 35. In this, the voltage drop across each of the emitter resistors is monitored by the base-emitter diodes of the protection transistors VT10 and 11. The value of R_E is designed so that at the maximum normal current of the output stage, there is just not quite enough voltage across R_E to turn the protection transistor on. Under overload condition, there will be sufficient drop across R_E to turn the protection transistor on. This removes base current from the driver transistors (VT5 or 6) and prevents the output current from rising any further. One disadvantage of this method is that output swing is reduced by the inclusion of resistor R_C which is required to limit the current sunk by transistor VT3 when the current limit is working. In this situation the output swing of the amplifier is limited by the current limit circuit, but the negative feedback loop on the amplifier tries to restore full output swing by turning VT3 hard on. Without R_C, both VT3 and the current limit transistor could be damaged by excess current. The actual reduction of output swing depends on the current gain of the upper output transistor and its driver. Under minimum gain conditions and if the current through VT3 is limited to 150mA, giving a value for R_C of about 150Ω, there will be a drop of 5.2V across the resistor in the 30W into 8Ω version of the amplifier. This problem does not arise on the other half of the amplifier as the current supplied by resistor R11 is limited and hence an equivalent of R_C is not required. With this circuit, distortion, especially at high output levels, is worse than without protection. This is because the current gain of the output stages is reduced by the protection circuit, partially working under normal operating conditions.

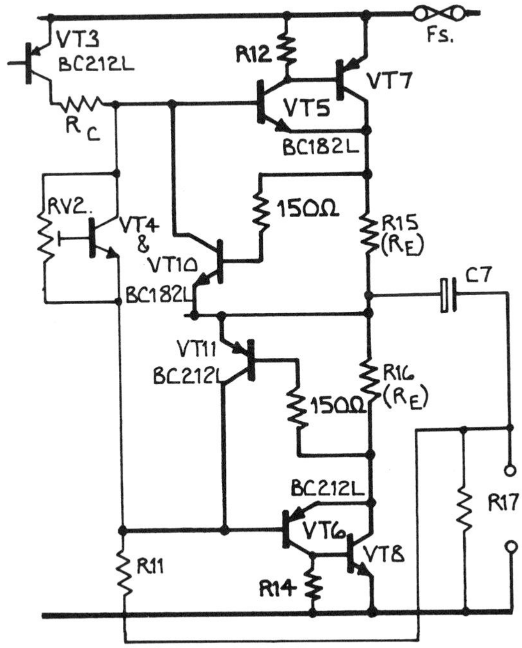

FIGURE 35

The main disadvantage of this circuit is that even with it working excessive dissipation can occur in the output transistors under overload condition because the output transistors have to pass a current in excess of the normal maximum output current with up to half the supply voltage across them.

The power supply fuse, however, tends to prevent this prolonged over-dissipation.

Dissipation in the output transistors under limit conditions may be reduced by modifying the characteristics of the overload detection circuit so that the current limit level is reduced when the voltage across the relevant output transistors increases. This may be done by using the circuit, shown for one half of the output stage, in Figure 36.

Table 2.6

Amplifier P_{OUT}	Load Ω	Norm. Pk. Current A.	Value of R15/R16 Ω	Limit Current A	Rating of Fuse A
10	8	1.58	0.33	1.96	0.75
10	15	1.15	0.47	1.38	0.50
15	8	1.94	0.27	2.40	0.75
15	15	1.41	0.39	1.67	0.75
20	8	2.24	0.27	2.40	0.75
30	8	2.76	0.22	2.96	1.0

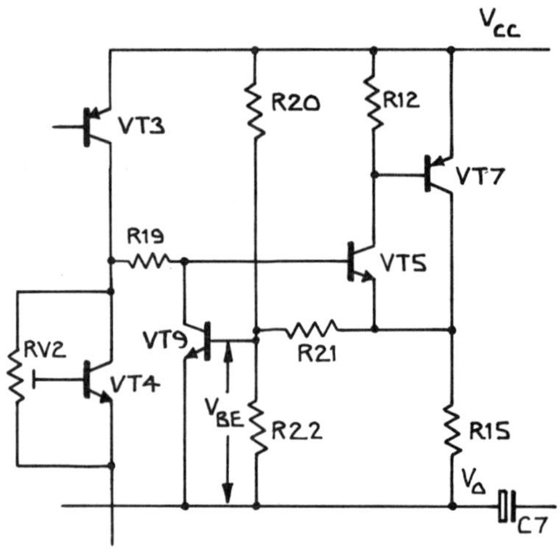

FIGURE 36

The bias voltage for the base of the limit transistor VT9 is made up of two components: a voltage due to the output current through resistor R15, divided down in the ratio R22/(R21 + R22) and a voltage due to the difference between the supply voltage and the output mid point voltage (this difference is almost equal to the collector-emitter voltage of the output transistor), divided down in the ratio R23/(R20 + R21). This gives the relation:

$$V_{BE} = \frac{(V_{CC} - V_o).R21.R22 + I_o.R15.\ R20.R21}{R20.R21 + R21.R22 + R20.R22}$$

which for a given value of V_{BE} gives a linear relation between the current output I_o (limit) and voltage output V_o (limit) which means that the transistor can be limited to operate below a given level of power dissipation.

Two further modifications to the circuit to enable it to work properly in the amplifier are needed.

1. When the amplifier output swings negative (i.e. the bottom output transistor is conducting), the voltage on the base of VT9 increases until it reaches a point where VT9 turns on, even though there is no current flowing through R15. This can be prevented by designing the value of R20 and R21 so that the voltage at their junction is less than V_{BE} even with the full supply voltage across the chain. This, unfortunately, reduces the effect of having the voltage contribution to the current limit. A better way of preventing VT9 turning on during the negative half cycle of the output swing is to split R20 into two components and to connect a zener diode between this mid-point and ground. Values are adjusted so that with half the supply voltage across the chain, the zener diode is just conducting. When the voltage across the chain increases beyond ½V_{CC}, the zener diode prevents the base-emitter voltage of VT9 from increasing.

2. At the extremes of the output swing the protection transistors can operate in the inverted mode and limit the amplifier with no overload present. This problem is best appreciated by referring to Figure 37, which

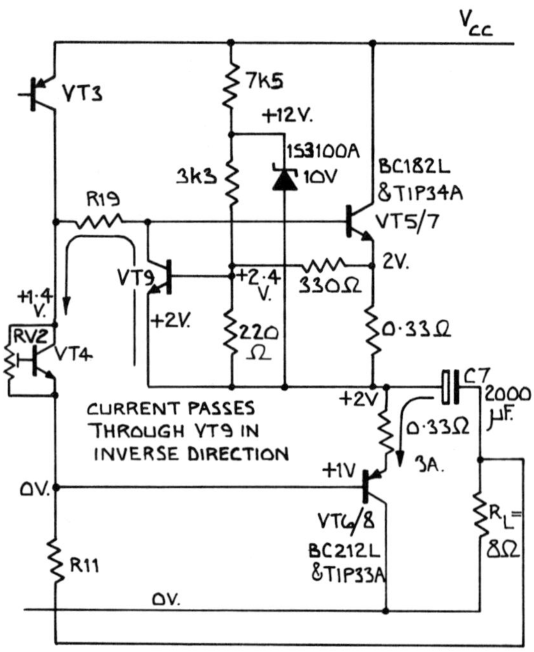

FIGURE 37

shows the voltages existing in the amplifier at the negative extreme of its output swing. It can be seen that the collector of VT9 is at about −1V with respect to its base. Its emitter is at about −0.4V with respect to its base. Under these conditions, the transistor will work in the inverted mode, passing current from emitter to collector and into the constant current sink formed by bootstrapped R11. This effectively steals base current from the lower output transistor and gives an unintentional limiting effect. To prevent this, VT9 must be made a truly unilateral device by placing a diode in its collector. The final protection circuit for the 30W 8Ω version of the amplifier circuit is shown in Figure 38.

This circuit, although effective, still suffers from three shortcomings:—

1. Due to the presence of R19, the output swing capability of the amplifier is reduced.
2. There is an increase in distortion compared with the unprotection circuit beginning to work during normal operation of the circuit.
3. Power dissipation in the output stage during overload is considerably higher than during normal operation. As the overload condition can, theoretically, last indefinitely the heatsink area must be increased to cope with this higher dissipation.

To overcome these disadvantages, a circuit has been devised which uses separate transistors for sensing the overload and limiting the amplifier. Between these transistors is placed a latching circuit so that once an overload is detected, the amplifier output stage is completely cut off, and is hence in a zero dissipation condition, until the amplifier is switched

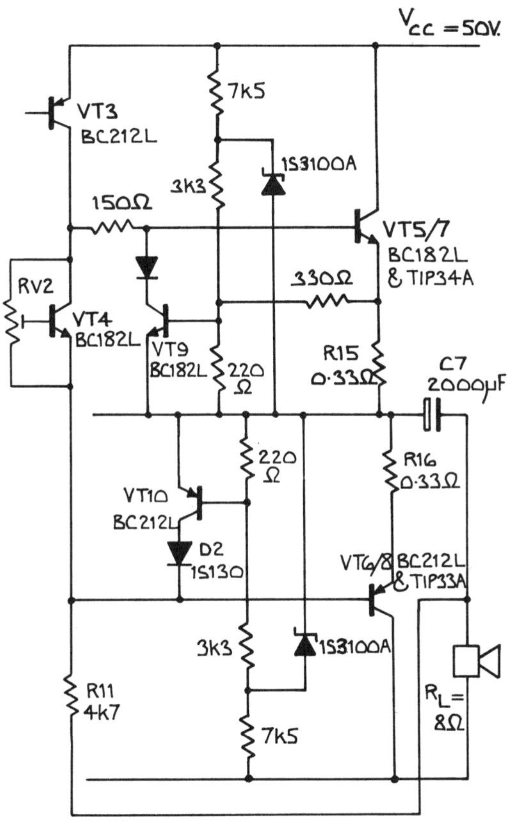

FIGURE 38

off, the fault corrected, and the amplifier switched on again. The extra complication of the protection circuitry is justified by the improved protection it gives (the output stage is left in a zero dissipation condition) and by the fact that until an over load condition arises, the protection circuit does not interfere with the normal operation of the circuit.

The basic amplifier described above does not readily lend itself to this form of protection. The amplifier is, therefore, re-arranged to work with split voltage rails. This rearrangement is a little more costly than the original version, since three extra small signal transistors are used as follows:–

1. To couple the output of the first stage into the voltage amplifier stage in an arrangement which gives a considerable reduction in the amount of supply ripple appearing on the output.
2. As a constant current sink in the collector of the voltage amplifier stage, this configuration is similar to that shown in Figure 23.
3. Connected as a diode in the resistor chain providing the base bias for the constant current sink. This enables the devices to be easily placed in thermal contact to ensure that the value of the constant current sink does not alter. The tail resistor of the long-tail pair provides a convenient constant current source from which the base bias voltage for the constant current sink can be derived, thus saving the use of an extra resistor.

The single potentiometer previously used to set up the collector-emitter voltage on the quiescent current setting transistor VT4, has now been replaced by a resistor R11 and a variable resistor RV1. This protects the output stage in the event of an open circuit occurring in the variable resistor. This failure mode, which was formerly most likely, would have resulted in the output stage trying to pass a very large quiescent current. An open circuit variable resistor now results in zero quiescent current which allows the amplifier to survive such an occurrence. However, the resulting audible cross-over distortion indicates a fault has occurred. Before switching on for the first time RV1 should be set to its maximum resistance and then, after switch on, its value reduced to set the required quiescent current.

On the other hand, a saving may be made on electrolytic capacitors as the power supply capacitors need only be 40V types instead of 64V types required for the power supply and output capacitors in the single supply version. The revised circuit is shown in Figure 39.

The limiter circuit is added to the bases of the transistors at the top and bottom of the voltage amplifier stage. When the limiter transistors turn on, they remove base drive completely from the voltage amplifier stage, which in turn removes all drive from the output stage, rendering the latter completely passive.

One half of the complete protection circuit is shown in Figure 40.

When the amplifier is switched on, because of the greater drive available at its base, transistor VT18 turns on rather than transistor VT16. Once turned on, VT18 holds both VT16 and the limiter transistor VT12 off. Currents due to the amplifier output current and the collector-emitter voltage of the output transistor VT5/7 are summed together on the base of the sensing transistor VT14. When an overload occurs, the base voltage of VT14 becomes sufficiently large to turn on VT14 which then starts to turn on VT16. Once a certain point has been reached, VT16 turning on will start to turn off VT18 which will hasten the turn on of VT16 and the bistable will latch into the opposite state. In this state, VT12 is turned on removing all the base current from VT3, thus shutting the amplifier off. To reset the amplifier, the power supply must be switched off to allow the bistable to reset. A similar arrangement could be used for the other half of the output stage. It is not sufficient however, for the positive trip circuit to act alone as the output of the amplifier tries to swing fully negative when this is done and causing the negative protection circuit to act.

If the overload is marginal, the negative protection circuit may not function and the output could sit at full negative potential indefinitely. (The converse situation does not occur. If the negative trip circuit acts the output returns to zero because transistor VT3 is still operating and hence the normal d.c. feedback circuit still works.) This difficulty may be overcome by interconnecting the two circuits as shown in Figure 41.

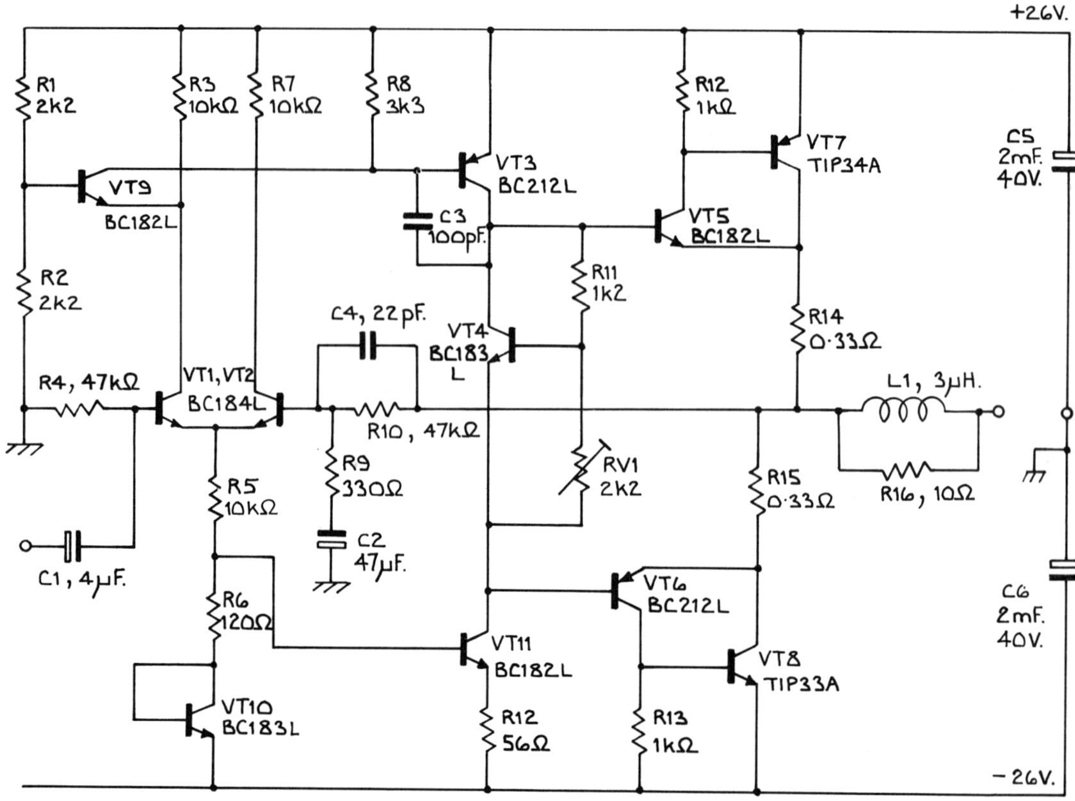

FIGURE 39

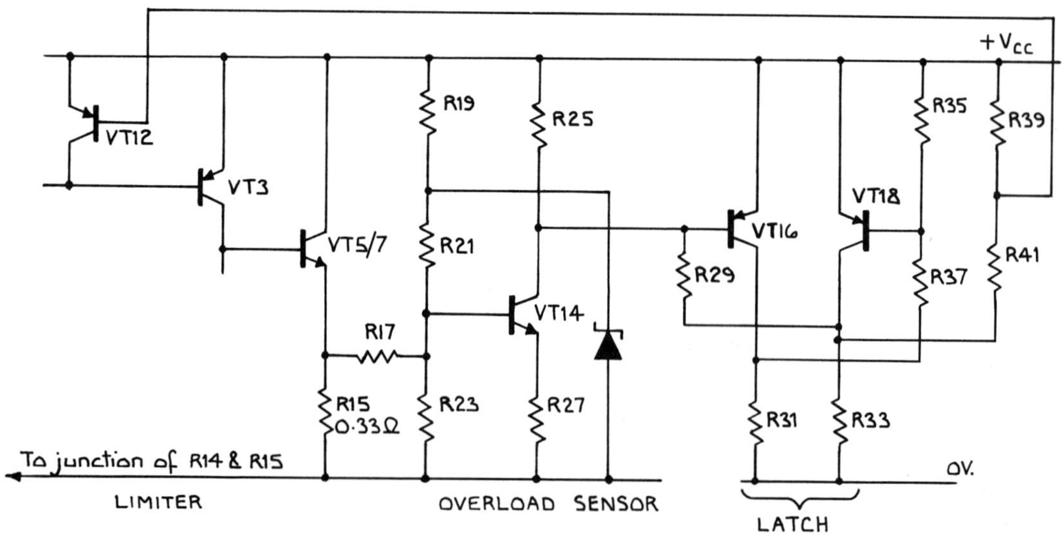

FIGURE 40

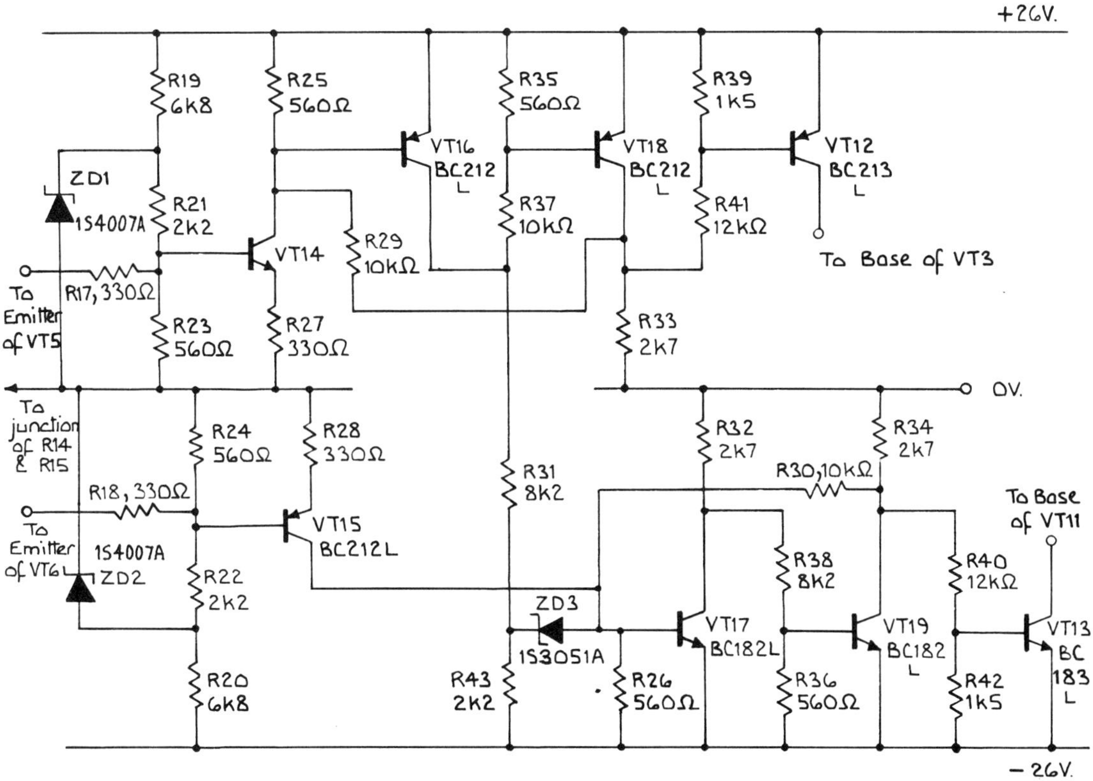

FIGURE 41

The similarity between the circuits shown in Figure 40 and 41 is obvious. The interconnection is achieved by connecting the collector load of VT16 to the negative supply instead of the earth. The feedback to VT18 is connected normally, but the base of VT17 is taken through a zener diode ZD3 to a tap on the collector resistor of VT16. When VT16 turns on, the zener diode conducts and turns VT17 on. This turns on VT12 and 13, which removes base current from both ends of the voltage amplifier stage.

For some applications instant cut-out is not desirable. The operation time of the cut out can be varied by adding capacitors between connector and emitter of transistors VT18 and 19.

It may also not be desirable to have to switch off the amplifier to reset the protection circuit. Transistors VT16 and 18 and VT17 and 19 may be coupled together to form monostables rather than bistables. When an overload occurs, the monostable trips and cuts off the amplifier. The amplifier stays cut off until the end of the monostable period, when it tries to restart. If the overload has cleared by then, the amplifier restarts successfully and continues to operate normally. If the amplifier is still overloaded, the monostable refires and cuts the amplifier off for a further period of time. This process repeats itself until the overload is cleared and although the amplifier operates briefly once per monostable period, the average power dissipation in the output stage is extremely low.

Using standard silect transistors as drivers and TIP33A/34A as output transistors, the maximum output power obtainable from the amplifiers described is limited to about 35W. By changing the load impedance to 4Ω and using TIP35A/36As about 70W could be obtained (this power would be limited by driver dissipation), but such a system would be uneconomic due to the high cost of TIP35A/36A relative to that of TIP33A/4A.

A more economical approach is to use the BFR39 – 41 super silect range and their pnp equivalents for the driver and voltage amplifier stages. These devices are available with breakdown voltages greater than 80V, current gains specified at 1A and over 800mW dissipation in free air. With these devices the rail voltage may be increased giving an output power of about 60W into 8Ω using TIP33B/34B output transistors. The amplifier configuration shown in Figure 39 is to be preferred to that shown in Figure 20 because of the easier capacitor voltage ratings.

PART 3

AMPLIFIERS FOR 100W AND ABOVE

When the output power of the circuit shown in Figure 39 is increased beyond 60W it is necessary to change both driver and voltage amplifier devices to the smallest TIP types – TIP29/30. Using these devices, the power may be increased to 100W if TIP35C/36Cs are used in the output stage. A complete 100W into 8Ω amplifier is shown in Figure 42.

The distortion performance of this amplifier is not as good as the 30W version mainly due to the larger swings involved and the lower loop gain.

Performance figures are given in Table 3.1.

This amplifier represents about the ultimate power which may be obtained from this configuration into 8Ω. Any increase in output power would require an increase in rail voltage which would necessitate a full range of complementary transistors with breakdown voltages exceeding 100V.

Such devices become comparatively expensive and unavailable in the pnp form. An alternative approach is to reduce the load impedance. However, unless the amplifier is to be built into a combination unit with its own speakers, one must stick to standard impedance values. The lowest common standard value is 4Ω which would, theoretically, allow one to obtain 200W from a modification of the circuit shown in Figure 39. The output transistors in such a circuit, however, would operate outside their safe area of operation, and hence such an amplifier would be grossly unreliable. The maximum practical power limit of the circuit in Figure 39 using TIP35/36 type devices is, therefore, about 150W. For powers above this, the current capability of the output transistors may be increased by using two or more in parallel. However, once extra power devices are to be used, it is profitable to look for a better way of using them than merely placing them in parallel. As an example for this exercise the design of an alternative form of 100W amplifier will be considered. The basic requirements for the output stage for 100W into 8 and 4Ω are given in table 3.2.

Table 3.1

1. P_{in} to 8Ω for THD = 1%, THD @ 100W

Frequency kHz	P_{out} W for 1% T.H.D.	T.H.D. % @ 100W
0.02	101.3	0.62
0.1	106.6	0.52
1.	108.0	0.52
10	108.0	0.52
20	108.0	0.65

2. P_{out} into 8Ω for various THD @ 1kHz

T.H.D. %	P_{out} W
1	108.0
3	111.7
5	119.4
10.	132.0

3. THD versus P_{out} @ 1kHz and 10kHz

P_{out} W	T.H.D. %	
	@ 1kHz	@10kHz
100	0.52	0.52
30	0.13	0.14
10	0.09	0.09
3	0.08	0.07
1	0.09	0.09

Frequency response

Output Voltage is –1dB @ 5Hz and 28kHz
Signal/Noise ratio 72dB

The Load Drive Configuration

There is no way in which the current and voltage requirements into the load can be reduced. The current taken from the output stage can be reduced by using an 8Ω load, but this presents the problem of providing the large voltage swing into the load, which requires high supply voltages. The total supply voltage required in all the configurations described so far is about three times the r.m.s. output voltage which can be placed across the load is limited to plus or minus half the supply voltage. This difficulty may be overcome by using the configuration shown in Figure 43.

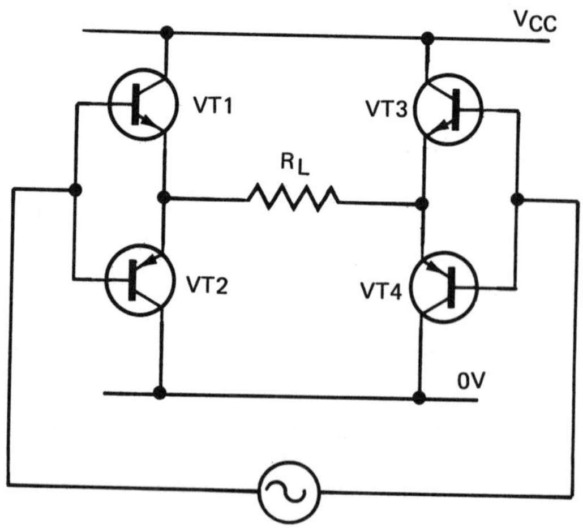

FIGURE 43

Table 3.2

Load Impedance Ω	R.M.S. Voltage V	Peak Voltage V	Pk–Pk Voltage V	R.M.S. Current A	Peak Current A	Power Supply V
8	28.3	40.1	80.2	3.53	5.0	90
4	20	28.3	56.6	5	7.06	64

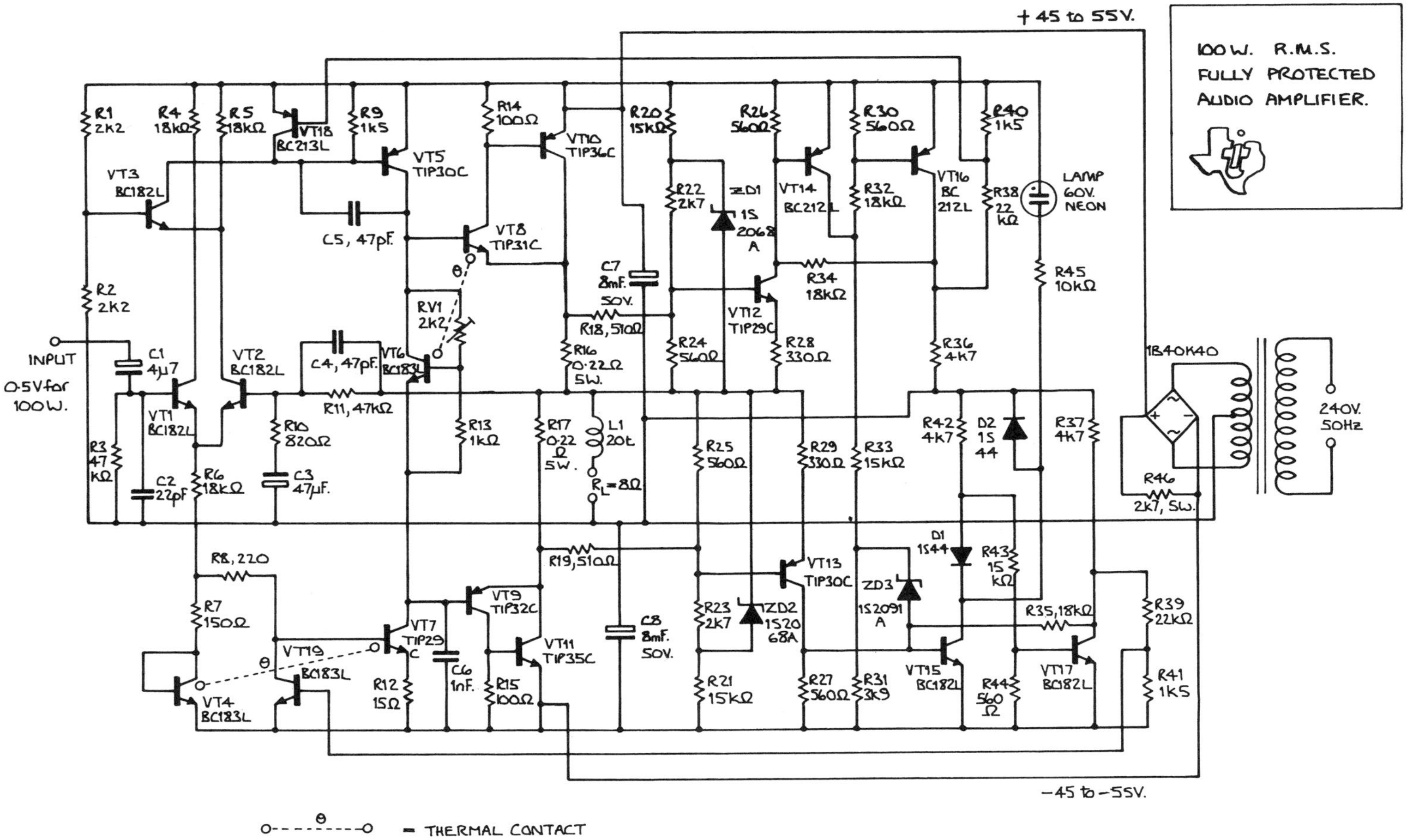
100 W. R.M.S.
FULLY PROTECTED
AUDIO AMPLIFIER.
+ 45 to 55V.
−45 to −55V.
INPUT
0·5V for 100W.
240V. 50Hz
1B40K40
LAMP 60V. NEON
R1 2K2
R2 2K2
R3 47 kΩ
R4 18kΩ
R5 18kΩ
R6 18kΩ
R7 150Ω
R8, 220
R9 1K5
R10 820Ω
R11, 47kΩ
R12 15Ω
R13 1kΩ
R14 100Ω
R15 100Ω
R16 0·22Ω 5W.
R17 0·22 Ω 5W.
R18, 510Ω
R19, 510Ω
R20 15kΩ
R21 15kΩ
R22 2K7
R23 2K7
R24 560Ω
R25 560Ω
R26 560Ω
R27 560Ω
R28 330Ω
R29 330Ω
R30 560Ω
R31 3K9
R32 18kΩ
R33 15kΩ
R34 18kΩ
R35, 18kΩ
R36 4K7
R37 4K7
R38 22 kΩ
R39 22kΩ
R40 1K5
R41 1K5
R42 4K7
R43 15 kΩ
R44 560 Ω
R45 10kΩ
R46 2K7, 5W.
RV1 2K2
C1 4μ7
C2 22pF
C3 47μF.
C4, 47pF.
C5, 47pF.
C6 1nF.
C7 8mF. 50V.
C8 8mF. 50V.
L1 20t
RL = 8Ω
D1 1S44
D2 1S 44
ZD1 1S 2068 A
ZD2 1S20 68A
ZD3 1S2091 A
VT1 BC182L
VT2 BC182L
VT3 BC182L
VT4 BC183L
VT5 TIP30C
VT6 BC183L
VT7 TIP29 C
VT8 TIP31C
VT9 TIP32C
VT10 TIP36C
VT11 TIP35C
VT12 TIP29C
VT13 TIP30C
VT14 BC212L
VT15 BC182L
VT16 BC 212L
VT17 BC182L
VT18 BC213L
VT19 BC183L
θ
= THERMAL CONTACT

FIGURE 42

In this arrangement, the two halves of the output stage are driven in anti-phase. When there is no signal, there is no voltage across the load. At positive peak signal (say) transistors VT1 and VT4 are both hard on (the other two transistors are off) and the full supply voltage appears across the load. At peak negative signal, transistors VT2 and VT3 are both hard on (transistors VT1 and VT4 are off) and full supply voltage again appears at the load, but in this case current flows through the load in the opposite direction to its previous one. Hence, with this system, the effective voltage swing across the load for any given power supply voltage is doubled and the output power obtainable from any given supply voltage quadrupled ($P \propto V^2$). This amplifier configuration shows considerable promise, therefore, for obtaining higher output powers without the need for high supply voltages or low load impedances.

Re-examination of Table 3.2 shows that an r.m.s. output voltage of 28.3V is required for 100W into 8Ω. With this configuration each half of the output stage is required to swing half the required voltage, i.e. 14.15V r.m.s. = 20.05V peak = 40.10 volts peak to peak. This output voltage swing can easily be obtained from a 50V power supply which enables one to use readily available, low cost 50V transistors.

Furthermore, the relatively low power supply voltage required now means that there is opportunity for increase if higher power versions are required. Using an 8Ω load and 60V power supply, an output power of about 175W is possible. If an output stage with a higher current capability than the one to be described here and a 4Ω load is used, power in excess of 350W should be attainable.

A block diagram of the system to be used is shown in Figure 44 and a diagram of the practical circuit is shown in Figure 45.

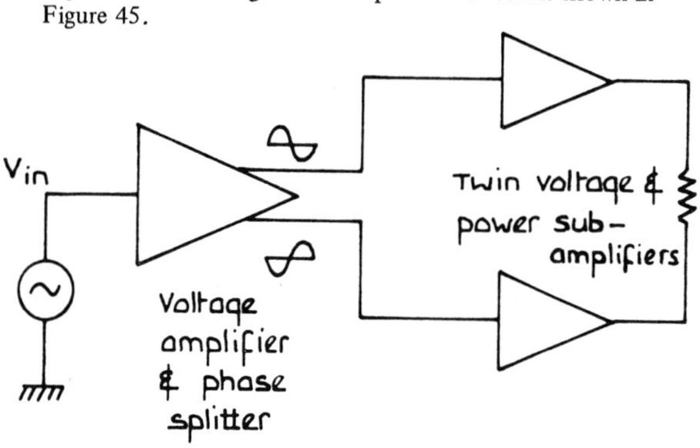

FIGURE 44

The Power Sub-Amplifiers

The left-hand power sub-amplifier in Figure 45 will now be described. A similar argument may be applied to the right hand power sub-amplifier.

The output stage configuration is similar to the used in the sub-amplifier which was described in Part 2 of this chapter The TIP35A and TIP36A are used for the output transistors in the 100W amplifier circuit.

These transistors have a minimum current gain of about 20 at 5A and thus require a peak base current of 250mA. This current requirement is beyond the power dissipation capabilities of the normal silect transistors previously used to drive the output transistors. However, the TIP29A and TIP30A will perform this function well within their ratings, (no heat sink is required for them) and, as their current gain at this peak current is at least 40, their peak base current is about 6mA. Their rather low current gain at low collector currents necessitates the use of the 150Ω resistors, R21 and R23, to shunt the base emitter diodes of the output transistors. This ensures that under quiescent conditions, the collector current of the TIP29A and TIP30A is about 45mA, at which their current gain is typically about 100. The 150Ω resistors also provide a path for the output transistors' collector-base leakage current, ensuring thermal stability.

To set the quiescent current of the output transistors, the usual transistor/potentiometer bias voltage setting arrangement is used. Sufficient current gain will be obtained from the output transistors if the quiescent current through them is set in the 20 to 50mA range.

The Voltage Amplifier Stage

The high current gain of the output stage of each sub-amplifier and the low supply voltage used, enables the voltage amplifier stage to be operated at 10mA, which is well within the current and power handling capabilities of normal silect transistors. The manner is which the two transistors are used in the voltage amplifier stage is shown in Figure 23. As well as the improved performance which this configuration was shown to have in Part 2, there are further reasons for using it in preference to the bootstrap resistor configuration also described in Part 2. In this circuit there is no output capacitor to transfer the output voltage from the output mid-point to earth. Hence an extra capacitor and resistor would have to be used in each output amplifier to provide a suitable point for connecting the bootstrap resistor. The expense of this lessens the cost premium one has to pay for the extra performance of the transistor current sink arrangement. This circuit also uses a common potential divider chain to provide the base voltage for the constant current sink in both sub-amplifiers, further reducing the cost premium and making the use of this particular configuration justifiable. A diode-connected transistor is used in the potential divider chain to ensure constancy of current in the voltage amplifier stage with cahnges in ambient temperature. It should be placed in thermal contact with either VT9 or VT10.

The Input Stage

So far the configuration of the output sub-amplifiers have been similar to that described in Part 2. The long-tail pair first stage has, however, been replaced by a single transistor used in the configuration described in Part 1, Figure 11. This transistor (VT3) has its collector connected directly to the base of the driver transistor VT5, whose base emitter junction is shunted by the 2k7Ω resistor R13 to reduce variation in VT3's collector current due to variations in the

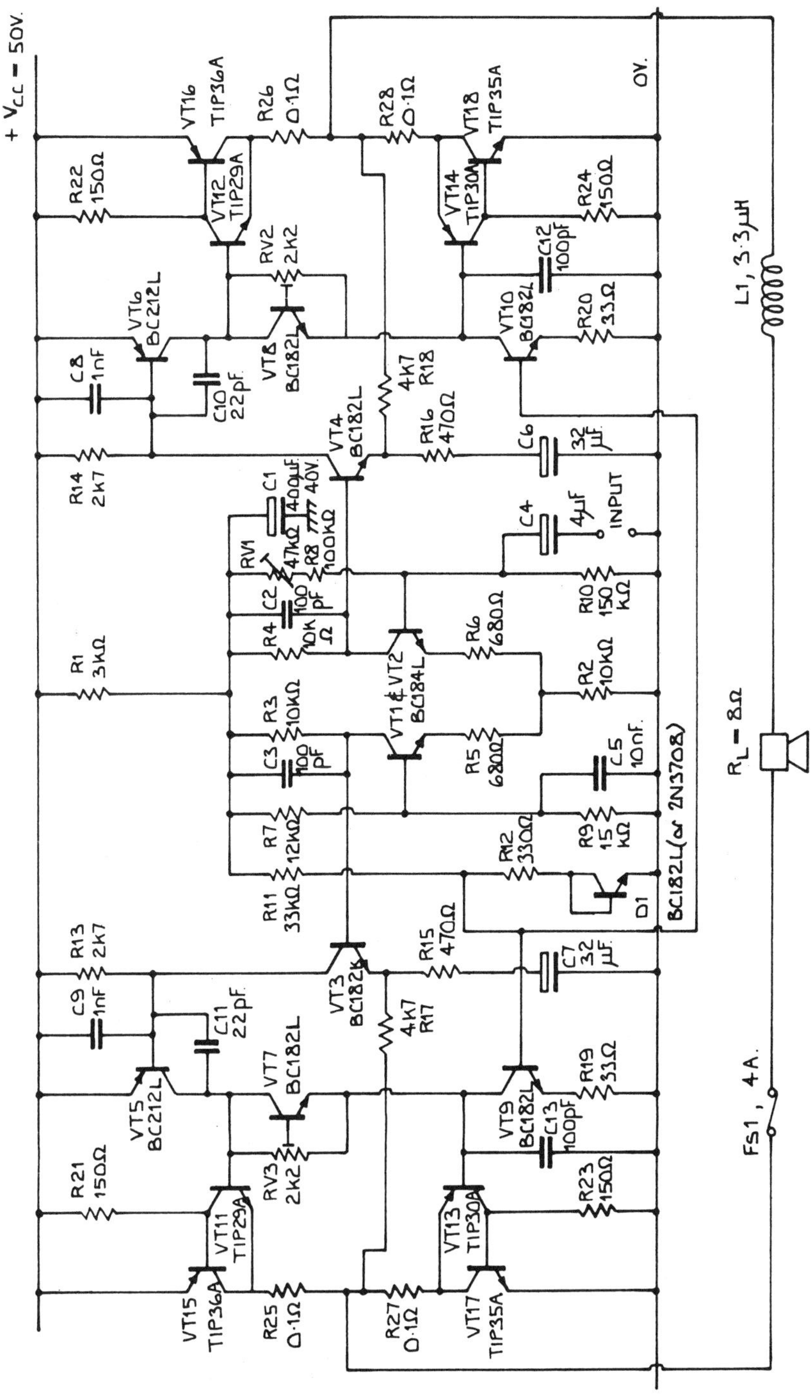
+ V_{CC} − 50V.
0V.
VT16 TIP36A
R26 0·1Ω
R28 0·1Ω
VT18 TIP35A
R22 150Ω
VT12 TIP29A
VT14 TIP30A
R24 150Ω
RV2 2K2
C12 100pF
VT6 BC212L
VT8 BC182L
VT10 BC182L
R20 33Ω
L1, 3·3µH
C8 1nF
C10 22pF
4K7 R18
VT4 BC182L
R16 470Ω
C6 32µF
R14 2K7
C1 400µF 40V.
RV1 47kΩ
R8 100kΩ
C4 4µF
INPUT
C2 100 pF
R4 10K Ω
R10 150 kΩ
R6 680Ω
R1 3kΩ
R3 10kΩ
VT1 & VT2 BC184L
R2 10kΩ
R5 680Ω
C3 100 pF
C5 10nF.
R_L − 8Ω
R7 12kΩ
R12 330Ω
R9 15 kΩ
R11 33kΩ
D1
BC182L (or 2N3708)
R13 2K7
VT3 BC182L
R15 470Ω
C7 32 µF.
C9 1nF
C11 22pF
VT7 BC182L
4K7 R17
VT5 BC212L
R19 33Ω
VT9 BC182L
C13 100pF
FS1, 4A.
R21 150Ω
RV3 2K2
VT11 TIP29A
VT13 TIP30A
R23 150Ω
VT15 TIP36A
R25 0·1Ω
R27 0·1Ω
VT17 TIP35A

FIGURE 45

current gain of the driver transistor. Its emitter is connected to the output mid-point by a 4k7Ω resistor R17 and to earth by the 470Ω resistor R15 and capacitor C7. This sets the a.c. closed loop gain to 11. This simplification can be justified by the performance benefits which were claimed as a result of using a long-tailed pair configuration in the first stage of the amplifiers described in Part 2 being realised in this total system using a single transistor. These benefits were:–

a) Stability of the DC level of the output mid-point. The stability of the DC level of the two output mid-points of these amplifiers is worse than that of the amplifiers described in Part 2, as changes of both the base emitter voltage of transistor VT3 and the current gain of transistor VT5 change the DC output level. These changes are, however, small, e.g. a temperature change of 20°C will cause a change in the output level of 40mV; a gain change in transistors VT5 and VT6 from 60 to 300 will cause an output change of 400mV. However, the change in voltage across the load, the important factor as large DC currents through the load must be avoided, is much less than these amounts. This is because the parameter changes, causing these output voltage drifts, tend to change similarly in both output sub-amplifiers, giving only a small differential voltage change across the load.

b) High input impedance to the input and feedback circuits. As the feedback circuit is connected to the emitter of transistors VT3 and VT4, component values need to be low. The main advantage of high impedance in the feedback circuit is that low value capacitors may be used to decouple it. In this circuit, because both the phase splitter and output stages contribute to the voltage gain, the output stage voltage gain (defined by resistors of 4k7Ω and 470Ω) is only 11. It is the 470Ω which has to be decoupled and a −3dB frequency of 10Hz may be obtained with a decoupling capacitor of 32μF. In a more normal application of this amplifier configuration, where the gain would have to be, say, 100, a capacitor value of ten times this amount would have to be used to get the same −3dB frequency. The impedance of this configuration, looking into the bases of transistors VT3 and VT4, is high because the signal on their emitters is in phase with that on their bases.

To set the DC level of the output voltage, a DC reference voltage must be fed onto the bases of transistors VT3 and VT4. This could be achieved by feeding the a.c. signal from a phase splitter stage through capacitors and using a pair of potential divider chains. (Two would have to be used so that the DC levels of the two output mid-points could be equalised despite differences in the V_{BE}s of transistors VT3 and VT4 and the voltage drops across the resistors R17 and R18). However, considerable simplification may be obtained by directly coupling the output sub-amplifiers to the phase splitter and using the latter to set the voltage reference level for the former.

The Phase Splitter Stage

The phase splitter consists of a conventional long tail pair formed by transistors VT1 and VT2. This was chosen in preference to using a single transistor with equal collector and emitter resistors for the following reasons:

a) It gives voltage gain, enabling the power amplifier gain to be reduced, which both simplifies the design of the power amplifier and reduces distortion by allowing a greater ratio of open loop to closed loop gain.

b) It provides identical drive conditions for both power amplifiers

c) It is capable of being directly connected to the power amplifiers which saves several components.

The base voltages for the two transistors are provided by two potential divider chains formed by resistors R7 and R9 and resistors RV1, R8 and R10. It can be seen that the impedance of the resistors R7 and R9 is about one tenth of that of resistors R8 and R10, since:–

a) R8 and R10 need to be large to provide a high input impedance for the amplifier. For a.c. they are shunted by capacitor C4 and the source impedance of the input signal, which means that transistor VT2 usually sees a source impedance for which its noise output is low.

b) Transistor VT1 would see a source impedance too high for minimum noise output if a similar impedance network to R8 and R10 were used to define the voltage on its base. Also this point would easily pick up radiated interference.

By using the values shown for resistors R7 and R9, a favourable source impedance for low noise is obtained. The capacitor C5 shunts R9 to reduce this source impedance at higher frequencies to prevent interference pick-up. The only disadvantage of this method of reducing noise is that because of the different D.C. source impedances to each half of the long-tail pair, any gain change in the transistors arising from a change of temperature is likely to result in a drift in the relative D.C. voltage levels of the two bases. This voltage drift manifests itself as a differential output between the collectors of the long-tail pair, and hence between the two outputs of the amplifier system, amplified by the D.C. gain of the complete amplifier. This effect is extremely small due to high gain of the transistors specified for transistors VT1 and VT2. If minimum gain devices are used for VT1 and VT2 and their gains double due to a change in temperature (and this is highly unlikely), a differential change in base voltage of about 120μV can be expected. The resulting differential change in output voltage is likely to be about 1.7mV.

The gain of the phase splitter stage is set to 15.3 by the ratio of the collector resistors R3 and R4 to the emitter resistors R5 and R6. Each collector resistor is shunted by the input impedance of the sub-amplifier it drives, but the large input impedance of the sub-amplifiers allows one to ignore its effect in the gain calculations.

The complete phase splitter stage is connected to the positive rail by a single resistor R1 which is decoupled to ground by capacitor C1. This arrangement is intended to prevent ripple and noise on the power supply from appearing on the inputs to the sub-amplifiers. (By similar arguments applied to the amplifiers in Part 2, the sub-amplifiers themselves are completely immune to noise and ripple on the power supply). This precaution is not theoretically necessary because any power supply noise voltage will appear equally and in phase on the two phase splitter outputs. Hence, there will be no differential noise voltage across the load, provided the sub-amplifier gains are exactly equal. Unless special care is taken exact equality of gain is unlikely in practice. As the extra decoupling has low cost, it has been included and gives reduced noise level and improved high frequency stability.

The resistor values in the phase splitter stage are designed to give a voltage of 28.5V on each collector. This sets the mean D.C. level of the power amplifier output mid-points to about +26.2V, which allows for the fact that the output voltage of the power amplifiers cannot swing as close to earth as it can to the positive rail (because the emitters of the current sink transistors VT5 and VT6 are set at 0.3V). Because of component tolerances, a considerable difference in the D.C. voltage levels of the output mid-points of the two power amplifiers can arise. This condition is most undesirable as it can give rise to a large D.C. current through the load. It is avoided by adjusting the base voltage of transistor VT2 by means of potentiometer RV1 until the two D.C. output voltages are equal. Resistor R7 should be reduced to 10kΩ and a 4k7Ω potentiometer placed in series with it so that the two output voltages can be equalised and set to about +26V if loose tolerance resistors are used in the amplifier.

High Frequency Stability

As in all the other amplifiers described in this chapter, this amplifier has the gain in the megahertz region and is therefore unstable. With the straight resistive loads, stability may be obtained by placing 100pF capacitors between collector and base of transistors VT5 and VT6. If a capacitor is placed across the load, the amplifier again becomes unstable and capacitors C2 and C3 have to be added and the value of capacitor C10 and C11 increased to 330pF. This stabilises the amplifier but restricts its undistorted output capability to about 10kHz. To stabilise the amplifier and achieve an undistorted frequency response beyond 20kHz the capacitors C8, C9, C12 and C13 have to be added. The small inductance L1, which may be made by winding 25 turns of about 26 s.w.g. enamelled copper wire in a single layer on a high value 1W resistor, completed the array of stabilisation components. The use of such a large number of components is necessitated by a large open loop gain of the amplifier and the large current and voltage swings involved.

Construction and Heat Sinks

Provided that the leads carrying large currents are kept well away from the input, the layout of this amplifier is not particularly critical. The following rules for connecting up the output stages should be observed, however:

(a) The two stages consisting of transistors VT11, VT12, VT13 and VT14 should be mounted in close proximity to one another on the circuit board.

(b) Connections should be made with short wires between these two stages and the four output transistors.

(c) The two npn output transistors should be mounted fairly close together, and their emitters jointed together with thick wire and connections made from this point to the negative terminal of the power supply output capacitor and to the negative line on the circuit board.

The pnp output transistors should also be mounted fairly close to one another, connected first together and then to the power supply and the board in a similar manner to that used for the npn transistors.

Each output transistor has to dissipate a total of about 12.5W and, therefore, requires heat sinking. The transistors may be mounted in pairs on heat sink having a thermal resistance of less than about 2°C/W. In the prototype Radiospares heat sinks were used for this.

Setting-up Procedure

After the circuit has been constructed it should be thoroughly checked, preferably by someone else, and the following points observed when it is switched on for the first time:

(a) Terminate the input with a 10kΩ resistor. Do not connect the load to the output. Connect a voltmeter across the output terminals. Connect the amplifier to its power supply.

(b) Set potentiometer RV2 and RV3 so that their wipers are at the end of their tracks connected to the collectors of the bias transistors VT7 and VT8.

(c) Set the wiper of potentiometer RV1 mid-way along its track.

(d) Turn the power supply on.

(e) Adjust potentiometer RV1 until the meter across the output reads 0±50mV;

(f) Switch off, connect in the load, switch on and re-check the differential output voltage. Re-adjust if necessary, switch off.

(g) Place a 100mA current meter in series with the power supply. Switch on the power supply and adjust the potentiometer RV3 until the meter reading increases by about 30mA.

(h) Adjust RV2 until the meter reading increases by a further 30mA. Switch off and remove the meter.

(i) Switch on the power supply and re-check the D.C. differential output voltage. The amplifier is now ready for use.

Overload Protection

It is strongly recommended that the fuse Fs1 be included in any constructed version of this amplifier. The D.C. nature of the connection of the load to the output of the amplifier means that if one of the output transistors should fail, a large current can flow through the load which may well damage it. The fuse will protect the load in this eventuality. Overload protection circuitry of the type described in Part 2 of this report can be used with this amplifier, a current trip on the power supply is a much simpler answer to protection. This will be described fully in the next section.

The Power Supply

The large variation in current drawn by this amplifier from no load to full load means that the use of a stabilised power supply is preferable. (An unregulated power supply can be used, provided that its no load voltage does not rise beyond about 60V and its full load voltage does not fall below about 50V.) The suggested power supply design is shown in Figure 46.

The 50V r.m.s. winding, which needs to have a current rating of about 10mA, when rectified and smoothed forms a constant voltage source. When passed through the 1kΩ resistor R28 this forms the constant current source for the zener diode ZD1. The two transistors VT18 and VT19 are used to provide current gain in order to minimise the effect of the swing of the base current of the output transistors VT20 and VT21 on the current through the zener diode. The potentiometer RV4 is used to set the output voltage to 50V, allowing for the tolerance in the zener diode voltage. The output transistor has to pass a peak current of about 5A and must be able to withstand the peak otuput voltage of the power supply (about 65V). The normal dissipation in the transistor is fairly high, but a TIP35B could be used. A lower cost solution, however, is to use two TIP 3055s in parallel. Two of these transistors are necessary to ensure reliable operation at the power levels involved. The 0.22Ω resistors placed in the emitter of each device ensure that the current is shared equally between the two transistors, both of which should be mounted on the same heatsink. The leads from capacitor C16 to the emitters of the four output transistors of the amplifier should be kept short.

Overload protection is provided by transistor VT22 which limits according to the power dissipation in the output transistors. The base-emitter voltage of VT22 depends on the current (approximately half the output current) flowing through resistor R34 and the difference in voltage between C15 and C16. It is defined by these quantities, resistors R35, R36 and R37 and zener diode ZD2. When the output current becomes sufficient to forward bias the base of VT22, the transistor starts to turn on and removes base current from transistor VT18. This causes the voltage across VT20 and hence that across resistors R35 and R37 and zener diode ZD2 to increase. Thus VT22 is turned harder on and hastens the fall of the output voltage towards zero. Here the output voltage stays until the mains supply is switched off and a while is allowed for the circuit to discharge. This, therefore, enables the limiter circuit to reset so that the circuit will function normally when the mains are reconnected – provided that the overload condition no longer persists.

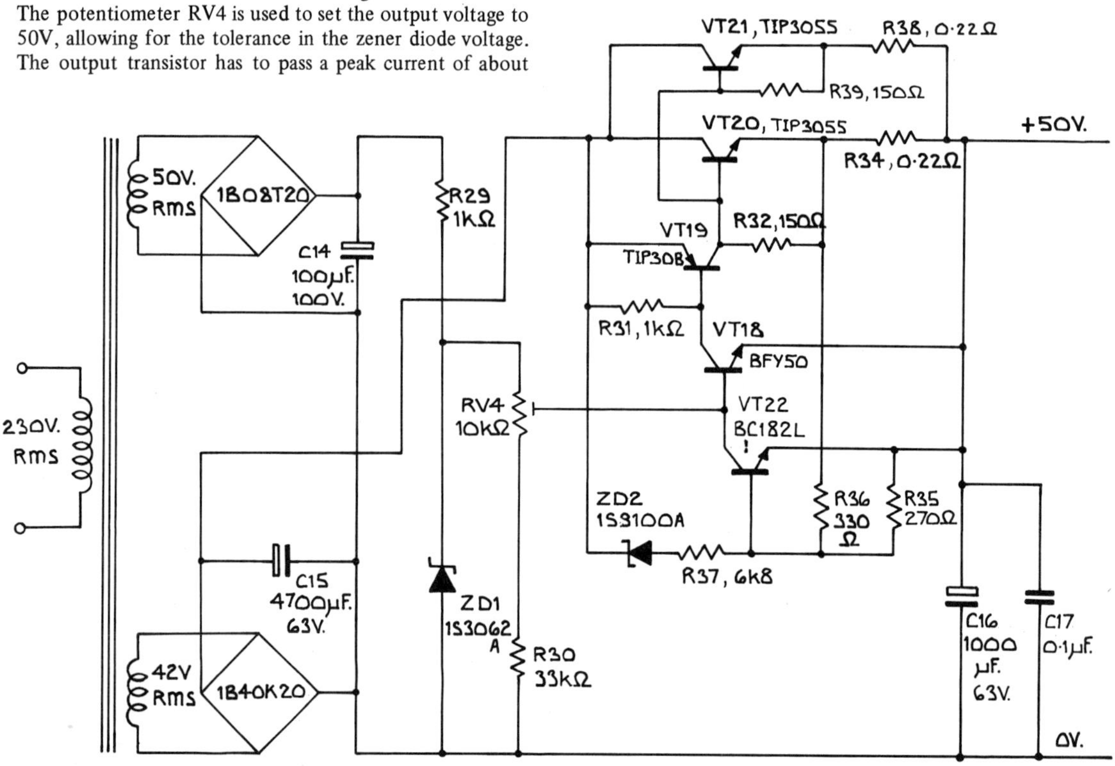

FIGURE 46

Performance

The performance of this amplifier has been fully characterised and performance figures are given in tables 3.3 and 3.4. and Figure 47.

Table 3.3

Power Output	100W
Power Output (T.H.D. = 10%)	150W
Frequency Response (± 1dB)	10Hz – 20kHz
Signal - Noise ratio (source resistance = 600Ω)	89dB
Signal - Noise ratio (source resistance = 10kΩ)	84dB
Input Impedance at 1kHz	56kΩ
Output Impedance at 1kHz	0.08Ω
Input Voltage (P_{out} = 100W at 1kHz)	180mV
Power Supply Current	3.5A r.m.s.

Table 3.4

Output Frequencies and Amplitudes				Intermodulation Distortion
Hz	V	Hz	V	%
1100	14.4	900	14.4	0.16
11k	14.4	9k	14.4	0.23
10k	24.0	1k	6.0	0.14
10k	6.0	1k	24.0	0.14
10k	24.0	120	6.0	0.15
10k	6.0	120	24.0	0.15
1k	24.0	120	6.0	0.135
1k	6.0	120	24.0	0.145

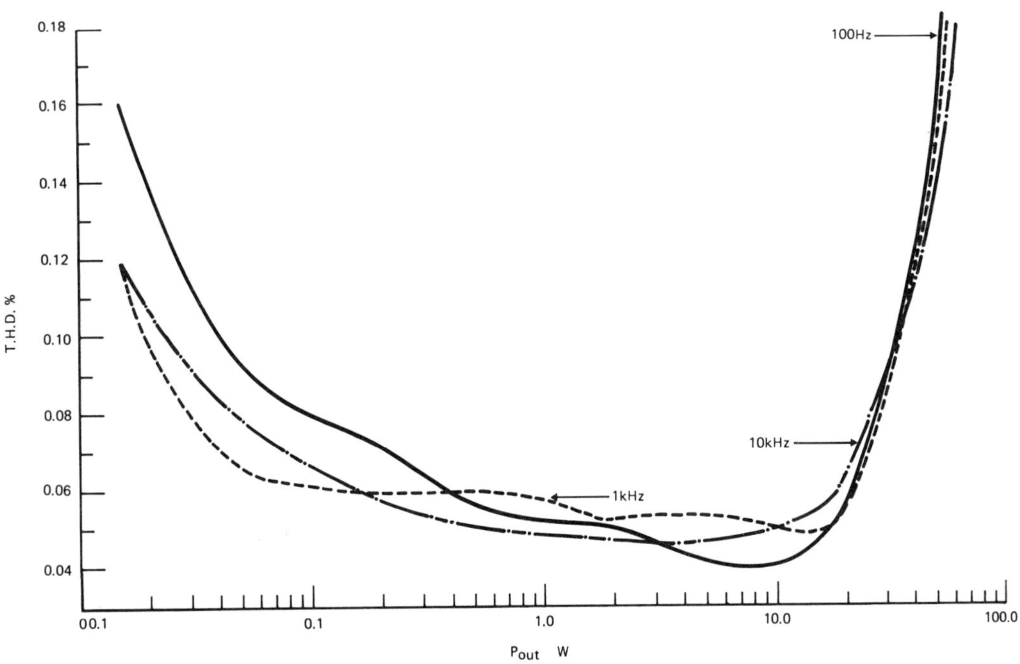

FIGURE 47 Total Harmonic Distortion VS Output Power for 100W/8Ω Amplifier

PART 4

MEASUREMENT METHODS

In this section block diagrams of the methods used to measure the various performance parameters quoted in the preceding sections are given.

Harmonic Distortion

Distortion and Noise is directly read as a percentage on the distortion meter.

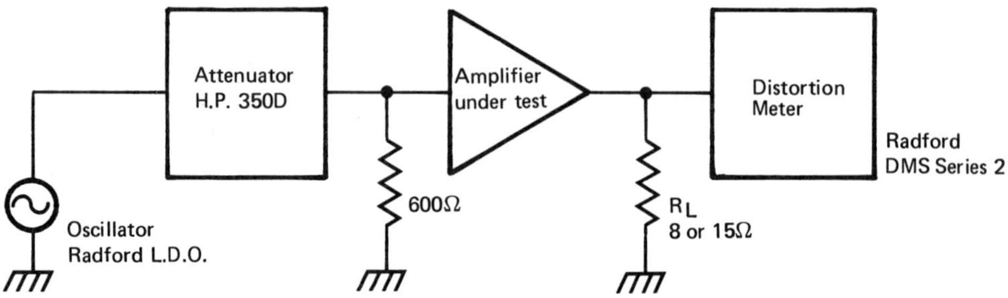

FIGURE 48

Intermodulation Distortion

Output of oscillators P and Q are adjusted to give amplifier output waveform equal in pk – pk amplitude to pk – pk amplitude of full power sine wave. Relative amplitudes of P. and Q. are defined in results table. Amplitudes of these intermodulation products at such frequencies as (P + Q), (P – Q), (2P – Q), (3P – Q), etc. (P + 2Q), (P – 2Q), (P + 3Q), (P – 3Q) etc. (3P + 2Q), (3P – 2Q), etc. are measured on the Wave Analyser.

Intermodulation % distortion is defined as

$$\frac{\sqrt{\text{sum of squares of intermodulation products}} \times 100\%}{\text{amplitude of P + amplitude of Q}}$$

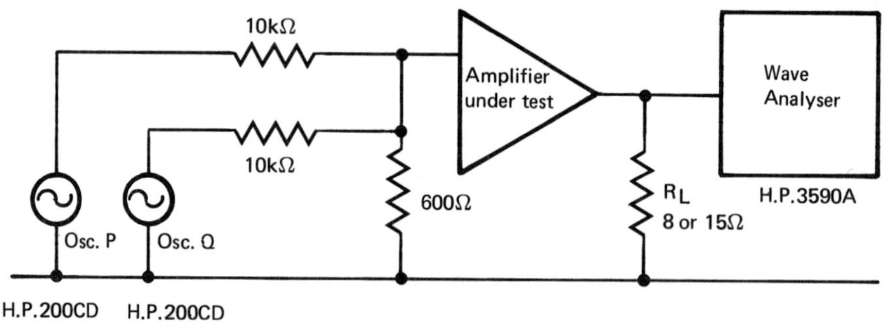

FIGURE 49

Input Impedance

Oscillator amplitude is adjusted for full amplifier power output. Voltages are both ends of 10kΩ resistor are measured.

$$\text{Input Impedance} = \frac{V2}{(V1 - V2)} \times 10k\Omega$$

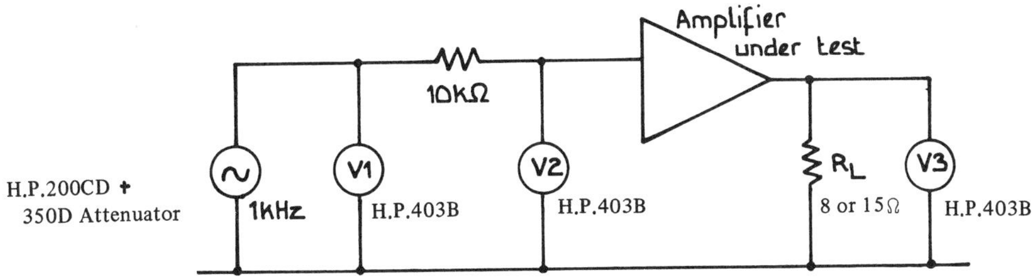

FIGURE 50

Output Impedance

Output of oscillator is adjusted to give amplifier output of IV rms (V1) into 15Ω. Load resistor is changed to 1Ω and new otuput voltage (V2) is measured.

$$\text{Output Impedance} = \frac{(V2 - V1)}{(V2 - V1/15)}$$

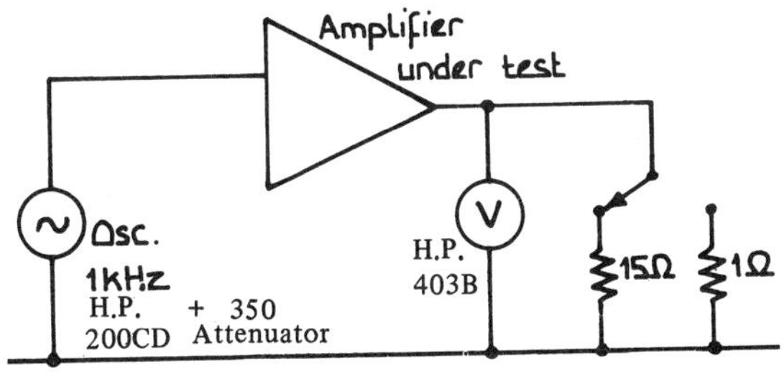

FIGURE 51

Signal/Noise Ratio and Frequency Response

Oscillator output amplitude is set to give amplifier full power output at 1kHz. Frequency is then decreased to give amplifier output voltage 1dB and then 3dB down on 1kHz values. Frequency is then increased to give amplifier output voltage 1dB and then 3dB down.

Noise output is measured with oscillator deconnected (S1 open) with S2 set to give 600Ω source impedance. S2 is then set to 10kΩ source impedance and a new noise voltage read.

Signal to Noise ratio is calculated by:–

$$\text{S/N ratio} = 20 \log_{10} \frac{\text{Full power output voltage}}{\text{Noise voltage}} \text{ dB}$$

The bandwidth of the meter is 5Hz to 2MHz. An ideal 10kΩ resistor could be expected to contribute a noise output of 15μV to the input of the amplifier – i.e. most of the reduction of the signal to noise ratio with a 10kΩ source is due to noise from the resistor itself.

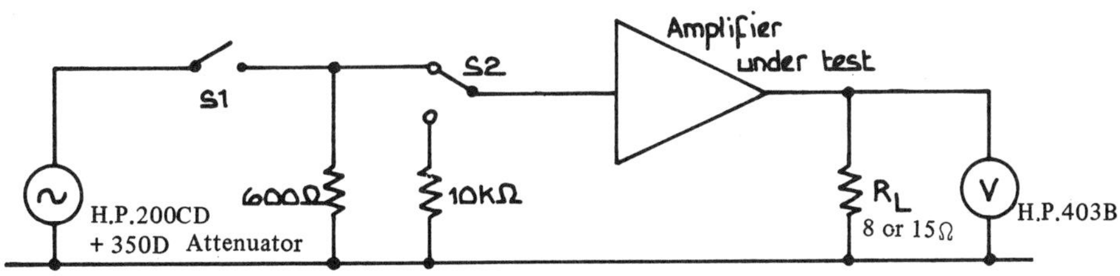

FIGURE 52

XIII SUPER SILECT TRANSISTORS IN AUDIO AMPLIFIERS

In the previous chapter a number of audio amplifiers designed to fulfil a wide range of requirements are described. The first section uses Silect † transistors in simple circuitry to give up to 2.5W into the load. The second section describes how Silect 'drivers' feeding plastic power output devices may be used in again a relatively simple circuit, to give 10 to 30W.

The introduction of the Super Silect complementary transistor range (NPN – BFR39/40/41 and PNP – BFR 79/80/81) gives even greater flexibility. These devices have a higher current, voltage and high power dissipation capability and using them in the circuit designs of Chapter XII means that, into a given load, considerably more power than before can be provided.

5W AMPLIFIERS

General

Assuming that the requirement is 5W into the load, Table 1 gives the voltage and current requirements for various values of load impedance (R_L).

Table 1

R_L Ω	V_{out} rms V	V_{out} pk. V	V_{out} pk-pk V	I_{out} rms A	I_{out} pk.A
50	15.8	23.3	46.6	0.32	0.48
35	13.2	18.7	37.4	0.38	0.53
16	8.94	12.6	25.2	0.56	0.79
8	6.32	8.94	17.9	0.79	1.12
4	4.47	6.32	12.6	1.12	1.58

Load Impedances of 50, 35 or 16Ω

For these load impedances a very simple amplifier, as shown in Figure 1, may be used. (A detailed description of the circuit operation is given on page 164 in the previous chapter.) A resistor R6 could be included between base and collector of transistor VT2, rather than just using potentiometer RV2 (base to wiper) as shown, to make the setting of quiescent current easier and provide a safety precaution (i.e. should the potentiometer go open circuit, its most likely failure mode, transistor VT2, would be held hard 'on' and damage to the output transistors prevented.)

Potentiometer RV2, should be adjusted to its maximum resistance condition before the amplifier is switched on for the first time, i.e. so that its wiper is hard up to the collector of transistor VT2.

Table 2 lists the component values for the three load impedances and Figure 2 gives a suggested circuit layout

† Trademark of Texas Instruments

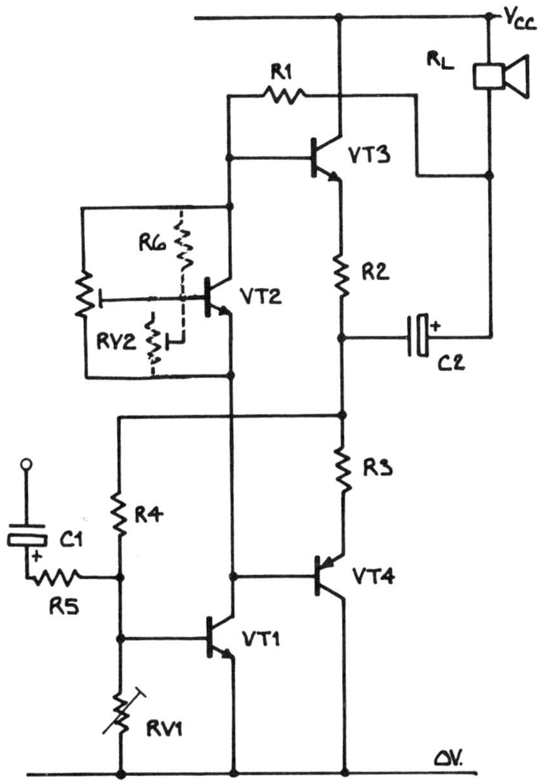

FIGURE 1

Table 2

R_L Ω		50	35	16
V_{CC}	V	52	45	34
C1	μF	22	22	22
C2	μF	330	500	1000
R1	Ω	1k2	1k2	560
R2	Ω	4.7	4.7	3.3
R3	Ω	4.7	4.7	3.3
R4	Ω	27k	27k	15k
R5	Ω	1k	1k	1k
RV1	Ω	1k	1k	1k
RV2	Ω	1k	1k	1k
VT1		BFR40	BFR41	BFR41
VT2		BC183L	BC183L	BC183L
VT3		BFR40	BFR41	BFR41
VT4		BFR80	BFR81	BFR81

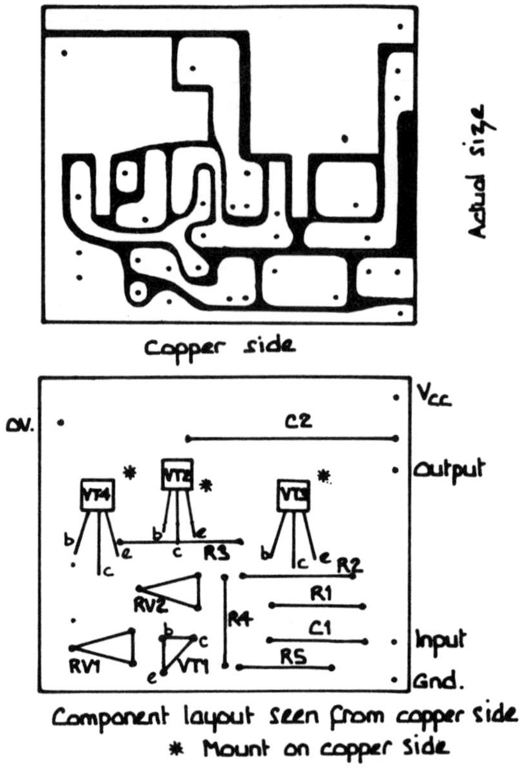

FIGURE 2

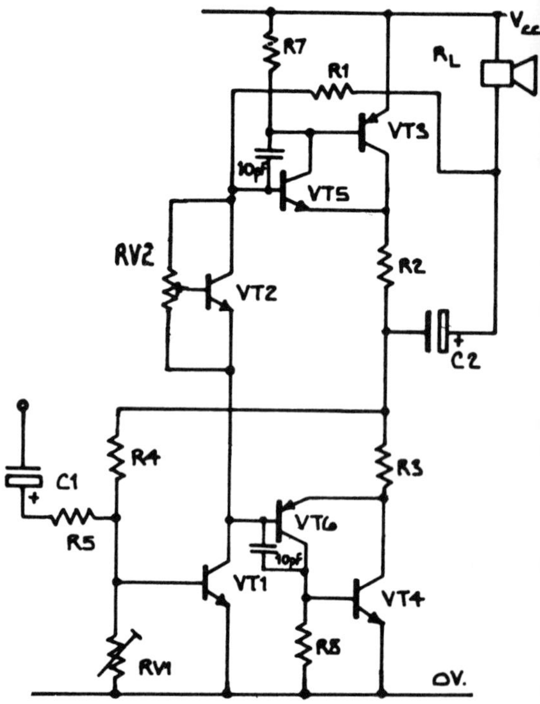

FIGURE 3

and printed board design. Transistors VT2 and VT4 are mounted on the same heatsink to ensure thermal contact between them.

Performance figures for this circuit with a 35Ω load are given in Table 3.

Table 3

	Distortion %				
Power W / Frequency kHz	0.01	0.1	1	2.5	5
0.12	0.22	0.40	0.44	0.75	3.2
1	0.05	0.14	0.35	2.03	5.1
10	0.05	0.12	0.31	1.64	4.0

Sensitivity 0.6V for 5W

Input Impedance 1kΩ

Load Impedances of 8 and 4Ω

With an 8 or 4Ω load Class B driver transistors VT5/6 must be used in order to supply sufficient base current to the output devices now running at higher collector currents. Figure 3 gives the required circuit, and Table 4 the respective component values.

Transistors VT2, 3, 4, 5 and 6 should all be mounted on the same heatsink.

Table 4

R_L Ω	8	4
V_{CC} V	25	20
C1 μF	4	4
C2 μF	1500	2000
R1 Ω	12k	4k7
R2 Ω	1	1
R3 Ω	1	1
R4 Ω	220k	100k
R5 Ω	15k	12k
R7 Ω	1k	1k
R8 Ω	1k	1k
RV1 Ω	20k	10k
RV2 Ω	10k	4k7
VT1		BC183L
VT2		BC183L
VT3		BFR81
VT4		BFR41
VT5		BC183L
VT6		BC213L

With this circuit, slight instability can occur and the 10pF capacitors across the base to collector of the driver transistors VT5 and VT6 are included to prevent this.

Performance figures for both versions of the amplifier are given in Table 5 for 8Ω and Table 6 for 4Ω loads.

Table 5

Power W / Frequency kHz	Distortion % 0.01	0.1	1	2.0	5
0.12	0.28	0.13	0.20	0.57	1.8
1	0.15	0.05	0.44	1.00	1.7
10	0.16	0.14	0.54	1.09	1.9

Sensitivity 0.45V for 5W
Input Impedance 15kΩ

Table 6

Power W / Frequency kHz	Distortion % 0.01	0.1	1	2.0	5
0.12	0.07	0.09	0.5	0.23	3.04
1	0.04	0.03	0.47	1.18	3.03
10	0.06	0.16	0.51	1.18	4.12

Sensitivity 0.65V for 5W
Input Impedance: 12kΩ

Higher Voltage Rail Version

The need for potentiometer RV1 may be removed without losing efficiency by increasing the rail voltage by 3V, adding resistor R9 and capacitor C3 in the emitter of transistor VT1 and changing potentiometer RV1 to a resistor R10, as shown in Figure 4.

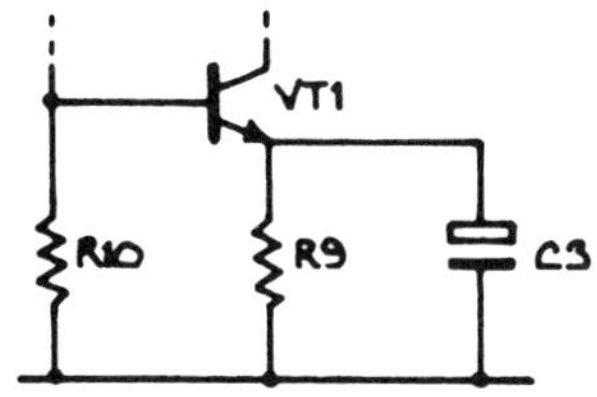

FIGURE 4

Table 7 gives the rail voltages and component values then required for all load impedance values.

Table 7

R_L Ω	50	35	16	8	4
V_{CC} V	55	48	37	28	23
R9 Ω	150	150	100	3k	1k5
R10 Ω	4k7	4k7	4k3	75k	47k
C3 μF [6V]	250	250	250	100	100

Simple Preamplifier

To obtain a considerable increase in sensitivity and some increase of input impedance an a.c. preamplifier may be attached to the basic circuit of Figure 1, as shown in Figure 5. Table 8 gives the appropriate values of the resistors whose value depends on the load, i.e. R11 and R12, and

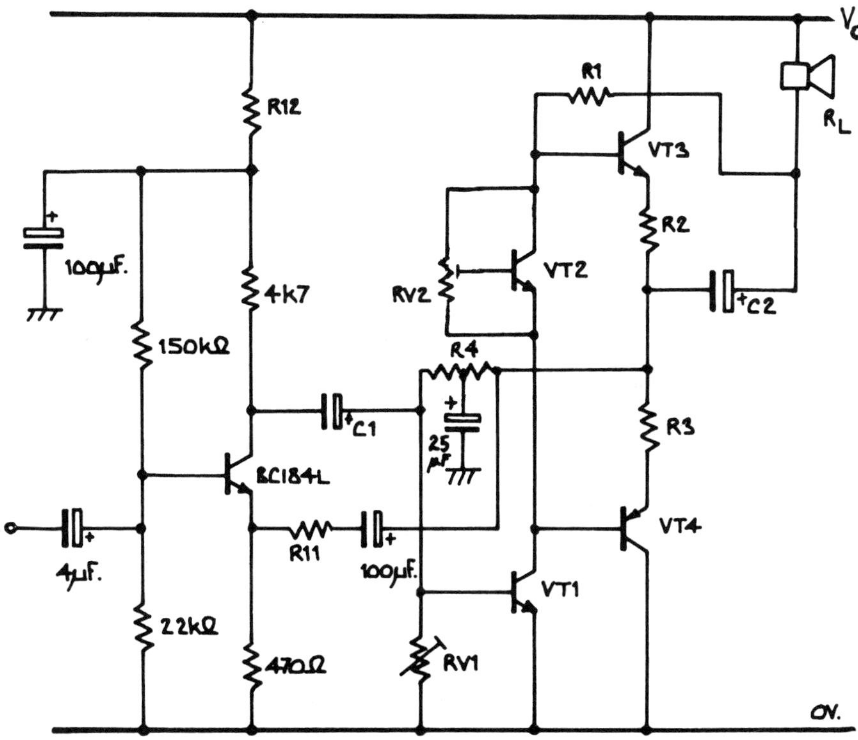

FIGURE 5

Table 8

R_L Ω	50	35	16
R4Ω in series	12k & 15k	12k & 15k	6k8 & 8k2
R11Ω	75k	62k	43k
R12Ω	39k	33k	22k

resistor R4 which is split and the centre point decoupled to earth with a 25μF capacitor.

Basic performance figures for the 35Ω version are given in Table 9.

Table 9

Sensitivity 0.1 for 5W
Input Impedance 15kΩ

Improved Amplifier Configurations

If the extra transistor can be afforded, Figures 6 and 8 show a more economical way to use it than the simple preamplifier previously described. The circuits have a higher closed loop gain and input impedance than the basic circuits of Figures 1 and 3 and, as the extra transistor, VT7, is D.C. coupled and used to give current gain, the open loop gain of the circuit is higher and hence its distortion lower. A further advantage is that it allows the speaker to be returned to 0V rather than the V_{CC} supply rail.

Load Impedances of 50, 35 and 16Ω.

Figure 6 shows the circuit diagram for 50, 35 and 16Ω loads and Figure 7 a suggested printed circuit board layout. Table 10 shows the component values for the 3 loads and Table 11 gives the performance figures for the 35Ω version.

Table 10

R_L Ω	50	35	16
V_{CC}, C2, R1, R2, R3, RV2	As in table 2		
C1 μF	4	4	4
C3 μF	150	150	100
C4 μF	100	100	100
R4 Ω	3k3	3k3	3k3
R10 Ω	560	560	560
R12 Ω	56k	33k	27k
R13 Ω	180k	150k	100k
R14 Ω	330k	270k	220k
R15 Ω	100	120	180
VT1	BFR80	BFR81	BFR81
VT2	BC183L	BC183L	BC183L
VT3	BFR40	BFR41	BFR41
VT4	BFR80	BFR81	BFR81
VT7	BC183L	BC182L	BC182L

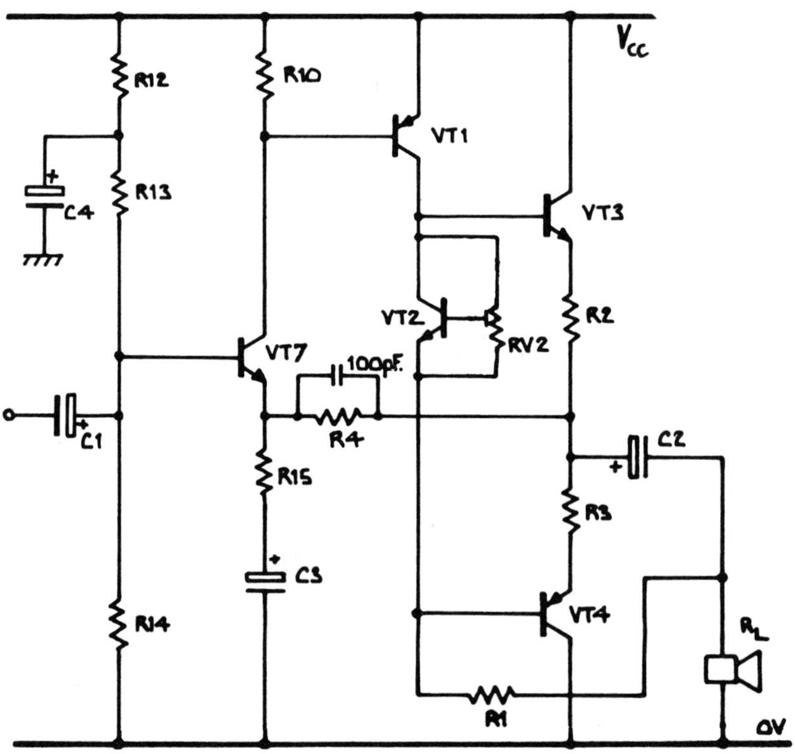

FIGURE 6

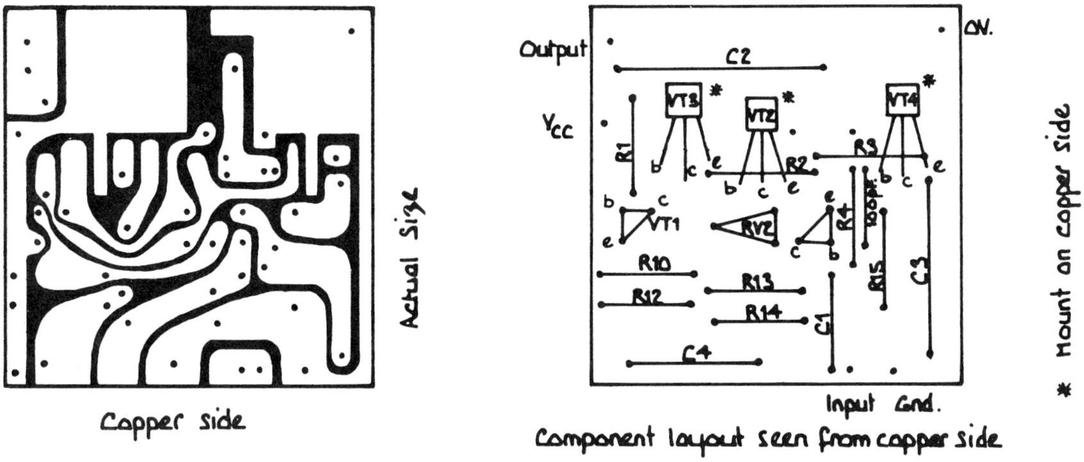

FIGURE 7

Table 11

Power W / Frequency kHz	Distortion % 0.01	0.1	1	2.5	5
.12	0.04	0.04	0.06	0.16	1.54
1	0.05	0.07	0.08	0.13	1.44
10	0.03	0.05	0.07	0.13	1.47

Sensitivity 0.5V for 5W
Input Impedance 80kΩ

When the output was loaded capacitively (0.1μ) some oscillation occurred and the 100pF phase advance capacitor across resistor R4 was included to prevent this.

Capacitor C4 should be of good quality as any leakage through it will upset the biassed arrangement of transistor VT5.

Load impedances of 8Ω and 4Ω

Figure 8 shows the circuit diagram for 8 and 4Ω loads. Table 12 gives the component values for both load values and Table 13 the performance figures for the 4Ω version.

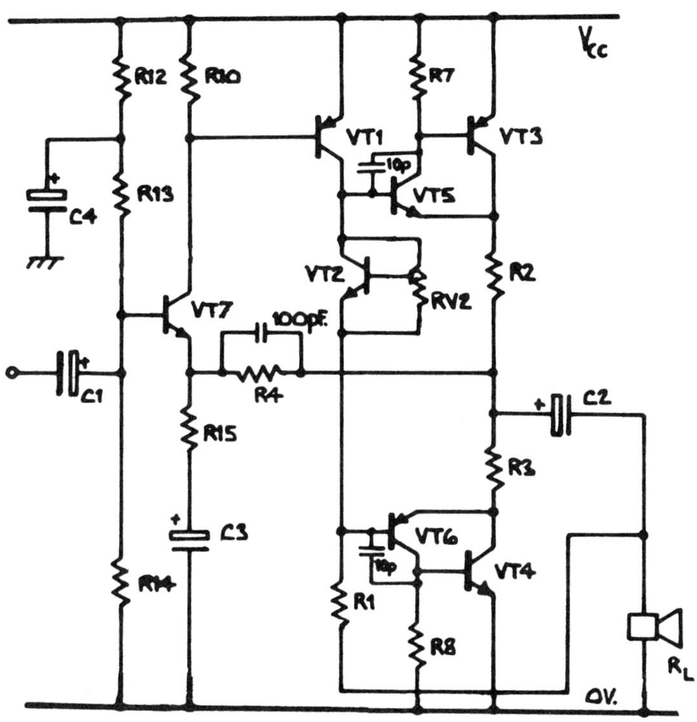

FIGURE 8

Table 12

R_L Ω	8	4
V_{CC}, C1, C2, R1, R2, R3, R7, R8, RV2.	As Table 4	
C3 μF	150	150
C4 μF	100	100
R4 Ω	6k8	6k8
R10 Ω	5k6	5k6
R12 Ω	39k	22k
R13 Ω	180k	150k
R14 Ω	270k	220k
R15 Ω	100	150
VT1	BC213L	
VT2	BC183L	
VT3	BFR81	
VT4	BFR41	
VT5	BC183L	
VT6	BC213L	
VT7	BC183L	

Table 13

Power W / Frequency kHz	Distortion % 0.01	0.1	1	25	5
0.12	0.09	0.06	0.10	0.14	0.61
1	0.10	0.06	0.10	0.25	0.60
10	0.10	0.09	0.18	0.30	0.59

Sensitivity 0.1V for 5W

Input Impedance: 47kΩ

Lower Voltage Supplies

The circuit of Figure 8 with a 4Ω load impedance could be run from a 12V car battery. To ensure correct biassing resistor R14 is raised to 270kΩ. With a V_{CC} of 13.5V the output power at 10% distortion is 3.0W. At low rail voltages a significant amount of voltage swing is dropped across the output transistors VT3/4 and emitter resistors R2 and R3. These resistors can be removed provided extreme care is taken to ensure that the transistor VT2 is in good thermal contact with transistors VT5 and VT6 to maintain quiescent stability. At 10% distortion the output power obtained is then 4.6W. With a 3Ω speaker the output is further raised to 5.3W.

HIGHER OUTPUT POWER AMPLIFIERS

General

Substituting Super Silect for 'normal' Silect transistors in simple audio amplifier circuits has been shown to considerably increase the available output power into the load. This principle of employing more powerful devices in previously designed circuits can be used to further extend their output power ratings.

10W into 8Ω

With the relatively simple circuit shown in Figure 6, but employing a super silect device as the driver (VT1) of plastic power output devices (VT3 & 4), a considerable increase in output power is available, e.g. 10W into 8Ω. The component values are given in Table 14 and brief performance details in Table 15.

Table 14

V_{CC}	32V	R12	82kΩ
C1	4μF	R13	82kΩ
C2	1500μF	R14	220kΩ
C3	160μF	R15	10Ω
C4	32μF	RV2	1kΩ
R1	220Ω 2W	VT1	BFR81 (with heatsink)
R2	0.47Ω	VT2	BC184L
R3	0.47Ω	VT3	TIP41
R4	1k	VT4	TIP42
R10	1k	VT7	BC184L

Table 15

% Distortion @ 1kHz	– 5W out	0.5
	– 10W out	0.8

Sensitivity 0.11V for 10W

Input Impedance 22kΩ

Lower Voltage Supply

By changing some of the above component values, (i.e. R1 = 68Ω, R12 = 27kΩ, R13 = 47kΩ, R14 = 150kΩ, R2 and R3 = 0Ω) the circuit will drive a 3Ω load from a 12V car battery (i.e. V_{CC} = 13.5) giving an output power at 10% distortion of 5.5W.

100W into 4Ω

General

The high performance amplifier shown in Figure 20 and described extensively in Part 2 of the previous chapter (XII), was basically limited in its power range by the Silect driver transistors. Using Super Silect transistors as the drivers the power output can be considerably increased. The following section gives the calculations for obtaining the component values which will allow the amplifier circuit to give 100W into 4Ω.

Calculations

Requirement 100W into 4Ω

$\therefore$ R.M.S. $V_{out} = (100 \times 4)^{1/2} = 20V$

Peak $V_{out} = 1.414 \times 20 = 28V$

Peak to peak $V_{out} = 56.56V$

R.M.S. $I_{out} = (100/4)^{1/2} = 5A$

$\therefore$ Peak $I_{out} = 1.414 \times 5 = 7.07A$

Voltage dropped across emitter resistors R15 and R16
= 2 x 7.07 x 0.22 = 3.1V

∴ V_{CC} must be > 56.56 + 3.1 + $V_{CE(sat)}$ s of VT7 and VT8

i.e. 64V*, say.

∴ Output TIP device must be a B type device.

At 7A, the gain of TIP33/34Bs would be very low.

∴ Devices necessary – TIP35B for VT8* and TIP36B for VT7*

$V_{CE(sat)}$ @ 7A is <2V

∴ V_{CC} of 64V is sufficient

h_{FE} min @ 7A = h_{FE} typ. @ 7A x h_{FE} min @ 15A/h_{FE} typ. @ 15

∴ For TIP35 h_{FE} min @ 7A = 45 x 10/21 = 21

and for TIP36 h_{FE} min @ 7A = 38 x 10/20 ≃ 19

∴ Maximum base current for them to be supplied by driver transistors VT5 and VT6 collector current.

I_C (VT6) = 7.07/21 = 340mA
I_C (VT5) = 7.07/19 = 370mA

Suitable devices for these currents and voltage requirement

VT5 – BFR 39* VT6 – BFR 79*

Maximum base current needed ≃ 340 to 370/50 x 120/105
≃ 6 to 6.5m A

∴ Run voltage amplifier stage VT3 @ 2 x 6 = 12mA

Suitable transistor for VT3 is BFR79*

∴ Resistor R11 ≑ (V_{CC}/2) /12 = 32/12 = 2.66
i.e. 2k7Ω* say.

The long-tailed pair VT1/2 is run @ 2.5mA tail current by base voltage of VT1/2 = V_{CC}/4 = 64/4 = 16V

∴ Emitter voltage = 16 – 0.7 = 15.3V

∴ Resistor R5 = 15.3/2.5 = 6.1, i.e. 5k6Ω* say.

The collectors should sit @ ≃ V_{CC}/3

∴ Resistors R4 and R6 = (V_{CC}/3) x 2 x 5.6/15.3
= 15.6 i.e. 15kΩ* say.

∴ Voltage drop across them = 15 x 15.3/5.6 x 2 = 20.5

Typical base current of VT3 = 12/115 = 0.105mA

Current through resistor R8 = I_B (VT3) + I_{R7}
= 0.105 + 7/4.7 = 0.253mA

∴ R8 = (20.5 – 0.7)/0.253 = 78k i.e. 82kΩ* say.

Value of resistor R9 depends on sensitivity required.

As V_{out}/V_{in} = (100k + R9)/R9

where V_{out} ≃ 20V,

if V_{in} required is 100mV, R9 x 20/0·1 – R9 = 100k

∴ R9 ≃ 100k/200 i.e. 560Ω* say.

Suitable output capacitor C7 4000μF* (64V) say.

With the 4Ω speaker, resistors R14/R16 should be lowered from the 1kΩ shown in Figure 20 to 560Ω to allow correct turn 'off' of the output devices.

Summary of component values not given (*) in Figure 20 are given in Table 16 and brief performance figures in Table 17.

Table 16

V_{CC} V	64
C7 μF	4000
R4 Ω	15k
R5 Ω	5k6
R6 Ω	15k
R8 Ω	82k
R9 Ω	560
R11 Ω	2k7
VT3	BFR79
VT5	BFR39
VT6	BFR79
VT7	TIP36B
VT8	TIP35B

Table 17

Power W	1W	4W	10W	25W	50W	100W
Distortion % (@ 1kHz)	0.25	0.35	0.4	0.45	0.45	1.8

Sensitivity 120mV for 100W
Input Impedance 40kΩ

CONCLUSIONS

It has been shown that using Super Silect transistors in simple circuits means that an output power of 5W into a wide range of load impedances is readily achievable with low distortion figures.

It has also been illustrated that using them to drive plastic power output transistors enables high power high fidelity amplifiers to be realised. The 100W into 4Ω amplifier described was taken as a specific example to show this realisation but obviously the permutations possible are numerous. A split rail configuration, as shown in Figure 39 of chapter XII, would cut down the expense of obtaining 100W, when compared to Figure 20 as the voltage ratings of the circuit and power supply electrolytic capacitors could be reduced. Alternatively, if, say, 60W was an adequate output power TIP33B/34B transistors could be used again cutting cost.

It should be noted that although all the circuits in this chapter have been built and tested in the laboratory, (the latter being, of course, the source of the performance figures), the designs are intended to show the possibilities using Super Silect devices. Circuits for specific applications may require minor modifications which should be tested under intended operating conditions.

SECTION 4

USING TTL INTEGRATED CIRCUITS

XIV TTL COUNTERS AND REGISTERS

The series 74 TTL family of integrated circuits is one of the fastest and most complete line of saturated, digital logic available today. In addition to the familiar basic TTL circuits, most gating and flip-flop functions are available in a high-speed and a low-power circuit. This combination of compatible circuits and the largest selection of complex functions, (sometimes referred to as "medium scale integration" or MSI) offers the design engineer considerable flexibility in the selection of speed versus power dissipation best suited for his system.

This chapter consists of the functional analyses required to construct shift-or storage-registers and various types of counters using Texas Instruments Series 74 transistor-transistor logic (TTL) circuits. The information is primarily intended to assist in the utilization of complex Series 74 circuits; however, some examples require only the multi-function flip-flop or latch circuits. No detail, other than functional symbols, are provided for the gating functions unless special notations are required.

The devices specified throughout this section are basic Series 74 circuits in the familiar TTL configuration. These monolithic TTL circuits feature typical d-c noise margins of one volt, and full fan-out of 10 is available from all outputs. As some of the Series 74 complex functions require 16 functional pins, circuits packaged in the molded dual-in-line package are used in the examples.

The rapid advance in high complexity MSI devices has meant that some of the circuits shown may no longer present the best or easiest method of achieving a particular function. However, although some circuits can, for example, be replaced by a single new MSI package they have still been included. This is because they could be useful for specific applications, e.g. more intermediate outputs are available or they can be assembled using military F pack packages. (Some newer MSI devices only being available in Dual in Line (DIL) packages). The references at the end of the chapter indicate where further information may be found, or where an alternative approach is given.

Although the whole section refers mainly to Series 74 devices, which are characterised for operation over the temperature range 0° to 70°C, all the devices are also available in Series 64 and 54 which have temperature ranges –40° to 85°C and –55° to 125°C respectively.

REGISTER AND COUNTERS DESCRIPTIONS

Quadruple Bistable Latch SN7475

This quadruple, bistable latch is composed of four storage elements, each having complementary Q and $\bar{Q}$ outputs. See Figure 1. Information present at a data (D) input is transferred to the Q output if the clock is high, and the Q output will follow the data input as long as the clock remains high. When the clock goes low, the information (that was present at the data input at the time this transition occurred) is retained at the Q output. Pin assignments were selected to coincide with the physical placement of logical functions of other series 74 circuits which are likely to be used as inputs to, or outputs from, the SN7475.

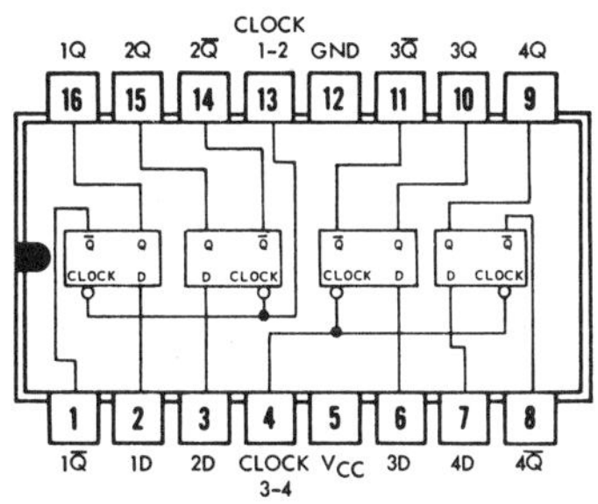

PROPAGATION DELAY	30 ns
POWER DISSIPATION	160 mW

FIGURE 1. Quadruple Bistable Latch SN7475N

This device is designed specifically for use as a temporary storage element between the SN7490 decade counter and the SN7441A binary-coded-decimal (BCD) to decimal decoder. See Figure 2. However, the simplicity of this particular storage function, and availability of complementary Q and $\bar{Q}$ outputs makes the SN7475 readily adaptable to other storage register applications.

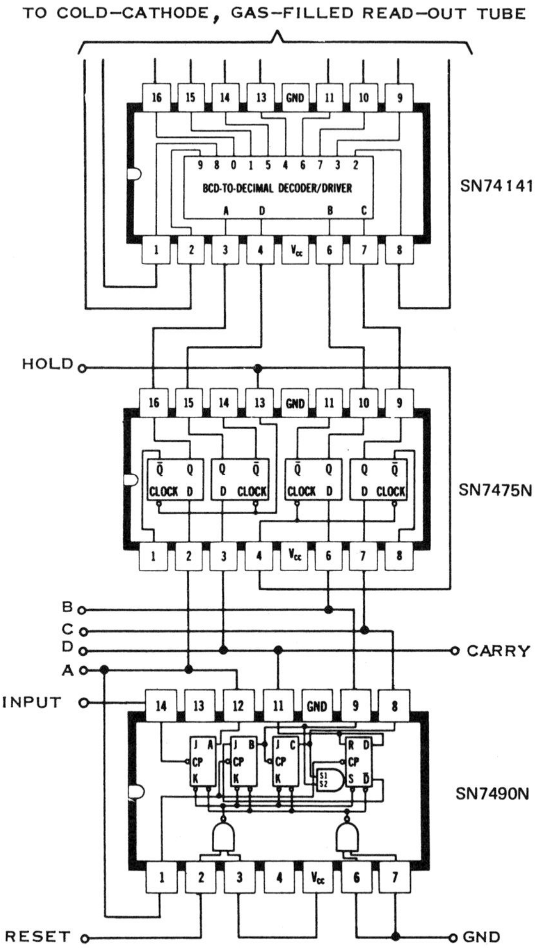

FIGURE 2. Storage and Readout of BCD Data From High-Speed Counter

8-Bit Shift Register SN7491A

This serial-in, serial-out, 8-bit shift register is composed of eight R-S master-slave flip-flops, input gating, and an inverting clock driver. See Figure 3. The register is capable of storing and transferring data at clock rates up to 18 MHz.

Single-rail data and input control are gated through inputs A and B and an internal inverter to form the complementary inputs to the first bit of the shift register. Drive for the internal common clock line is provided by an inverting clock driver. Each of the inputs (A, B, and CP) appears as only one TTL input load.

The clock pulse inverter/driver causes the SN7491A to shift information to the output on the positive edge of an input clock pulse. † 1

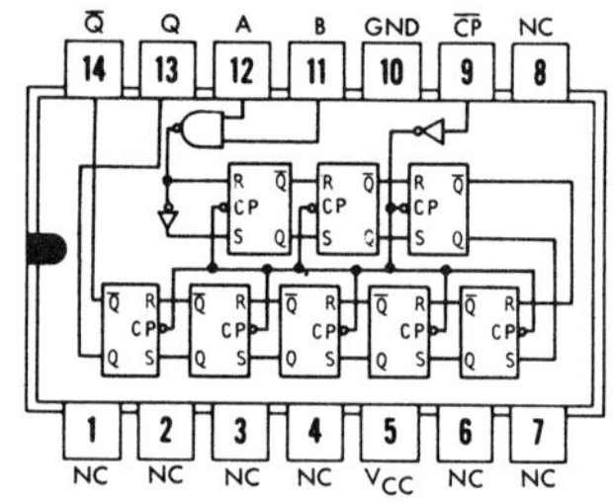

PROPAGATION DELAY 25 ns
POWER DISSIPATION 175 mW

FIGURE 3. 8-Bit Shift Register SN7491AN

4-Bit Shift Register SN7494

This serial shift register is composed of four R-S master Slave flip-flops, four AND-OR-INVERT GATES, and four inverter-drivers. See Figure 4. Internal interconnections of these functions provide a versatile register which performs right-shift operations as a serial-in, serial-out register with parallel load capability or as a dual-source, parallel-to-serial converter. A number of these registers may be connected in series to form an n-bit register.

All flip-flops may be simultaneously set to the logical state by applying a logical 1 voltage at the clear input. The register may be cleared independent of the state of the clock input.

The flip-flops may be independently set to the logical 1 state from either of two, gated, preset input sources. Preset inputs 1A through 1D are enabled during the time that a positive pulse is applied to preset 1, if preset 2 is at a logical level. When the logic levels at preset 2 are reversed, preset inputs 2A through 2D are active, enabling the register to store a binary number or its complement.

Transfer of information to the output occurs on the positive edge of an input clock pulse or on the negative edge of a clock pulse. The proper information must appear at the R-S inputs of each flip-flop prior to the leading edge of the clock input waveform. The serial input provides this information to the first flip-flop while the outputs of each flip-flop provide information for the remaining R-S inputs.

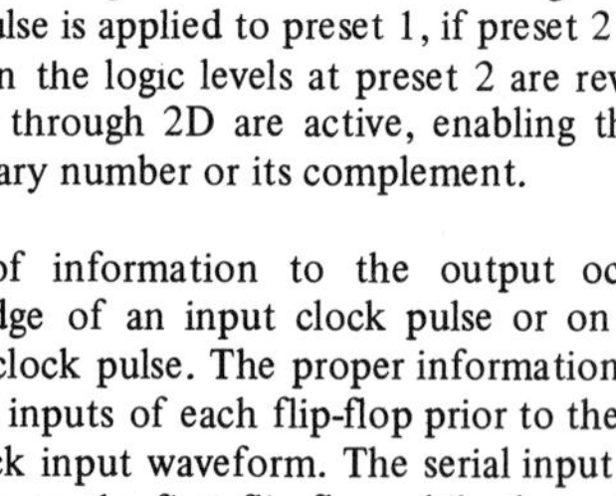

PROPAGATION DELAY 25 ns
POWER DISSIPATION 175 mW

FIGURE 4. 4-Bit Shift Register SN7494N

4-Bit Right-Shift Left-Shift Register SN7495

This parallel or serial shift register is composed of four R-S master-slave flip-flops, four AND-OR-INVERT gates, one AND-OR gate, and six inverter-drivers. See Figure 5. Internal interconnections of these functions provide a versatile register which will enter data serially or parallel dependent upon the logical input to the mode control. A number of these registers may be connected in series to form an n-bit right-shift or left-shift register by externally interconnecting each Q output to the parallel input of the preceding bit.

When a logical 0 level is applied to the mode control input, the AND gates numbered "1" are enabled and the AND gates numbered "2" are inhibited. In this mode the Q output of each flip-flop is coupled to the R-S inputs of the succeeding flip-flop and right-shift operation is performed by clocking at the clock 1 input. In this mode, serial data is entered at the serial input. Clock 2 and parallel inputs A through D are inhibited by the AND gates numbered "2".

When a logical 1 level is applied to the mode control input, the AND gates numbered "1" are inhibited (decoupling the Q outputs from the succeeding R-S inputs to prevent right-shift) and the AND gates numbered "2" are enabled to allow entry of data through parallel inputs A through D and clock 2. This mode permits parallel loading of the register or, with external interconnection, shift-left operation.

Clocking for the shift register is accomplished through the AND-OR gate "E" which permits separate clock sources to be used for the right-shift and parallel-shift modes. If both modes are to be clocked from the same source, the clock input may be applied commonly to clock 1 and clock 2. Information must be present at the R-S inputs of the master-slave flip-flops prior to the leading edge of the clock pulse.

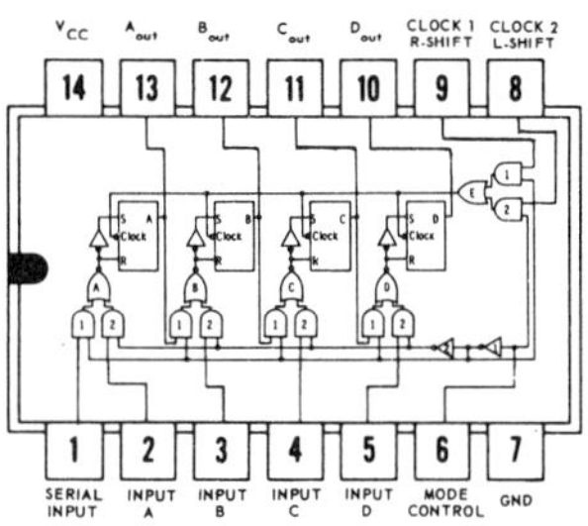

PROPAGATION DELAY 25 ns
POWER DISSIPATION 250 mW

FIGURE 5. 4-Bit Right-Shift Left-Shift Register SN7495N

5-Bit Shift Register SN7496

This register consists of five R-S master-slave flip-flops connected as a shift register to perform parallel-to-serial or serial-to-parallel conversion of binary data. Since both inputs and outputs to all flip-flops are accessible, parallel-in/parallel-out or serial-in/serial-out operation may be performed. Figure 6.

All flip-flops are simultaneously set to the logical 0 state by applying a logical 0 voltage to the clear input. This condition may be applied independent of the state of the clock input.

The flip-flops may be independently set to the logical 1 state by applying a logical 1 to both the preset input of the specific flip-flop and the common preset input. The common preset input is provided to allow flexibility of either setting each flip-flop independently or setting all flip-flops simultaneously. Preset is also independent of the state of the clock input.

Transfer of information to the output pins occurs when the clock input goes from a logical 0 to a logical 1. Since the flip-flops are R-S master-slave circuits, the proper information must appear at the R-S inputs of the first flip-flop prior to the rising edge of the clock input voltage waveform. The serial input provides this information to the first flip-flop, while the outputs of the subsequent flip-flops provide information for the remaining R-S inputs. The clear input must be at a logical 1 and at least one preset input of each flip-flop must be at a logical 0 when clocking occurs.

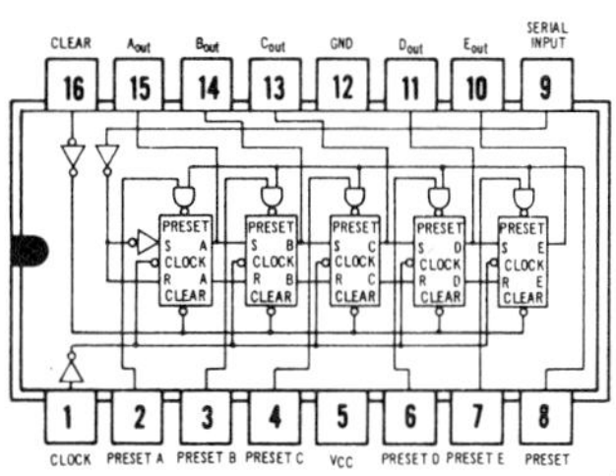

PROPAGATION DELAY 25 ns
POWER DISSIPATION 245 mW

FIGURE 6. 5-Bit Shift Register SN7496N

†1 Currently available 8 bit Shift Registers include:—

Serial in, Parallel out, SN74164
Parallel in, Serial out, SN74165
Parallel in, Serial out, Synchronous load SN74166
Parallel Access, Right-Left, Mode Control SN74198
Parallel Access, J—K inputs, Mode Control SN74199

Decade Counter SN7490

This decade counter consists of four master-slave flip-flops internally interconnected to provide a divide-by-two counter and a divide-by-five counter. See Figure 7. Gated, direct-reset lines are provided to inhibit count inputs and return all outputs to a logical zero or to a binary-coded-decimal (BCD) count of nine. As the output from flip-flop A is not internally connected to the succeeding stages, the count may be separated in three independent count modes:

1. When used as a binary-coded-decimal decade counter, the BD input must be externally connected to the A output. The A input receives the incoming count, and a count sequence is obtained in accordance with the BCD count sequence truth table. See Figure 8. In addition to a conventional zero reset, inputs are provided to reset a BCD 9 count for nine's complement decimal applications.
2. If a symmetrical divide-by-ten count is desired for frequency synthesizers, or other applications requiring division of a binary count by a power of ten, the D output must be externally connected to the A input. The input count is then applied at the BD input and a divide-by-ten square wave is obtained at output A. See Figure 8.
3. For operation as a divide-by-two counter and a divide-by-five counter, no external interconnections are required. Flip-flop A is used as a binary element for the divide-by-two function. The BD input is used to obtain binary divide-by-five operation at the B, C, and D outputs. See Figure 8. In this mode, the two counters operate independently; however, all four flip-flops are reset simultaneously.

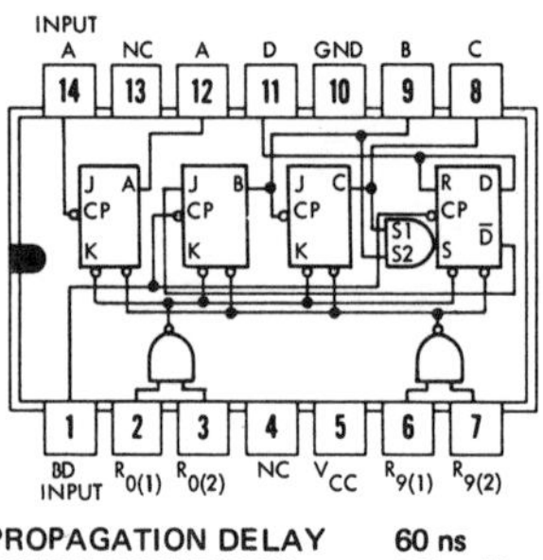

PROPAGATION DELAY 60 ns
POWER DISSIPATION 160 mW

FIGURE 7. Decade Counter SN7490N

MODE 1 (BCD)				MODE 2 (SYMMETRICAL DIVIDE-BY-TEN)				MODE 3 (DIVIDE-BY-FIVE)		
A	B	C	D	A	B	C	D	B	C	D
0	0	0	0	0	0	0	0	0	0	0
1	0	0	0	0	1	0	0	1	0	0
0	1	0	0	0	0	1	0	0	1	0
1	1	0	0	0	1	1	0	1	1	0
0	0	1	0	0	0	0	1	0	0	1
1	0	1	0	1	0	0	0			
0	1	1	0	1	1	0	0			
1	1	1	0	1	0	1	0			
0	0	0	1	1	1	1	0			
1	0	0	1	1	0	0	1			

FIGURE 8. Decade Counter Truth Tables

Divide-by-Twelve Counters SN7492

This 4-bit binary counter consists of four master-slave flip-flops which are internally interconnected to provide a divide-by-two counter and a divide-by-six counter. See Figure 9. A gated, direct-reset line can inhibit the count inputs and simultaneously return the four flip-flop outputs to a logical 0. As the output from flip-flop A is not internally connected to the succeeding flip-flops, the counter may be operated in three modes:

1. When used as a divide-by-twelve counter, output A may be externally connected to input BC. Then input count pulses are applied to input A. Simultaneous divisions of 2, 6, and 12 are performed at the A, C, and D outputs as shown in the mode 1 truth table. See Figure 10.
2. When used as a divide-by-six counter, the input count pulses are applied to input BC. Simultaneous frequency divisions of three and six are available at the C and D outputs. See Figure 10. Independent use of flip-flop A is available if the reset function coincides with reset of the divide-by-six counter.
3. Another divide-by-twelve code is available if output D is externally connected to input A. The input count pulses are then applied to input BC. Simultaneous frequency divisions of 3, 6, and 12 are available at the C, D, and A outputs as shown. See Figure 10.

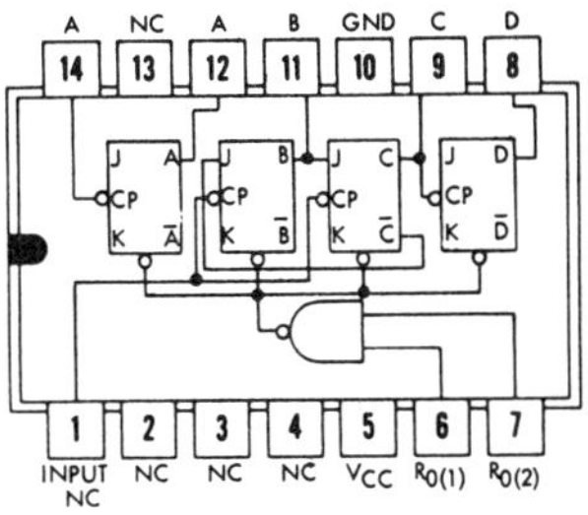

PROPAGATION DELAY 60 ns
POWER DISSIPATION 160 mW

FIGURE 9. Divide-by-12 Counter SN7492N

Figure 11. A gated, direct-reset line returns the four flip-flop outputs to a logical 0. As the output from flip-flop A is not internally connected to the succeeding flip-flops, the counter may be operated in two independent modes:

1. When used as a 4-bit ripple-through counter, output A must be externally connected to input B. The input count pulses are applied to input A. Simultaneous divisions of 2, 4, 8 and 16 are performed at the A, B, C, and D outputs as shown in the truth table. See Figure 12.
2. When used as a 3-bit ripple-through counter, the input count pulses are applied to input B. Simultaneous frequency divisions of 2, 4, and 8 are available at the B, C, and D outputs. See Figure 12. Independent use of flip-flop A is available if the reset function coincides with reset of the 3-bit ripple-through counter.

MODE 1 (DIVIDE-BY-12)				MODE 2 (DIVIDE-BY-6)			MODE 3 (DIVIDE-BY-12)			
A	B	C	D	B	C	D	A	B	C	D
0	0	0	0	0	0	0	0	0	0	0
1	0	0	0	1	0	0	0	1	0	0
0	1	0	0	0	1	0	0	0	1	0
1	1	0	0	0	0	1	0	0	0	1
0	0	1	0	1	0	1	0	1	0	1
1	0	1	0	0	1	1	0	0	1	1
0	0	0	1				1	0	0	0
1	0	0	1				1	1	0	0
0	1	0	1				1	0	1	0
1	1	0	1				1	0	0	1
0	0	1	1				1	1	0	1
1	0	1	1				1	0	1	1

FIGURE 10. Divide-by-12 Counter Truth Tables

MODE 1 (DIVIDE-BY-16)				MODE 2 (DIVIDE-BY-8)		
D	C	B	A	B	C	D
0	0	0	0	0	0	0
0	0	0	1	1	0	0
0	0	1	0	0	1	0
0	0	1	1	1	1	0
0	1	0	0	0	0	1
0	1	0	1	1	0	1
0	1	1	0	0	1	1
0	1	1	1	1	1	1
1	0	0	0			
1	0	0	1			
1	0	1	0			
1	0	1	1			
1	1	0	0			
1	1	0	1			
1	1	1	0			
1	1	1	1			

FIGURE 12. 4-Bit Binary Counter Truth Tables

4-Bit Binary Counter SN7493

This 4-bit binary counter consists of four master-slave flip-flops which are internally interconnected to provide a divide-by-two counter and a divide-by-eight counter. See

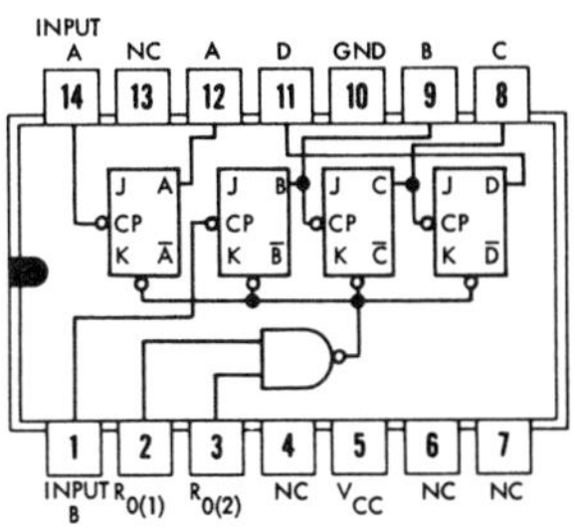

PROPAGATION DELAY 75 ns
POWER DISSIPATION 160 mW

FIGURE 11. 4-Bit Binary Counter SN7493N

†2 Currently available counters include:–

Decade, Synchronous load, Asynchronous Clear	SN74160
4–Bit Binary " " " "	SN74161
Decade, " " Synchronous Clear	SN74162
4-Bit Binary, " " " "	SN74163
Reversible Decade, Single Clock	SN74190
" 4-Bit Binary " "	SN74191
" Decade, Dual "	SN74192
" 4-Bit Binary Dual Clock	SN74193
50MHz Decade, Asynchronous	SN74196
50MHz 4-Bit Binary "	SN74197

REGISTER EXAMPLES

6-Bit Shift-Left/Shift-Right Register with Parallel Load Capability

The 6-bit left- or right-shift register illustrated in Figure 13 demonstrates how a versatile serial-in or parallel-in register may be constructed using SN7474 flip-flops.[1]

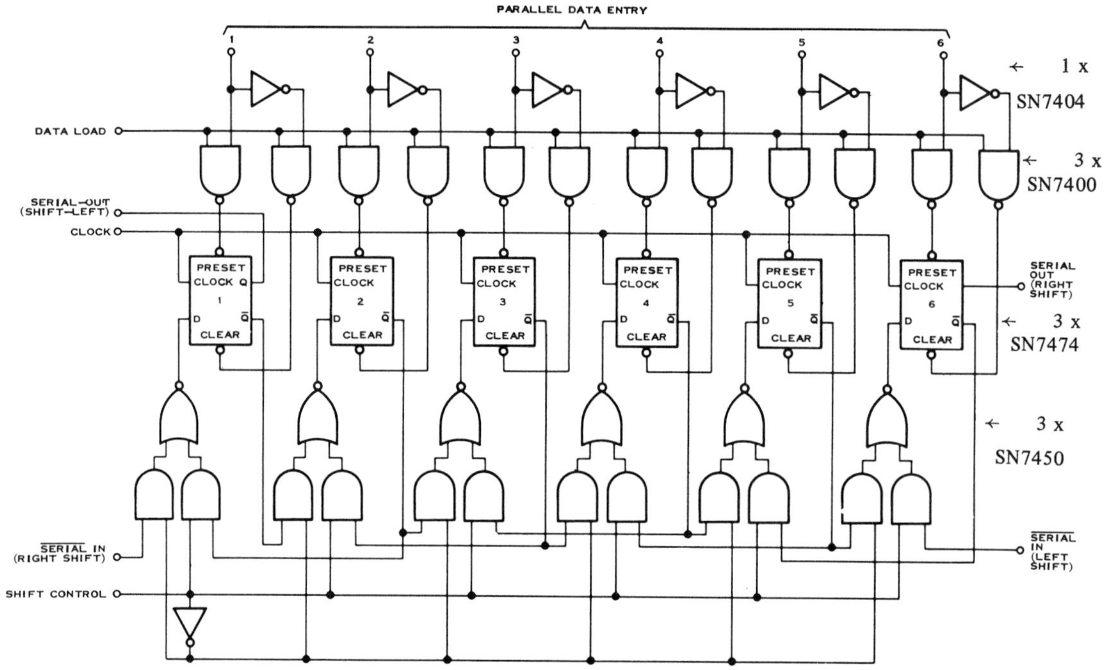

FIGURE 13. 6-Bit Left- or Right-Shift Register Using SN7474s

Parallel-To-Serial Converter Using SN7495

The parallel-to-serial converter in Figure 14 accepts parallel data in groups of 7 bits at the data inputs of the SN7495.[2] Together with the data, a logical "0" is entered into the first stage by applying a negative pulse at the INITIAL input. The INITIAL input has to be used only once at the beginning of the conversion to set the registers in a known state. Then the data is shifted out in series and logical "ones" are entered in series. After seven clock pulses, the data has all been shifted out, and the register is full of "ones." This state is decoded by the SN7430, which initiates a new 7-bit conversion cycle without requiring a new INITIAL pulse.

Addressable Register File Using SN7475

The SN7475 quad latch may be used as a four-bit-per-byte addressable storage register as shown in Figure 15. The addressing is achieved with the SN7442 three-to-eight or four-to-ten-line decoder. A 4-bit byte is entered into the input register and can then be transferred to any of eight output registers. More output registers may be added if additional SN7442s are used (2^n registers for n address lines). Furthermore, if additional input registers are included, the output registers may be extended to handle bytes in multiples of four bits. This example is very useful for decimal serial-by-character to parallel conversion.

Serial-To-Parallel Converter Using SN7495

Figure 16 shows SN7495 s in a serial-to-parallel converter.[3] By applying a positive pulse to the INITIAL line, the register is set to a known state (D1000000), where D is the first data bit. When the "one" is shifted to the last stage of the register, a new conversion cycle is initiated.

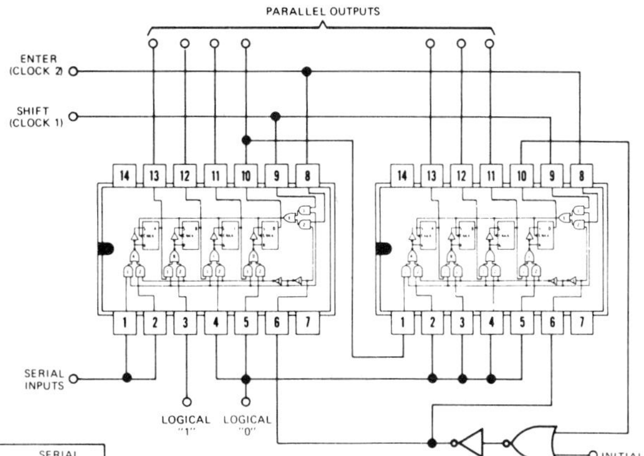

FIGURE 16.
7-Bit Serial-To-Parallel Converter Using SN7495 s

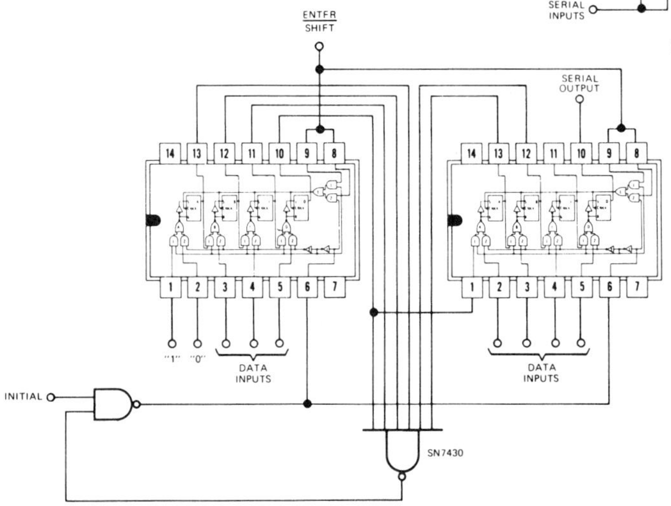

FIGURE 14. 7-Bit Parallel-To-Serial Converter Using SN7495 s

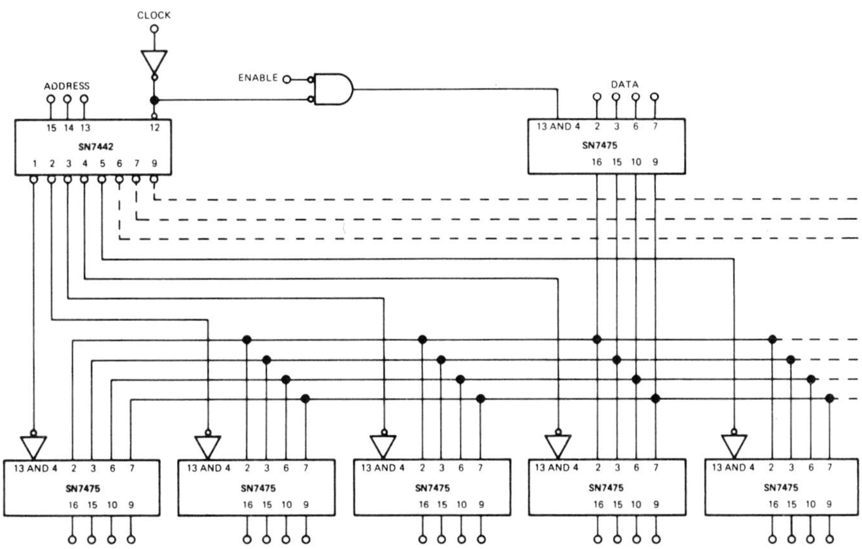

FIGURE 15. 4-Bit-Per-Byte Addressable Register File Using SN7475 s

DIVIDE-BY-N RIPPLE COUNTERS (N=CYCLE LENGTH)

Counters Using SN7490, SN7492, SN7493

For some counting applications, the SN7490, SN7492, SN7493 ripple counters may be modified to change the count cycle. By decoding any desired cycle length at the outputs of the A, B, C, D flip-flops and feeding this signal to the asynchronous clear inputs, the counter can be made to adopt a shortened cycle. The maximum frequency in this mode is determined by the restriction that the clock should not go low for approximately 40 nanoseconds after the counter has been reset to zero. See Figure 17. The outputs which are in a logical 1 state at the count (N) are fed into the $R_{0(1)}$ and $R_{0(2)}$ inputs. The counter will then reset to

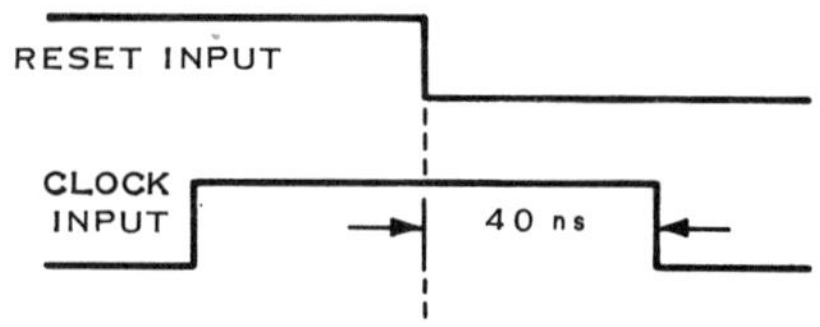

FIGURE 17. Typical Reset Timing Requirement

the all zero state when the count (N) appears at these outputs. According to the count cycle (N), voltage spikes may appear on some output lines. See output B waveform, Figure 18. When the outputs are to be decoded, a strobe gate should be provided to inhibit false output data. When the outputs are not equally and heavily loaded, timing difficulties may be encountered at the temperature extremes. Buffering outputs with the configuration shown in Figure 20 will prevent these difficulties.

CLOCK PULSE	OUTPUT			
	A	B	C	D
t_n	0	0	0	0
t_{n+1}	1	0	0	0
t_{n+2}	0	1	0	0
t_{n+3}	1	1	0	0
t_{n+4}	0	0	1	0
t_{n+5}	1	0	1	0
t_{n+6}	0/0	1/0	1/0	0/0

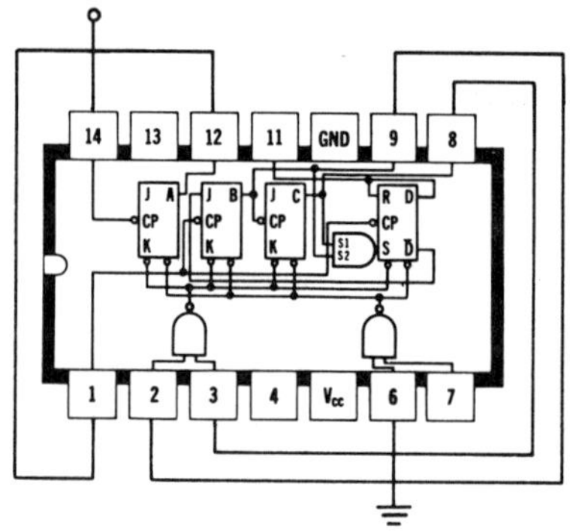

FIGURE 19. Binary Divide-by-6 Ripple Counter Using SN7490

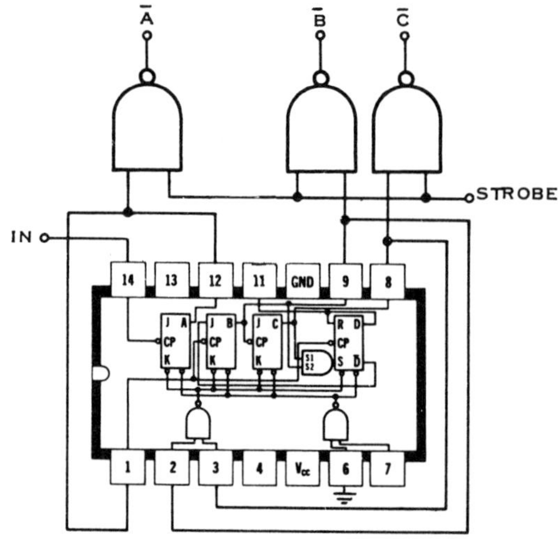

FIGURE 20. Binary Divide-by-6 Ripple Counter Using SN7490N

(Buffered Outputs)

For larger division ratios in BCD code, two or more SN7490 s may be cascaded. BCD numbers with no more than two "ones" in their sequence do not require external gates. See Figure 21. Some other numbers which do not require external gates are: 11, 12, 14, 18, 21, 22, 24, 28, 41, 42, 44, 48, 81, 82, 84, 88 . . . (and others).

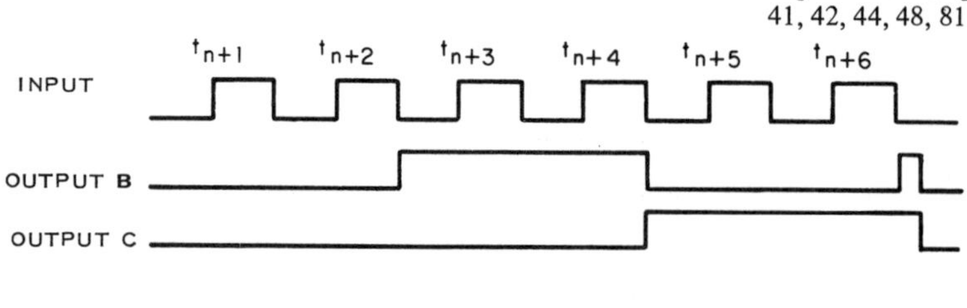

FIGURE 18. Binary Divide-by-6 Ripple Counter — Timing

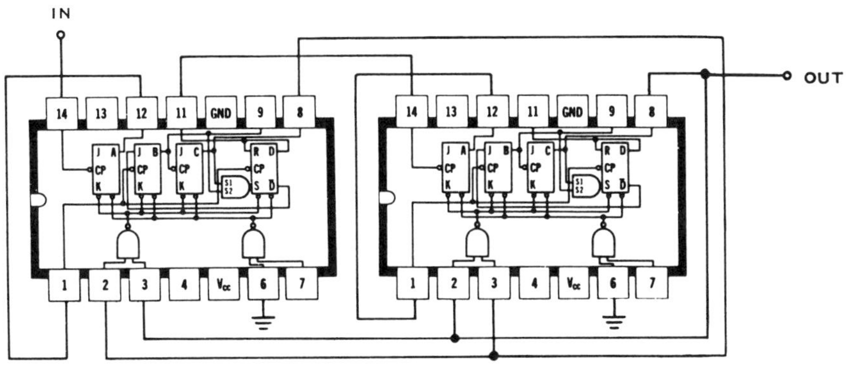

FIGURE 21. BCD Divide-by-44 Ripple Counter Using Two SN7490N

All other division ratios in BCD can be achieved with two SN7490 s, one NAND gate, and one inverter. See Figure 22.

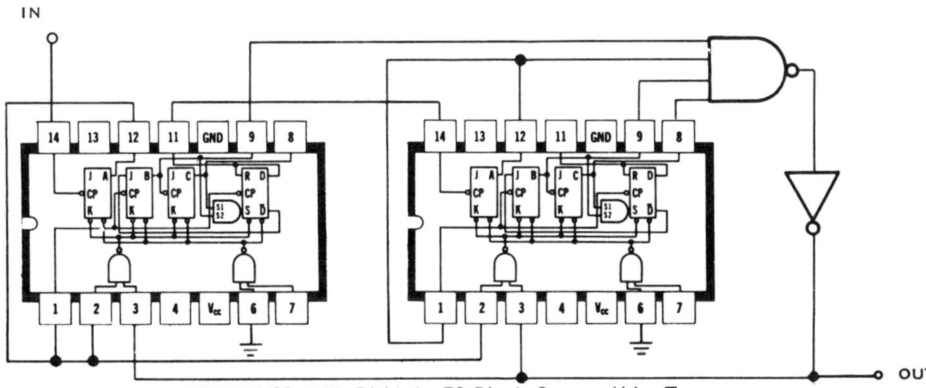

FIGURE 22. BCD Divide-by-73 Ripple Counter Using Two SN7490N

The following examples illustrate use of the SN7492 and SN7493 for various cycle length.

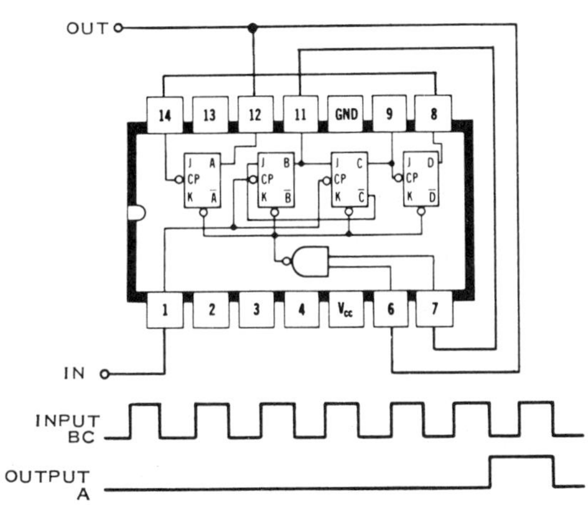

FIGURE 23. Divide-by-7 Ripple Counter Using SN7492

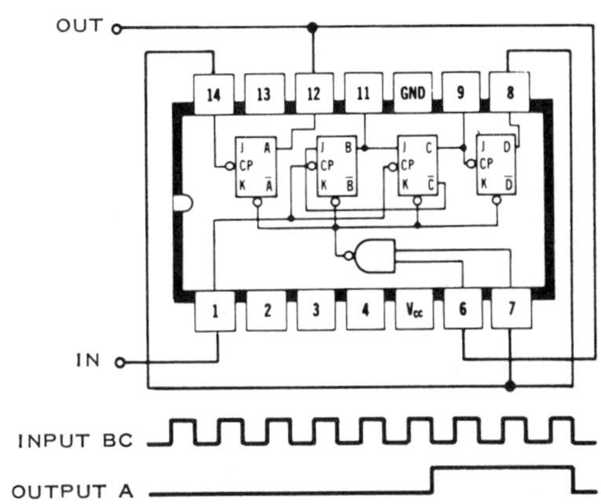

FIGURE 24. Divide-by-9 Ripple Counter Using SN7492

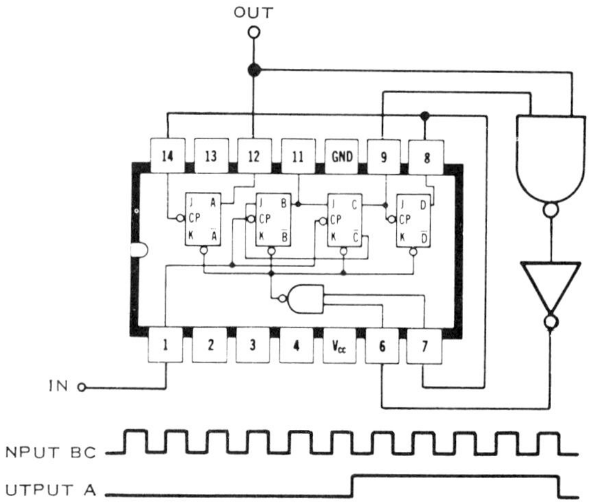

FIGURE 25. Divide-by-11 Ripple Counter Using SN7492

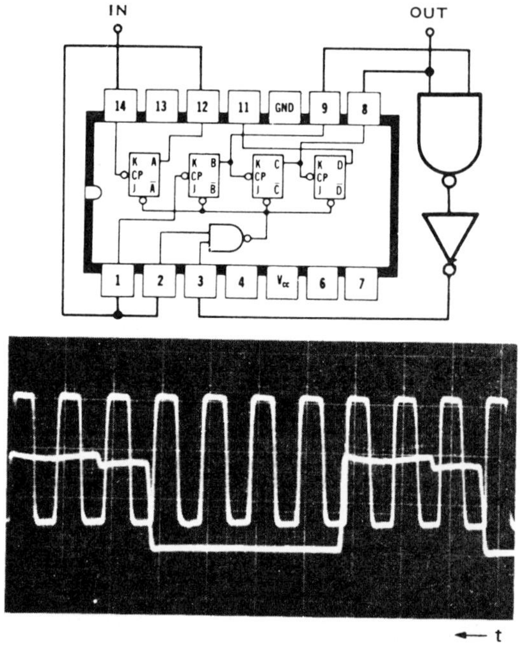

FIGURE 26. Binary Divide-by-7 Ripple Counter Using SN7493N

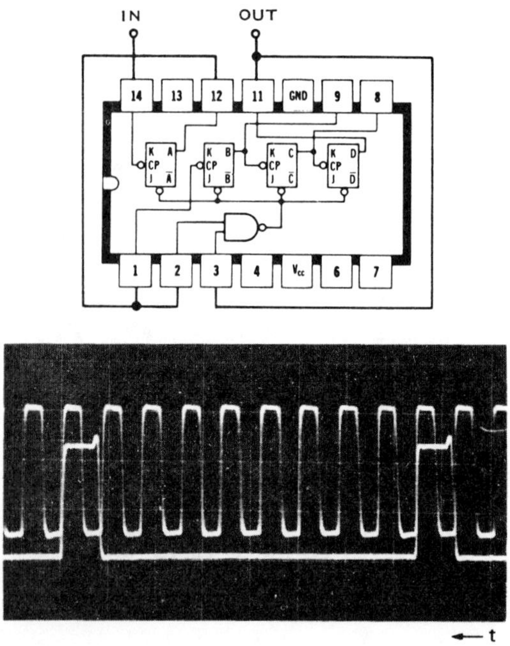

FIGURE 27. Binary Divide-by-9 Ripple Counter Using SN7493N

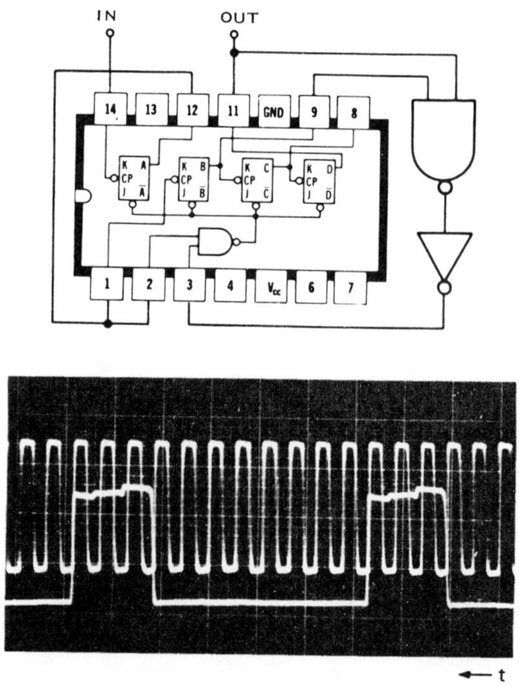

FIGURE 28. Binary Divide-by-11 Ripple Counter Using ***SN7493N***

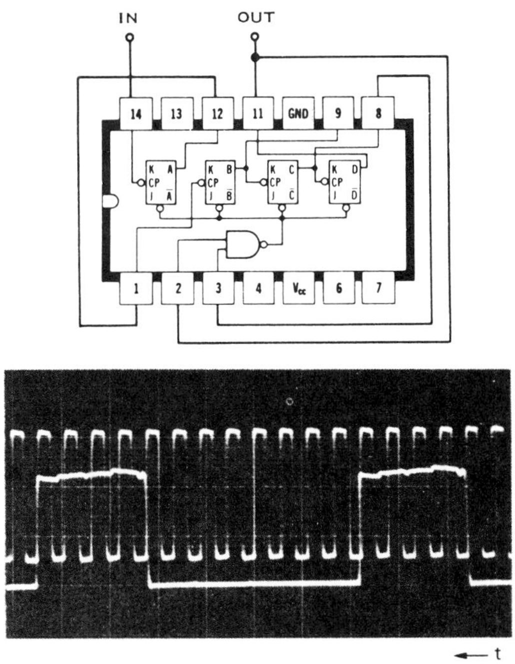

FIGURE 29. Binary Divide-by-12 Ripple Counter Using SN7493N

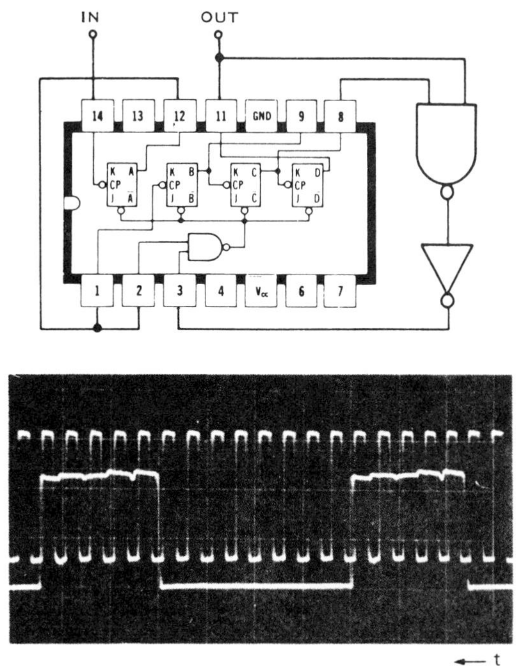

FIGURE 30. Binary Divide-by-13 Ripple Counter Using SN7493N

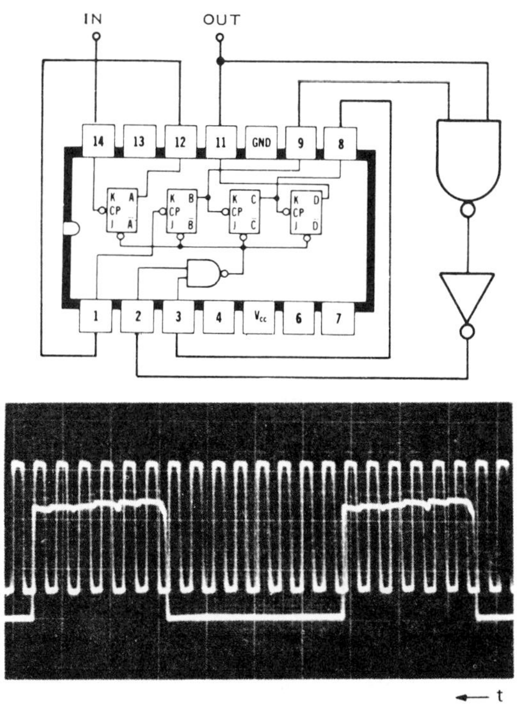

FIGURE 31. Binary Divide-by-14 Ripple Counter Using SN7493N

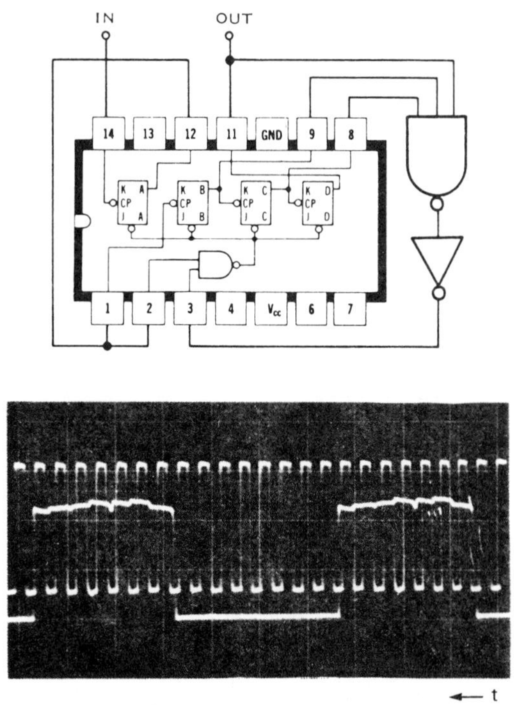

FIGURE 32. Binary Divide-by-15 Ripple Counter Using SN7493N

4-Bit Counter Using SN7490, SN7492 or SN7493

Another application using the SN7490, SN7492, SN7493 divides by any number (N) in binary code. This circuit will operate over the operating temperature range of the devices used under full loading conditions, and is not dependent on the internal delay of the counter. The circuit uses 1 to 1¾ external gate packages. See Figure 35.

In this configuration, the last state of a count cycle sets a latch which clears the counter to the all zero state. The positive edge of the following clock pulses resets the latch. The negative edge of the same pulse initiates the new cycle.

Binary Divide-By-14 Ripple Counter (Example):

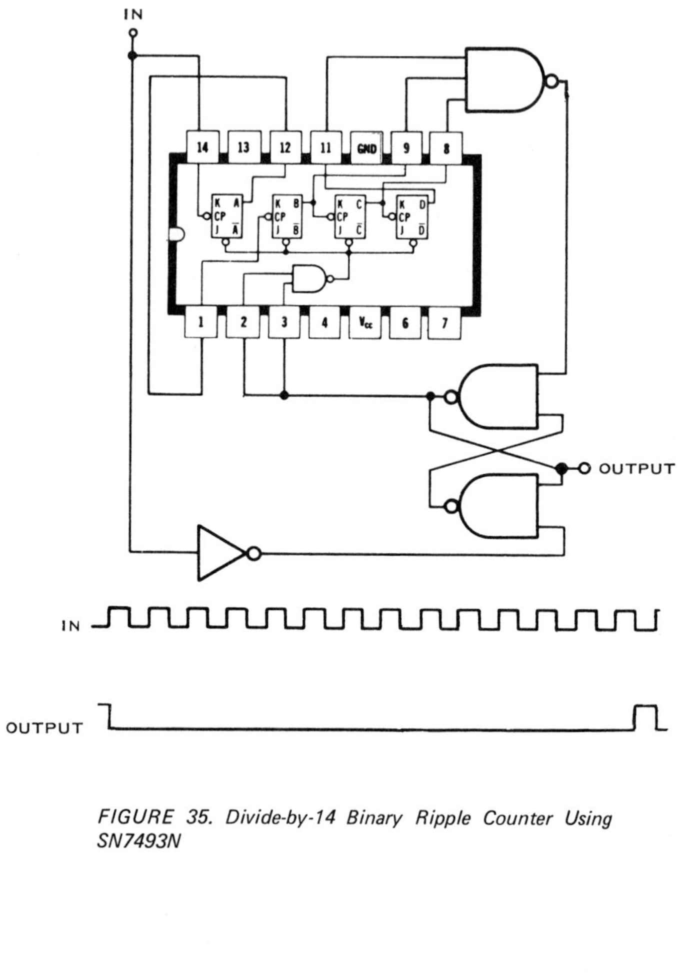

FIGURE 35. Divide-by-14 Binary Ripple Counter Using SN7493N

8-Bit Counter Using SN7493

A binary ripple counter which can count to any number[4] (N) up to 256 can be built with two SN7493 counters and 1-3/4 gate packages. In the following examples, the "ones" of the binary word (N), which is the last binary count of the desired cycle length, are fed into the NAND gate. This particular counter recycles for every 153 input pulses.

A binary divide-by-153 ripple counter example is illustrated in Figure 36.

BCD Divide-By-89 Counter Using SN7490s

Some extended counts in the BCD code are possible by using multiple SN7490 decade counters. The following example utilizes two SN7490 s and 1 external gate package. As in the two previous examples, the "ones" of the BCD word are fed into the NAND gate and the latch resets the two decade stages. This particular counter recycles for every 89 input pulses. See Figure 37.

Numerous other variations are possible especially when the symmetrical divide-by-10 mode of the SN7490 is considered.

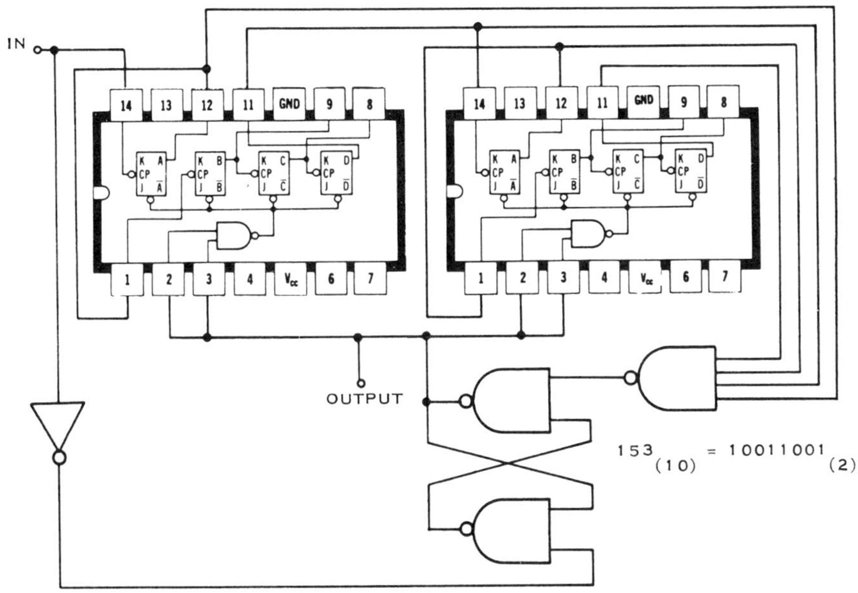

FIGURE 36. Divide-by-153 Binary Ripple Counter Using SN7493N

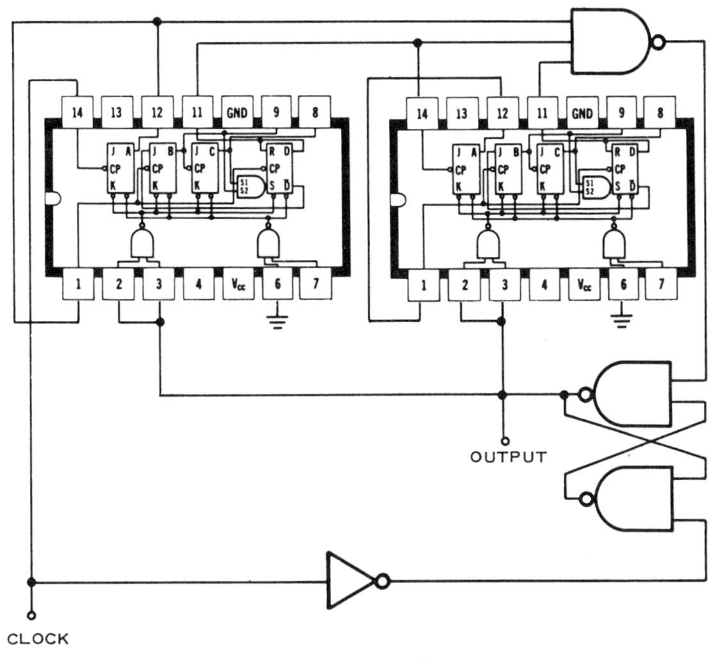

FIGURE 37. BCD Divide-by-89 Counter Using SN7490N

SYNCHRONOUS COUNTERS

Design Procedure for Synchronous Counters

The advantage of synchronous counters is, that they can produce any binary sequence of any desired length without long delays for propagating the clock pulse. However, the number of components is usually larger than for other types of counters and the design procedure is more difficult if the cycle length is not a power of two.

For straight, binary synchronous counters, for cycle lengths of $2, 2^1, 2^2, 2^3, 2^4, 2^5, \ldots 2^n$, the design pattern can be seen in the following example illustrating a divide-by-13 synchronous counter. Note that the toggle or inhibit condition of a particular flip-flop (n) is a function of all previous flip-flops.

1. Divide-By-13 Synchronous Counter
   a. Write a state table for the desired code and cycle length (figure 38):
      (1) A transition map shows the actual sequence of states more clearly. See Figure 39. For codes other than the straight binary code more efficient state assignments can be found using common stage assignment rules.
      (2) Next-Stage Karnaugh maps for flip-flop J & K input equations, figure 40.

Since each flip-flop has two inputs (J, K) and this counter requires four flip-flops (A, B, C, D), use eight 4-variable Karnaugh maps. If the transistion map is considered as the present-stage map, then the eight input maps can be considered next-state maps.

Example: From the stage 0 1 1 1 (transition map) the counter goes to 1000. Since the J inputs set the flip-flops to Q = 1 (the K-inputs set the flip-flops to Q = 0), a "one" is entered in the A_J-map and "one" in the B_K, C_K, D_K maps (next state will be 1000).

(3) Simplifying assumptions:

(a) Only 13 out of 16 possible states are used – three "don't care" conditions exist. (X = unassigned states)

(b) For all conditions for which a "one" is entered in a map and the particular flip-flop is in the "one" state already a no-change ("don't care") can be used to simplify.

ϕ = no change

DECIMAL	PRESENT STATE				NEXT STATE			
	A	B	C	D	A	B	C	D
0	0	0	0	0	0	0	0	1
1	0	0	0	1	0	0	1	0
2	0	0	1	0	0	0	1	1
3	0	0	1	1	0	1	0	0
4	0	1	0	0	0	1	0	1
5	0	1	0	1	0	1	1	0
6	0	1	1	0	0	1	1	1
7	0	1	1	1	1	0	0	0
8	1	0	0	0	1	0	0	1
9	1	0	0	1	1	0	1	0
10	1	0	1	0	1	0	1	1
11	1	0	1	1	1	1	0	0
12	1	1	0	0	0	0	0	0

FIGURE 38. State Table For Divide-by-13 Synchronous Counter

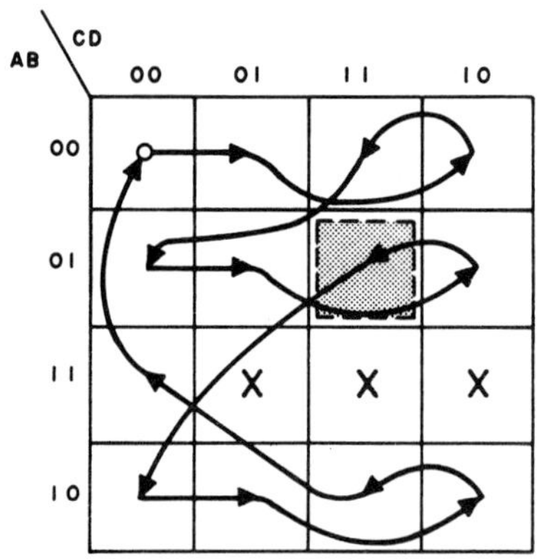

FIGURE 39. Transition Map For Divide-by-13 Synchronous Counter

Since a J-K flip-flop changes state for J-K = 1 a "one" in the J-Maps can be plotted as a toggle "don't care" (ϕ) in the K-maps.

ϕ = Toggle

b. With these simplifications, the input equations to the four flip-flops can be solved from the J-K input next-state maps:

$A_J = BCD \qquad C_J = D$

$A_K = B \qquad C_K = D$

$B_J = CD \qquad D_J = \bar{A} + \bar{B} = \overline{AB}$

$B_K = CD + A \qquad D_K = 1$

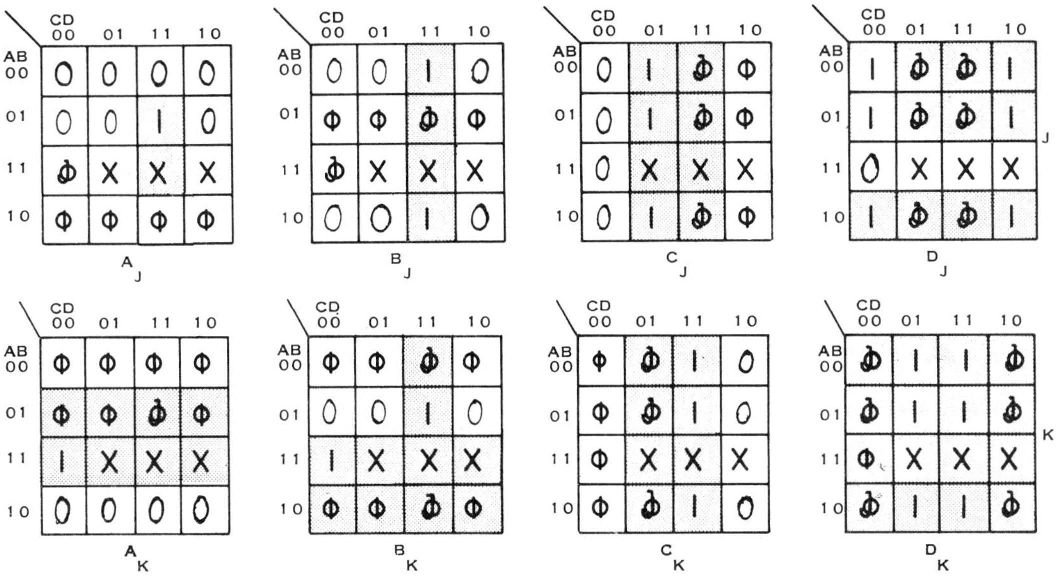

FIGURE 40. J and K Input Next-State Maps For Divide- by-13 Synchronous Counter

c. Logic Implementation of Divide-By-13 Synchronous Counter is shown in Figure 41.

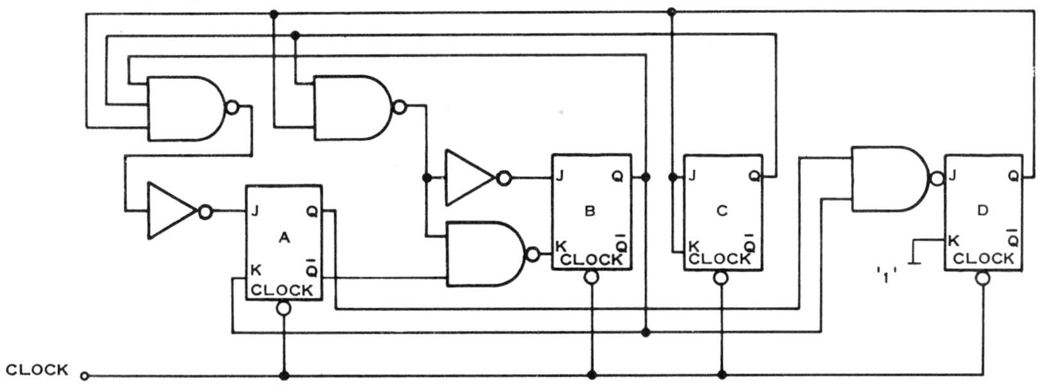

FIGURE 41. Divide-by-13 Sychronous Counter

2. 4-Bit Binary - UP-DOWN Counter:[5]

   a. Transition Table Figure 42:

   b. Logic Implementation of 4-Bit Binary Synchronous Counter

There are two different arrangements shown which represent a trade-off between speed and hardware especially for long cycle lengths.

(1) Serial-Carry method Figure 43.

(2) Parallel-Carry method Figure 44.

DECIMAL	PRESENT STATE				NEXT STATES X = 1				NEXT STATES X = 0			
	A	B	C	D	A	B	C	D	A	B	C	D
0	0	0	0	0	1	0	0	0	1	1	1	1
1	1	0	0	0	0	1	0	0	0	0	0	0
2	0	1	0	0	1	1	0	0	1	0	0	0
3	1	1	0	0	0	0	1	0	0	1	0	0
4	0	0	1	0	1	0	1	0	1	1	0	0
5	1	0	1	0	0	1	1	0	0	0	1	0
6	0	1	1	0	1	1	1	0	1	0	1	0
7	1	1	1	0	0	0	0	1	0	1	1	0
8	0	0	0	1	1	0	0	1	1	1	1	0
9	1	0	0	1	0	1	0	1	0	0	0	1
10	0	1	0	1	1	1	0	1	1	0	0	1
11	1	1	0	1	0	0	1	1	0	1	0	1
12	0	0	1	1	1	0	1	1	1	1	0	1
13	1	0	1	1	0	1	1	1	0	0	1	1
14	0	1	1	1	1	1	1	1	1	0	1	1
15	1	1	1	1	0	0	0	0	0	1	1	1

X = 1 (COUNT UP)
X = 0 (COUNT DOWN)

FIGURE 42. State Map For 4-Bit Synchronous Binary Up-Down Counter

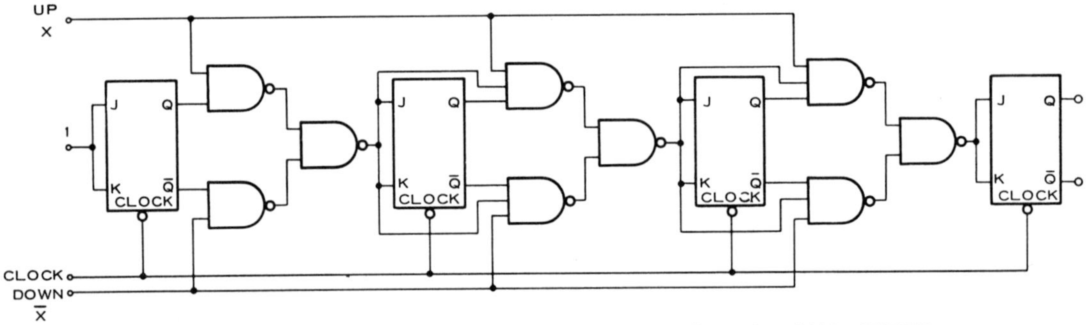

FIGURE 43. 4-Bit Synchronous Binary Up-Down Counter (Serial Carry) Using SN7473s

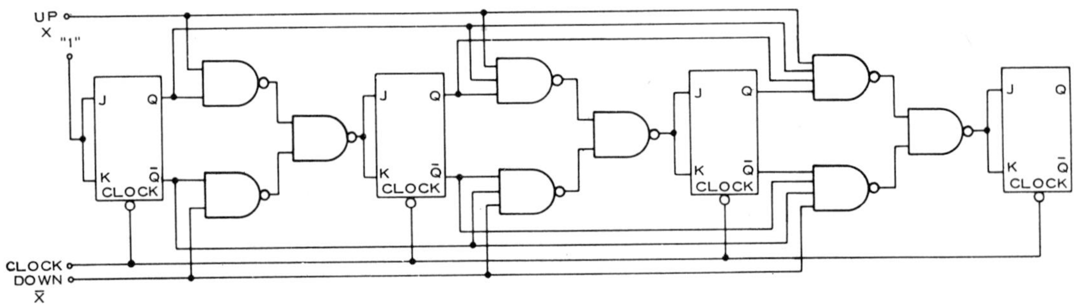

FIGURE 44. 4-Bit Synchronous Binary Up-Down Counter (Parallel Carry) Using SN7473s

3. Up-Down Decade Counter:[5]

   a. State-diagram (Figure 45)

DEC.	PRESENT STATE				NEXT STATES X = 1				NEXT STATES X = 0			
	D	C	B	A	D	C	B	A	D	C	B	A
0	0	0	0	0	0	0	0	1	1	0	0	1
1	0	0	0	1	0	0	1	0	0	0	0	0
2	0	0	1	0	0	0	1	1	0	0	0	1
3	0	0	1	1	0	1	0	0	0	0	1	0
4	0	1	0	0	0	1	0	1	0	0	1	1
5	0	1	0	1	0	1	1	0	0	1	0	0
6	0	1	1	0	0	1	1	1	0	1	0	1
7	0	1	1	1	1	0	0	0	0	1	1	0
8	1	0	0	0	1	0	0	1	0	1	1	1
9	1	0	0	1	0	0	0	0	1	0	0	0

X = 1 (COUNT UP)
X = 0 (COUNT DOWN)

FIGURE 45. State Map For Synchronous Up-Down Decade Counter

(1) Transition Map (Figure 46)

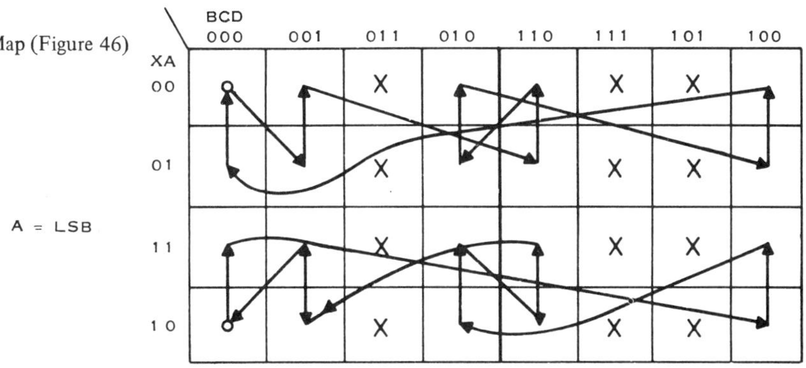

FIGURE 46. Transition Map For Synchronous Up-Down Decade Counter

(2) NEXT STATE MAPS for Flip-Flop Input Equations (Figure 47)

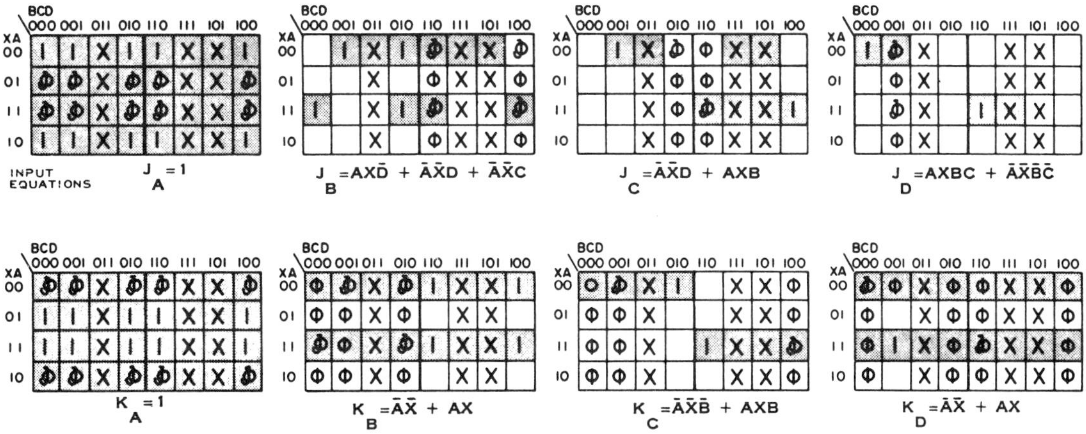

FIGURE 47. Next-State Maps For Synchronous Up-Down Decade Counter

b. **Logic Implementation of Synchronous Up-Down Decade Counter** (Figure 49)

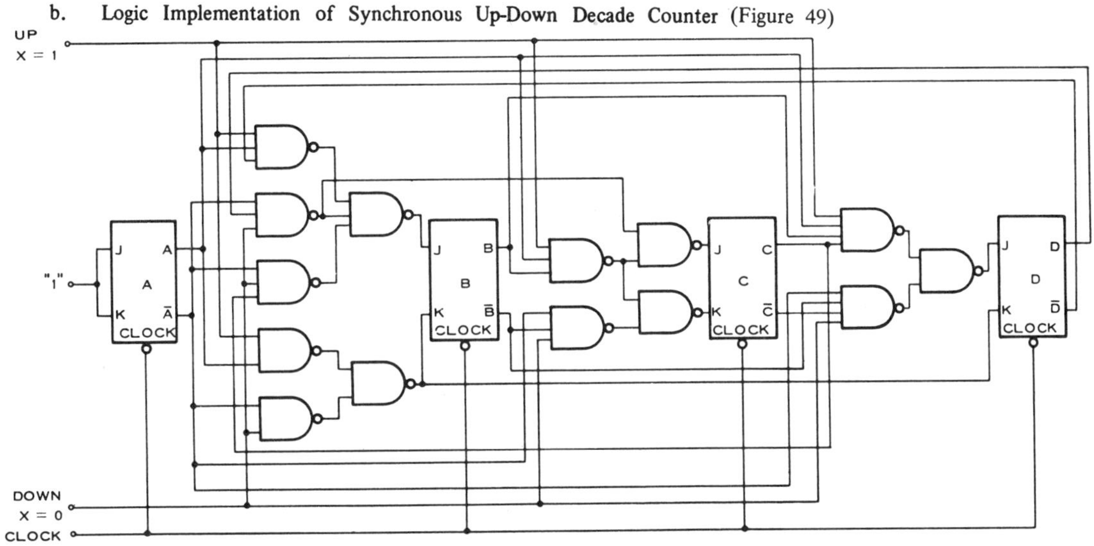

FIGURE 48. Synchronous Up-Down Decade Counter

JOHNSON COUNTERS AND RING COUNTERS

Johnson Counters For Even- And Odd-Cycle Lengths Using The SN7472, SN7473, SN7491, SN7494, SN7496

These registers can be used to build another counter type.[6] The counters to be discussed here are shift-register counters and are called Johnson Counters. These counters work synchronously and can be operated at typically 25 MHz using the SN7400 series. They can be very easily designed and the output decoding is very simple, especially where several counts need to be decoded. Using the SN7496 successive division ratios of 2, 4, 6, 8, 10 can be taken from the outputs, if the E output is inverted and fed back to the input. See Figure 49.

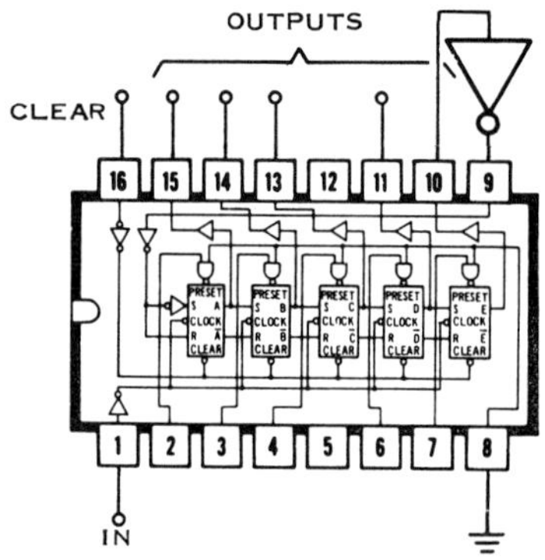

FIGURE 49. Johnson Counter For Even-Cycle Length Using SN7496N

For odd-cycle length, add gating to skip the all "1" state by feeding a "0" in the first stage:

1 1 1 1 0 – DETECT THIS STATE
1 1 1 1 1 – SKIP THIS STATE
0 1 1 1 1 – GO DIRECTLY TO THIS STATE

Division ratios of 3, 5, 7, 9 in the Johnson cycle can be taken from the outputs of the SN7496. See Figure 50.

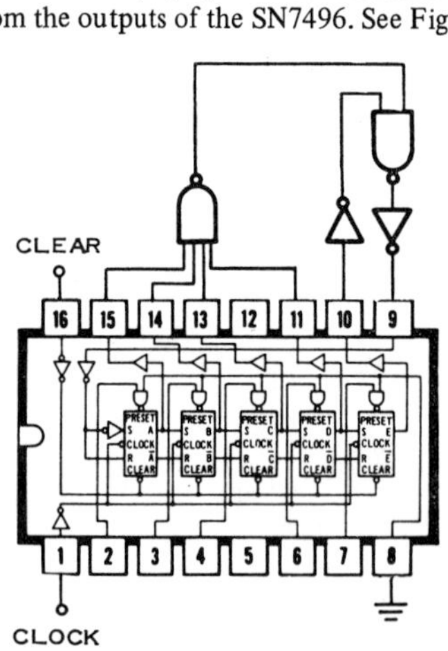

FIGURE 50. Johnson Counter For Odd-Cycle Length Using SN7496N

The cycle length of the Johnson counter is 2.n instead of 2^n. Therefore, more flip-flops are required for a given cycle length. On the other hand, a unique pattern is generated which allows any count to be decoded with a simple 2-input NAND gate. See Figure 51.

If asynchronous inputs (preset and clear) can be used to set the counter in a known state when it jumps out of the cycle or at the beginning of a count, the SN7491, SN7494, SN7496 may be used. Otherwise, gating must be provided to ensure that it will enter the correct cycle and count back into this cycle if it jumps out.

DECIMAL	A	B	C	D	E	OUTPUT DECODING	
0	0	0	0	0	0	$\bar{A}$	$\bar{E}$
1	1	0	0	0	0	A	$\bar{B}$
2	1	1	0	0	0	B	$\bar{C}$
3	1	1	1	0	0	C	$\bar{D}$
4	1	1	1	1	0	D	$\bar{E}$
5	1	1	1	1	1	A	E
6	0	1	1	1	1	$\bar{A}$	B
7	0	0	1	1	1	$\bar{B}$	C
8	0	0	0	1	1	$\bar{C}$	D
9	0	0	0	0	1	$\bar{D}$	E

FIGURE 51. Truth Table For Even-Cycle Length SN7496 Johnson Counter

The Johnson cycle is not self-starting without additional feedback. It has been found that the Johnson cycle is always self-starting (for n up to 25 stages) when feedback is provided from the last j stages of an n-stage shift register, where

$$j > \frac{n}{3}$$

Six-stage self-starting Johnson counters are illustrated in Figures 52 and 53. These counters can also be implemented using SN7472 dual flip-flops.

Ring Counters

Ring counters contain only one logical "1" or "0" and circulate this data. Total cycle length is equal to the number of flip-flops used.

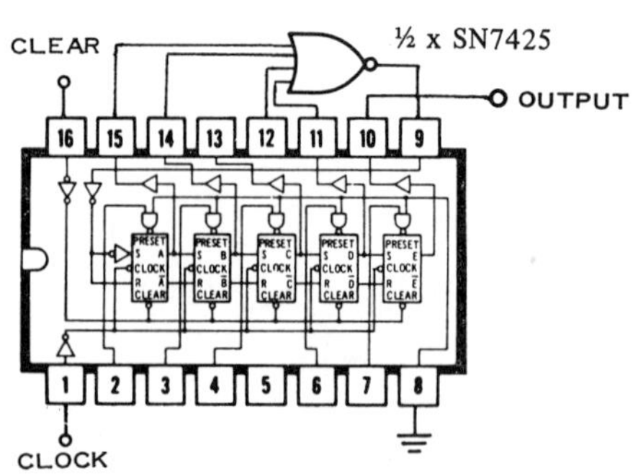

FIGURE 54. Divide-by-5 Ring Counter Using SN7496N

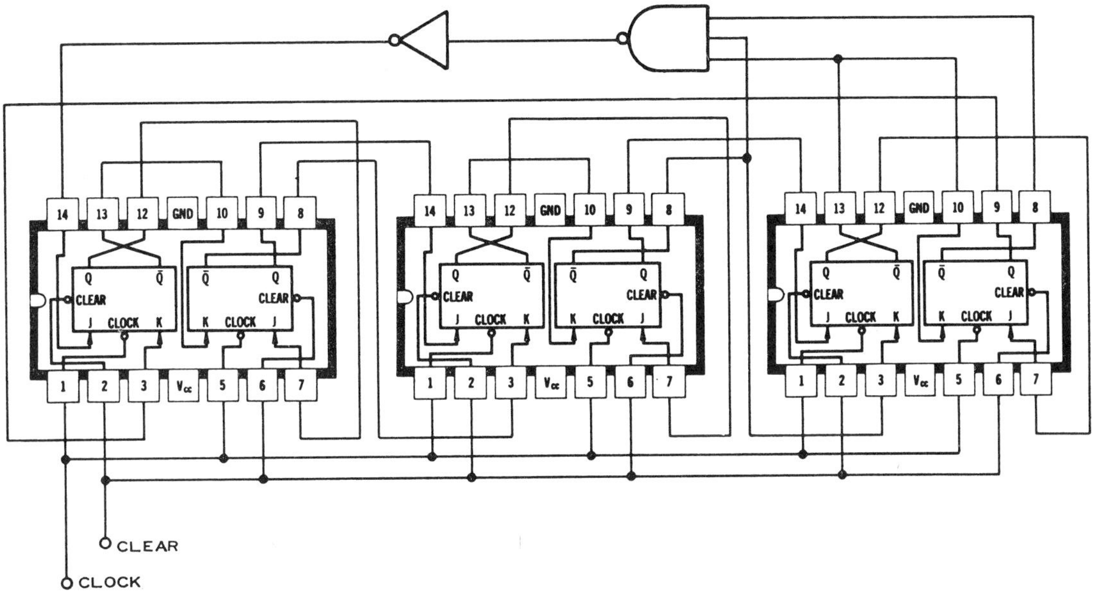

FIGURE 53. Self-Starting Six-Stage, Odd-Cycle Length Johnson Counter Using SN7473N

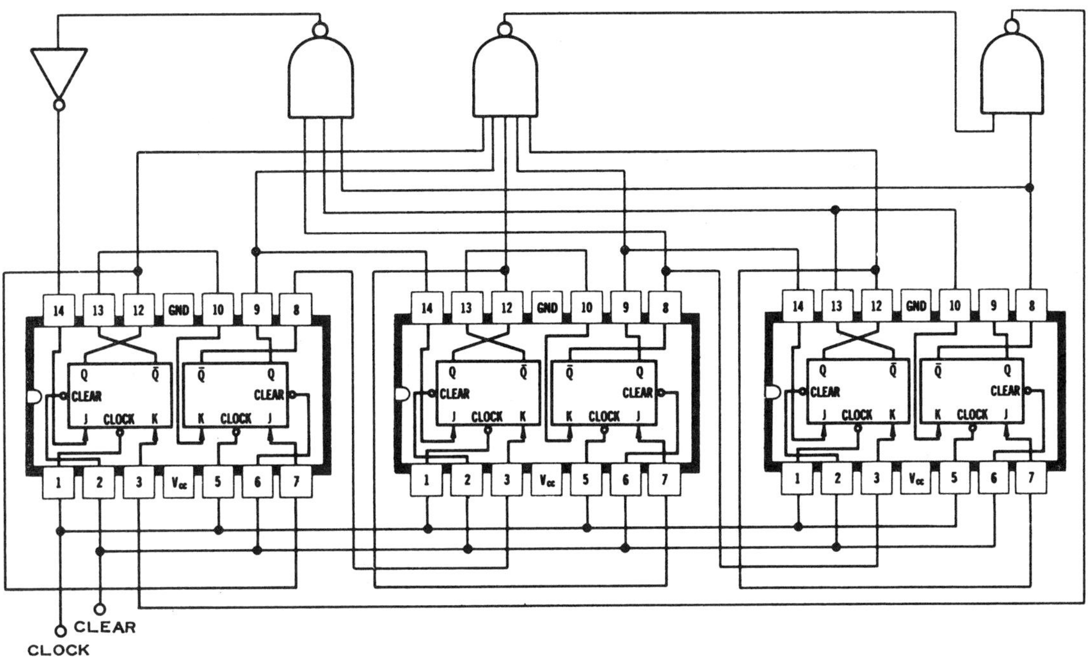

FIGURE 52. Self-Starting Six-Stage, Even-Cycle Length Johnson Counter Using SN7473N

SHIFT REGISTER GENERATOR COUNTERS

Other types of counters which can be built easier than synchronous binary counters and which become very economical for large cycle lengths are called linear shift register generator counters. These counters are shift register counters which use a mod. 2 feedback. These counters are synchronous counters and are divided into two classes:

A) Maximum length counters

B) Non-maximum length counters

Maximum Length Counters

Maximum length counters count to a cycle length of 2^n-1 (with n = number of stages in the shift register).

An exclusive – OR feedback term is provided from an even number of stages to the first stage. The stages to be fed back can be found in Table 1 for up to 12 stages.

Note: The feedback equation is the modulo 2 sum of the "1" terms in Table 1.

Example:

Use SN7496 to count to 31. The feedback term from above table for 5 stages is:

$$F = C\bar{E} + \bar{C}E$$

Since the modulo 2 sum of two zeros is zero, the all zero stages of the register have to be excluded from the counting sequence. This can be achieved by either additional gating or with the asynchronous preset inputs.

See Figure 55.

Another method of inhibiting the all zero state is to add the AND term of all zero outputs $(\bar{A}\cdot\bar{B}\cdot\bar{C}\cdot\bar{D}\cdot\bar{E})$ to the feedback equation:

$$F = C\bar{E} + \bar{C}E + \bar{A}\ \bar{B}\ \bar{C}\ \bar{D}\ \bar{E}$$

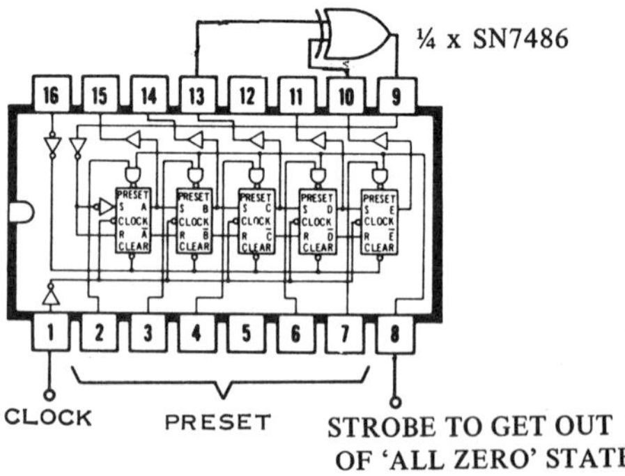

FIGURE 55. Divide-by-31 Shift-Register Generator Counter Using SN7496N

Another modification allows a count to 2^n instead of $2^n - 1$ (for n stages).

By adding a term to the feedback equation which inhibits the feedback (zero output of all stages, except the last, and the one output of the last stage $(\bar{A}\ \bar{B}\ \bar{C}\ \bar{D}\ E)$, the counter can be forced into the all zero state. At the next clock pulse the counter leaves the all zero state again on account of the term previously added.

Feedback equation for cycle length 2^n for n = 5:

$$F = (C\bar{E} + \bar{C}E)\cdot\overline{(\bar{A}\ \bar{B}\ \bar{C}\ \bar{D}\ E)} + \bar{A}\ \bar{B}\ \bar{C}\ \bar{D}\ \bar{E}$$

These counters can also be built with the SN7473, SN7474, SN7476, SN7494, SN7495.

Table 1

NUMBER OF STAGES IN THE SHIFT REGISTER	FEEDBACK STAGES											
	A	B	C	D	E	F	G	H	I	J	K	L
3	0	1	1									
4	0	0	1	1								
5	0	0	1	0	1							
6	0	0	0	0	1	1						
7	0	0	0	0	0	1	1					
8	0	0	0	1	1	1	0	1				
9	0	0	0	0	1	0	0	0	1			
10	0	0	0	0	0	0	1	0	0	1		
11	0	0	0	0	0	0	0	0	1	0	1	
12	0	0	0	0	0	1	0	1	0	0	1	1

Non-Maximum Length Counters

Here a technique referred to as the "jump technique" is described to shorten the cycle described above. Basically, another term is added to the technique described under (A) which forces, at a certain state, a 0 or 1 into the first stage, when a "1" or "0" is dictated by the mod. 2 feedback.

Effectively, the counter jumps over a determined number of stages (k) for $K < 2^n - 1$. Therefore, any desired cycle length can be generated.

Example:

Use the SN7496 to divide by 21. The maximum length cycle is $2^5 - 1 = 31$ (31-21 = 10) 10 stages must be jumped.

The design is the same as for MLS except that a jump term must be added to the feedback equation. See Figure 56.

Table I indicates the feedback equation:

$$F = \overline{C}E + C\overline{E}$$

Considering that the mod. 2 sum is "0" if C and E are equal and "1" if they differ, the counting sequence can be written. The jump term can be seen to be:

† 3

$$\begin{matrix} 1 & 0 & 1 & 0 & 1 \\ A & \overline{B} & C & \overline{D} & E \end{matrix}$$

because if the next clock pulse enters a "1" into the first stage, the count will be:

$$1 \quad 1 \quad 0 \quad 1 \quad 0$$

COUNT	A	B	C	D	E	
1	1	1	1	1	1	
2	0	1	1	1	1	
3	0	0	1	1	1	
4	0	0	0	1	1	
5	1	0	0	0	1	
6	1	1	0	0	0	
7	0	1	1	0	0	
8	1	0	1	1	0	
9	1	1	0	1	1	
10	1	1	1	0	1	
11	0	1	1	1	0	
12	1	0	1	1	1	
13	0	1	0	1	1	
14	1	0	1	0	1	
15	0	1	0	1	0	JUMP OF 10 (counts 15–24)
16	0	0	1	0	1	
17	0	0	0	1	0	
18	0	0	0	0	1	
19	1	0	0	0	0	
20	0	1	0	0	0	
21	0	0	1	0	0	
22	1	0	0	1	0	
23	0	1	0	0	1	
24	1	0	1	0	0	
25	1	1	0	1	0	
26	0	1	1	0	1	
27	0	0	1	1	0	
28	1	0	0	1	1	
29	1	1	0	0	1	
30	1	1	1	0	0	
31	1	1	1	1	0	

FIGURE 56. Divide-by-21 Shift-Register Generator Counter Count Sequence

The jump terms can be taken from Table 2 in connection with the maximum length feedback term given in Table 1.

Table 2

CYCLE LENGTH		A	B	C	D	E	F	(NOTE 1) J–K
	4	0	1	1				1
	5	1	0	0				1
	6	1	1	0				0
$2^n - 1$	7			MLS				
	8	0	1	1	0			0
	9	0	1	0	0			1
	10	1	1	0	0			1
	11	0	0	1	1			1
	12	1	0	0	0			1
	13	1	0	1	1			1
	14	1	1	1	0			0
$2^n - 1$	15			MLS				
	16	1	1	0	1	0		1
	17	1	0	0	0	1		0
	18	0	0	0	1	1		0
	19	0	1	1	0	1		1
	20	1	1	1	0	0		0
	21	1	0	1	0	1		1
	22	0	1	1	1	1		1
	23	1	1	0	0	1		0
	24	0	0	1	1	0		0
	25	1	0	0	1	0		1
	26	0	0	1	0	1		1
	27	1	0	1	1	0		0
	28	0	1	0	1	1		0
	29	0	1	0	1	1		
	30	1	1	1	1	0		0
$2^n - 1$	31			MLS				
	32	0	0	1	1	1	0	0
	33	1	1	0	0	1	0	0
	34	0	0	1	0	0	0	1
	35	0	1	0	1	1	1	1
	36	0	0	1	1	0	1	0
	37	1	1	1	1	0	1	0
	38	0	1	0	1	0	1	0
	39	0	1	1	1	1	1	1
	40	0	1	1	0	0	1	0
	41	1	0	1	0	0	0	1
	42	1	0	1	1	1	1	1
	43	0	0	0	1	0	0	1
	44	1	1	0	0	0	0	1
	45	1	1	0	1	0	0	1
	46	1	0	0	1	0	1	0
	47	1	0	0	1	1	1	1
	48	1	1	1	0	1	1	1
	49	1	0	0	0	1	1	1
	50	0	1	1	1	0	0	1
	51	0	0	0	1	1	0	0
	52	1	0	1	1	0	1	0
	53	0	1	0	0	0	0	1
	54	1	1	1	0	0	1	0
	55	0	0	1	0	1	0	0
	56	0	1	1	0	1	0	0
	57	0	0	0	0	1	1	1
	58	1	0	0	0	0	0	1
	59	1	1	0	1	1	0	0
	60	0	1	0	0	1	1	1
	61	1	0	1	0	1	1	1
	62	1	1	1	1	1	0	0
$2^n - 1$	63			MLS				

NOTE 1 THE TERM UNDER J–K SHOWS THE BIT TO BE ENTERED INTO THE FIRST STAGE WITH THE NEXT CLOCK PULSE.

Feedback equation for N = 21:

$$F = \underbrace{E\overline{C} + \overline{E}C}_{\text{MLS}} + \underbrace{A\,\overline{B}\,C\,\overline{D}\,E}_{\text{Jump Term}} = \underbrace{E\overline{C} + \overline{E}C + A\,\overline{B}\,C\,\overline{D}}_{\text{Simplified}}$$

See Figure 57.

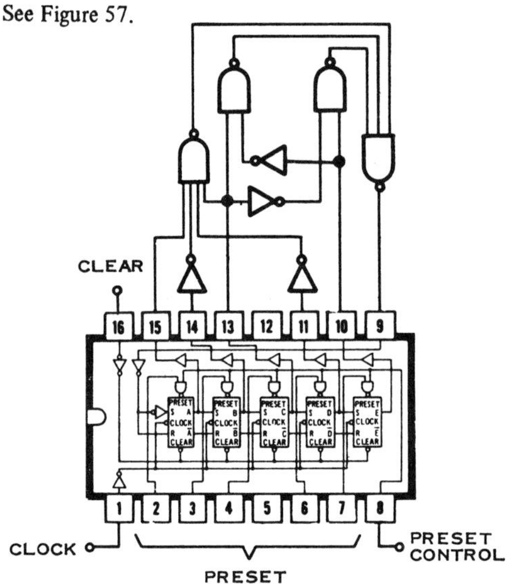

FIGURE 57. Divide-by-21 Shift-Register Generator Counter Using SN7496N

The all zero state can be inhibited by detecting that state and presetting the counter asynchronously to the all "1" state or adding the all "0" AND term to the feedback equation:

$$F = \underbrace{E\bar{C} + \bar{E}C}_{\text{MLS Feedback}} + \underbrace{A\ \bar{B}\ C\ \bar{D}\ E}_{\text{Jump Term}} + \underbrace{\bar{A}\ \bar{B}\ \bar{C}\ \bar{D}\ \bar{E}}_{\text{Prevent all zero state}}$$

which simplifies to:

$$F = C\bar{E} + A\ \bar{B}\ C\ \bar{D} + \bar{A}\ \ \bar{B}\ \ \bar{D}\ \ \bar{E}$$

†3
In any MLS counter there are always two states that differ only in the first bit and are separated by the required number of jump stages (k). Thus by inspecting the table as in Figure 56, these two stages may be found. (Practically, for example, if the table is drawn on an equally spaced grid, a template – the required jump number "high" – could be used to move down the table and the inspection made.) The change on the shift register input occurring ready for the next clock pulse is initiated from the state preceding the first of those to be jumped.

REFERENCES

1. A similar function – for 8 bits – may be performed using only one package with the SN74198. This has synchronous data loading.

2. A SN74165 may be used in place of the two SN7495s for this converter.

3. A SN74164 may be used in place of the two SN7495s for this converter.

4. Using techniques similar to those described in the text in conjunction with the load and clear functions in the SN74163, it is possible to make fully synchronous mod-N dividers – N counters, or N1 to N2 counters.

5. *Semiconductor Circuit Design,* Volume II, Chapter VI, pp 59-66, April 1973, describes the use of SN74190/3 synchronous 4 bit Up/Down counters.

6. Longer Johnson and Ring Counters can be made using the SN74164.

7. *Semiconductor Circuit Design,* Volume II, Chapter VII, pp 67-72, April 1973, describes a neater method of obtaining non MLS counters using the SN74164.

XV NUMEROUS TTL APPLICATIONS

This chapter has a number of assorted circuits which illustrate the use of digital integrated circuits.

PARALLEL INPUT SUMMING COUNTER

Figure 1 shows a circuit designed to combine a number of pulses arriving at random from separate sources into a single pulse train containing as many pulses as the sum of the input pulses from all sources. The minimum spacing between output pulses is defined by the frequency of the clock pulse oscillator, the width of the pulse being a function of the mark-space ratio of the oscillator. The system will accept coincident input pulses without error. Four inputs are shown, but the method may be extended to any number of inputs. In this example, the negative-going input pulses are derived from a microswitch. However, logic inputs may be used, thus increasing the operating frequency to several megacycles.

Brief Description of Circuit

Incoming information is stored in four pairs of cross-coupled gates, G1 to G8. During each strobe pulse cycle gates G9 to G12 open in sequence, only one gate being open at a time. The stored information then passes through these gates and via D-type flip-flops and gates G13 to G16, resets the input store and provides output pulses. G17 performs a NOR function on the four gate outputs, giving a serial output. An NPN transistor can be connected to output of G17 to give current gain and drive a load such as an electro-mechanical device.

The circuit shown in Figure 1 functions as follows:

A multivibrator provides a two phase clock pulse. If a standard cross-coupled multivibrator were used the positive going edges would be exponential due to C5 and C6 charging up to V_{CC} when VT1 and VT2 respectively were cut off. However, by splitting the collector load and inserting a diode to isolate the collector from the capacitor during a positive-going transition, R13 is used to charge up C6 and R14 to provide drive current for the logic stages when VT2 is turned off. The two outputs from the multivibrator are designated clock and $\overline{\text{clock}}$. With the resistor values shown, the clock output has a out of 5 and the $\overline{\text{clock}}$ output a fan out of 4. (R10 and R12 must be adjusted if a greater fan out is required so that transistors VT1 and VT2 can sink all the current from the inputs to the networks). The pulse repetition rate and mark-space ratio can be altered by varying capacitors C5 and C6.

The clock output is taken to a two stage ripple counter, consisting of two D-type flip-flops with their $\overline{Q}$ outputs taken to their D inputs as shown. Since $\overline{Q}$ goes to the complement of the D input state upon application of the clock pulse, with the $\overline{Q}$ output and the D input connected together the state of the flip-flop will change with each clock pulse. The ripple counter provides four outputs, A, $\overline{A}$, B and $\overline{B}$, which can be decoded to give AB, $\overline{A}$B, A$\overline{B}$, and $\overline{A}\,\overline{B}$. If these outputs are applied to gates G9 and G12 and an input pulse has been stored in each memory, then G9 and G12 will open and close in sequence, only one gate being open at any time.

In Figure 1 the inputs are shown to come from micro-switches. The resistors and capacitors are required to eliminate switch bounce.

Considering channel one only:

When the switch is normally closed capacitor C1 charges to ground via resistor R1. When the switch is opened and negative pulse appears across resistor R5 and the input of gate G2. This latches the gates G1 and G2 so that the output of G2 is a '1'. The capacitor C1 discharges rapidly through the base-emitter junction of the input gate and resistor R5 so that, in the event of switch bounce, the capacitor will be discharged on the first contact and no more pulses will be applied to G2 input.

With the output of G2 at a '1', when both A and B become a '1' the output of G9 goes to a '0', thus making the D input of flip-flop F1 a '0'. On the next clock pulse Q goes to a '0' and resets the store (G1 and G2) while $\overline{Q}$ goes to a '1' and opens gate G13. The clock pulse then goes to a '1' changing the output of G13 to a zero for the duration of the positive part of the clock pulse. Meanwhile, since G1 and G2 have been reset by the Q output of F1 going to a '0', D becomes a '1' and on the next clock pulse sets Q to a '1' and $\overline{Q}$ to a '0'. This unlocks gates G1 and G2 ready for a new input pulse and also locks gate G13 to inhibit any further clock pulses.

This is the sequence of events for each channel at one clock period intervals, so that after four clock pulses, all the inputs have been sampled and reset.

Gate 17 performs a NOR function, producing a serial output to drive transistor VT3.

Components

Capacitors	C1 to C4	0.1 µF
	C5	To suit timing
	C6	To suit timing N.B. For 10:1 mark-space ratio C6 is ten times C5.
Diodes	D1, D2	1S44

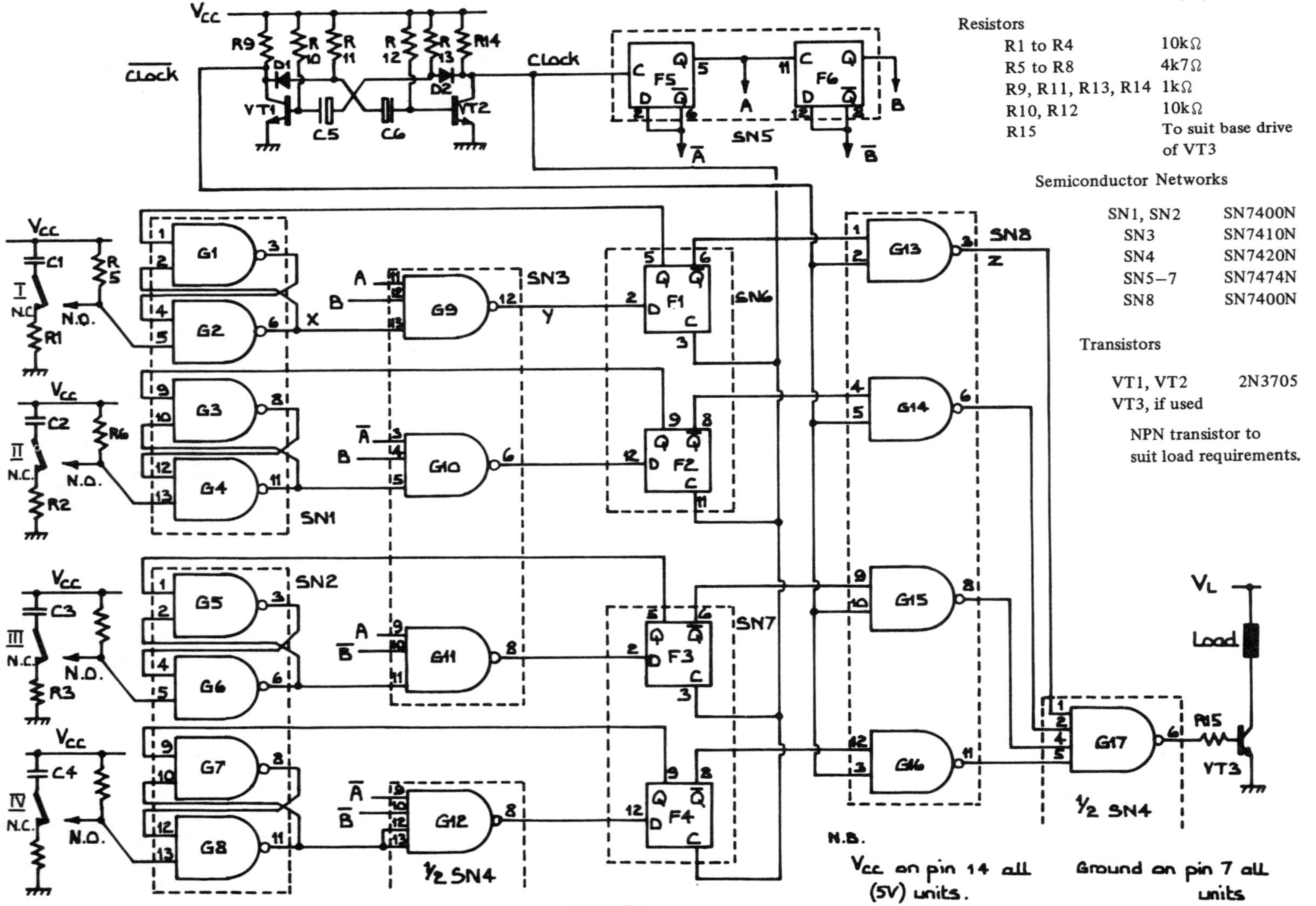

Microswitches are numbered I, II, III and IV.

Resistors

R1 to R4	10kΩ
R5 to R8	4k7Ω
R9, R11, R13, R14	1kΩ
R10, R12	10kΩ
R15	To suit base drive of VT3

Semiconductor Networks

SN1, SN2	SN7400N
SN3	SN7410N
SN4	SN7420N
SN5–7	SN7474N
SN8	SN7400N

Transistors

VT1, VT2	2N3705
VT3, if used	NPN transistor to suit load requirements.

FIGURE 1

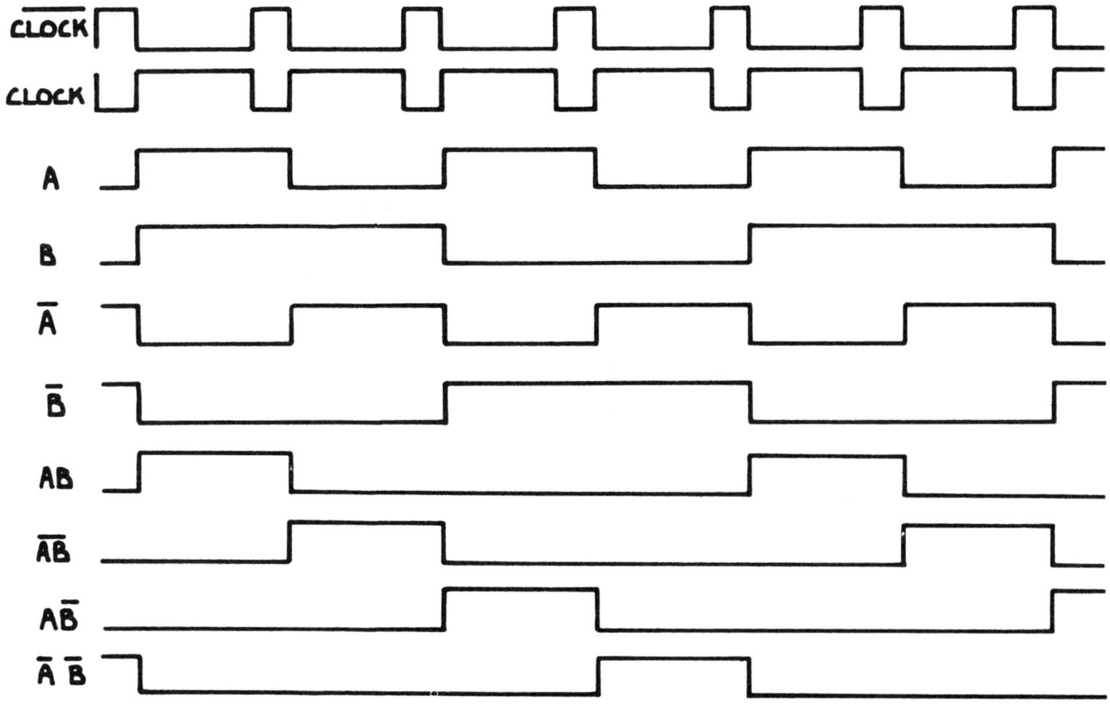

FIGURE 2 Timing Diagram for Sampling Gates

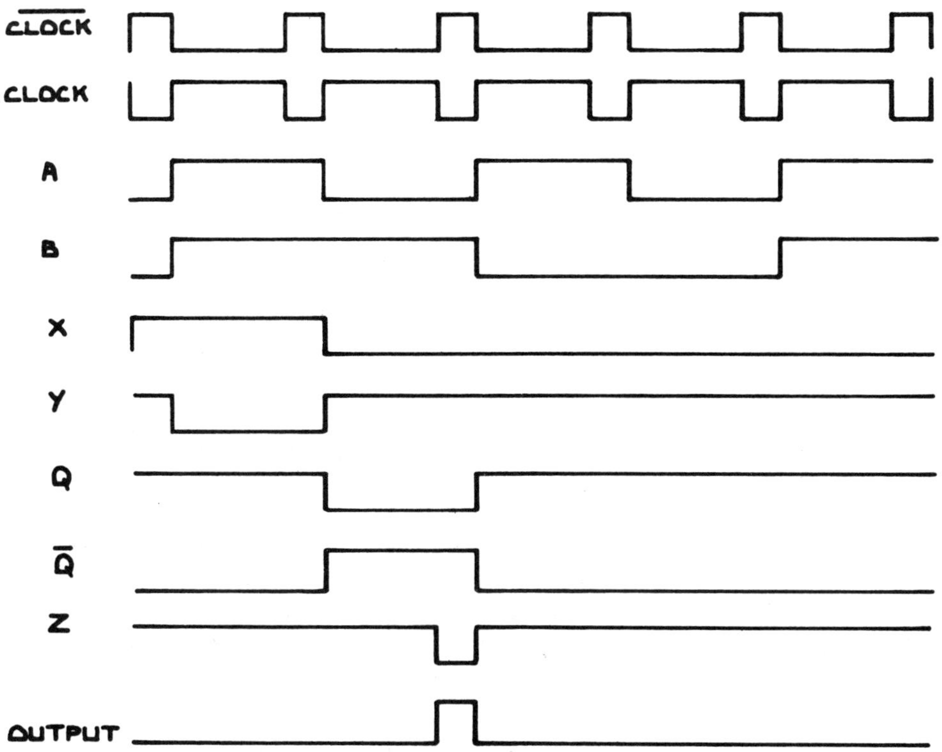

Figure 3 Timing Diagram for Circuit (One Channel Only)

TWENTY-FOUR HOUR CLOCK

This section describes a clock designed to give a digital readout of time for a twenty-four hour period. The unit uses series 74N networks, including four SN74141N BCD-decimal decoder/drivers which energise Nixie tubes indicating tens of hours, hours, tens of minutes and minutes.

Figure 4 is a block diagram of the clock and Figure 5 is the circuit diagram.

A 50Hz 5V square wave is buffered through one of the gates in quadruple two-input gate SN12 and is then divided down by three SN7490N decade counters. Counter SN1 is connected to divide by five and counters SN2 and SN3 divide by ten, so that clock pulses at a rate of one pulse per ten seconds are supplied to the divide-by-six section of SN4. The output of this unit clocks a further decade counter, SN5, which generates a BCD code. The outputs of SN5 are connected to the inputs of 'MINUTES' driver SN11, which decodes the BCD number to produce a count up to nine minutes in one minute increments. The 'D' output of SN5 clocks the divide-by-six section of the following counter, SN6, at a rate of one pulse per ten minutes. The output of SN6 is decoded by SN10 and the 'TENS OF MINUTES' tube indicates an increase of one digit every ten minutes. The 'C' output of SN6 supplies pulses to the input of SN7 at a rate of one pulse per hour, generating a BCD code which is decoded by SN9 to provide a count of hours. The divide-by-two sections of SN1 and SN4 are used to provide the drive for the 'TENS OF HOURS' unit. The 'D' output of SN7 supplies one pulse per ten hours to the 'A' input of SN4, so that one pulse twenty hours appears at the 'A' output to be fed to 'TENS OF HOURS' driver SN8. The same pulse is fed to SN1, where it is divided by two. A '1' now appears at Pin 12 of the quadruple 2 input NAND gate. SN12 and in the next cycle, when the 'C' input of SN9 goes to a '1' (i.e. at four hours) the 'A' output of SN12 goes to a '0'. This is inverted by the following gate to reset SN1 and SN7 outputs to '0'. The clock reading is reset to zero hours.

A manual 'RESET' switch is incorporated to allow all units to be set to zero. A 'SET HOURS' switch allows the 'D' output of SN2 to be routed directly to the 'A' input of SN7 so that hours are switched at a rate of one per second to any required hour. A more elaborate resetting mechanism may be employed if required.

Components

Semiconductor Networks	SN1	SN7490N
	SN2	SN7490N
	SN3	SN7490N
	SN4	SN7492N
	SN5	SN7490N
	SN6	SN7492N
	SN7	SN7490N
	SN8	SN74141N
	SN9	SN74141N
	SN10	SN74141N
	SN11	SN74141N
	SN12	SN7400N SN7413N is preferable

The 'SET HOURS' switch is a break-before-make switch.

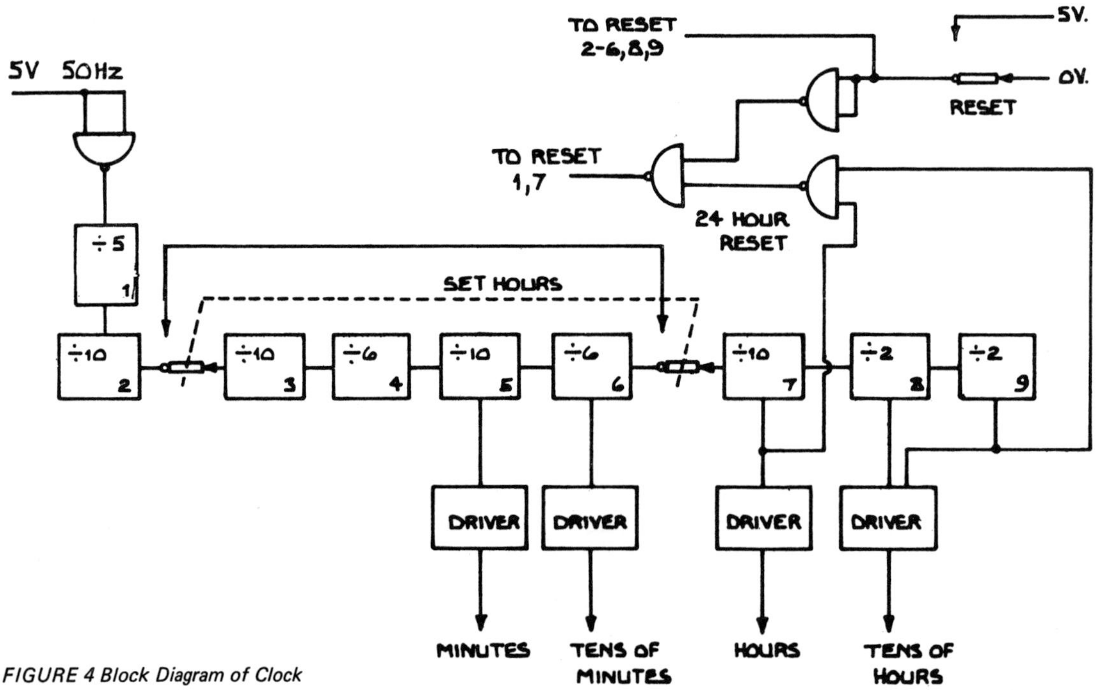

FIGURE 4 Block Diagram of Clock

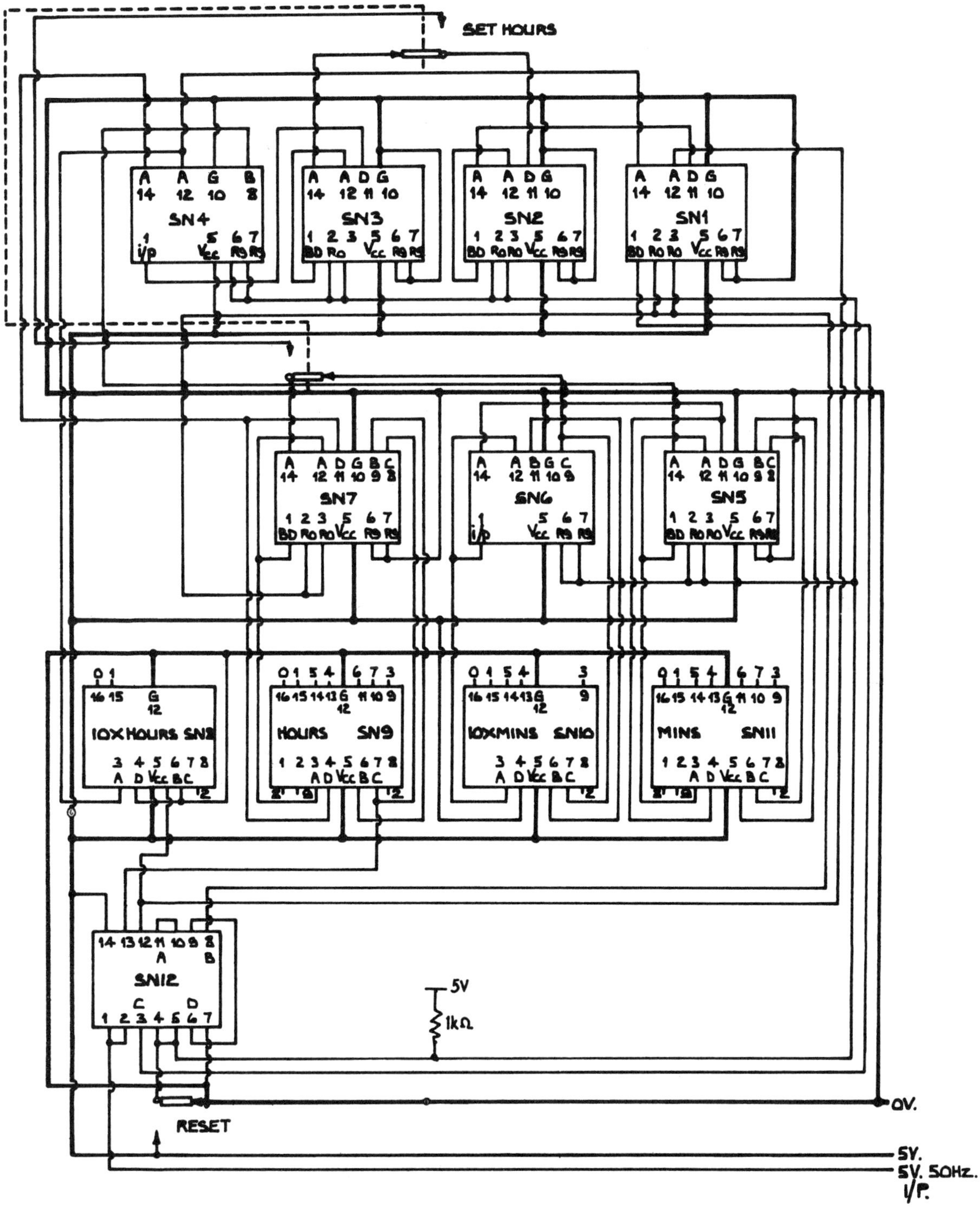

FIGURE 5 Circuit Diagram of Clock

RIPPLE COUNTER USING TTL D-TYPE FLIP-FLOPS

The SN7474 is a TTL dual D-type flip-flop with direct clear and preset inputs and complementary Q and $\overline{Q}$ outputs. Input information on D is transferred to the Q output on the positive edge of the clock pulse.

The truth table for the bistable is:–

	t_n	t_{n+1}	
CP	D	Q	$\overline{Q}$
1	0	0	1
1	1	1	0
0	X	Q_n	Q_n

Where t_{n+1} is bit time after the clock pulse and CP = 1 indicates that a '0' → '1' transition has taken place at the clock input.

The truth table required for a single counter stage is:–

CP	Q_n	Q_{n+1}
1	1	0
2	0	1
3	1	0

i.e. $Q_{n+1} = \overline{Q}_n$

From a simple Karnaugh map the input conditions on D to obtain this sequence can be derived. It is found that D = $\overline{Q}_n$.

The bistable divides by two and the Q output is used to drive a similar stage. The truth table for a three stage counter is shown below:–

	A	B	C
0	0	0	0
1	1	0	0
2	0	1	0
3	1	1	0
4	0	0	1
5	1	0	1
6	0	1	1
7	1	1	1

Figure 6 illustrates the circuit configuration.

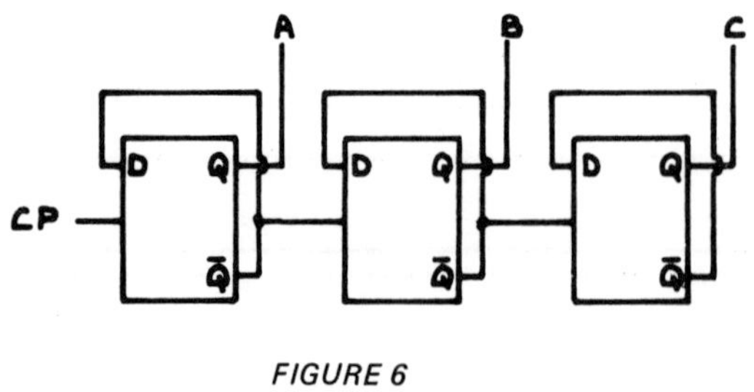

FIGURE 6

The count direction can be reversed by observing the $\overline{Q}$ outputs or by connecting Qs to clocks of succeeding stages.

16–BIT PARALLEL BINARY COMPARATOR

This section describes the use of full adder SN7483N to compare two binary numbers presented in parallel form. Each four-bit adder provides a sum output for each bit plus a resultant carry output which indicates the relative magnitude of the two numbers if a subtraction is being carried out.

If the two binary numbers are A and B, comparison is effected by subtracting A from B by the addition of 2's complement and observing if a borrow occurs. To obtain the 2's complement of A, the 1's and 0's are interchanged and a 1 is added to the least significant digit by feeding in a 1 at the carry input of the first adder. The 2's complement of A is therefore (2^{n+1} – A), where n is the number of bits. If this is added to B the result is (B – A) + 2^{n+1}. The 2^{n+1} carry output (at C_4) only occurs if B is equal to or greater than A. If B is smaller than A the carry output is a 0.

Example

For convenience four-bit numbers are used as an example.

(i) **A = 5 ≡ 0101 in binary**
B = 6 ≡ 0110 in binary
The 2's complement of A is 1011
added to B 0110
gives 1.0001
'1' carry output indicating B ⩾ A — **difference (B – A)**

(ii) **A = 5**
B = 5
The 2's complement of A is 1011
added to B 0100
gives 0.1111
'1' carry output indicating B ⩾ A — **difference (B – A)**

(iii) **A = 5**
B.= 4 ≡ 0100 in binary
The 2's complement of A is 1011
added to B 0100
gives 0.1111
'0' carry output indicating B <A — **difference (B – A)**

Figure 7 illustrates a 16-bit comparator. Fewer adders or parts of adders may be used to compare smaller numbers. The circuit must be modified, however, if part of an adder is used and the carry output can be obtained in two ways:–

1. **With unused inputs at 0 the 'carry' output will be found at the sum output following the last bit, e.g. if two inputs are used (A_1, A_2, B_1, B_2) the 'carry' output will be found at the Σ_3 output.**

Figure 7 illustrates a 16-bit comparator. Fewer adders or parts of adders may be used to compare smaller numbers. The circuit must be modified, however, if part of an adder is used and a carry output can be obtained in two ways:–

1. With unused inputs at 0 the 'carry' output will be found at the sum output following the last bit, e.g. if two inputs are used (A_1, A_2, B_1, B_2) the 'carry' output be found at the Σ_3 output.

2. '1's may be added in the remaining A inputs and 0's at the remaining B inputs to bring the carry output to the normal C_4 output. This method will provide a resultant output faster than the previous one.

By adding additional logic to sense when all the Σ outputs are '0', the condition A = B can be detected. The required 16 input AND gate is formed from 4, 4 input NAND gates and a 4 wide 2–2–2–2 input AND OR INVERT GATE.

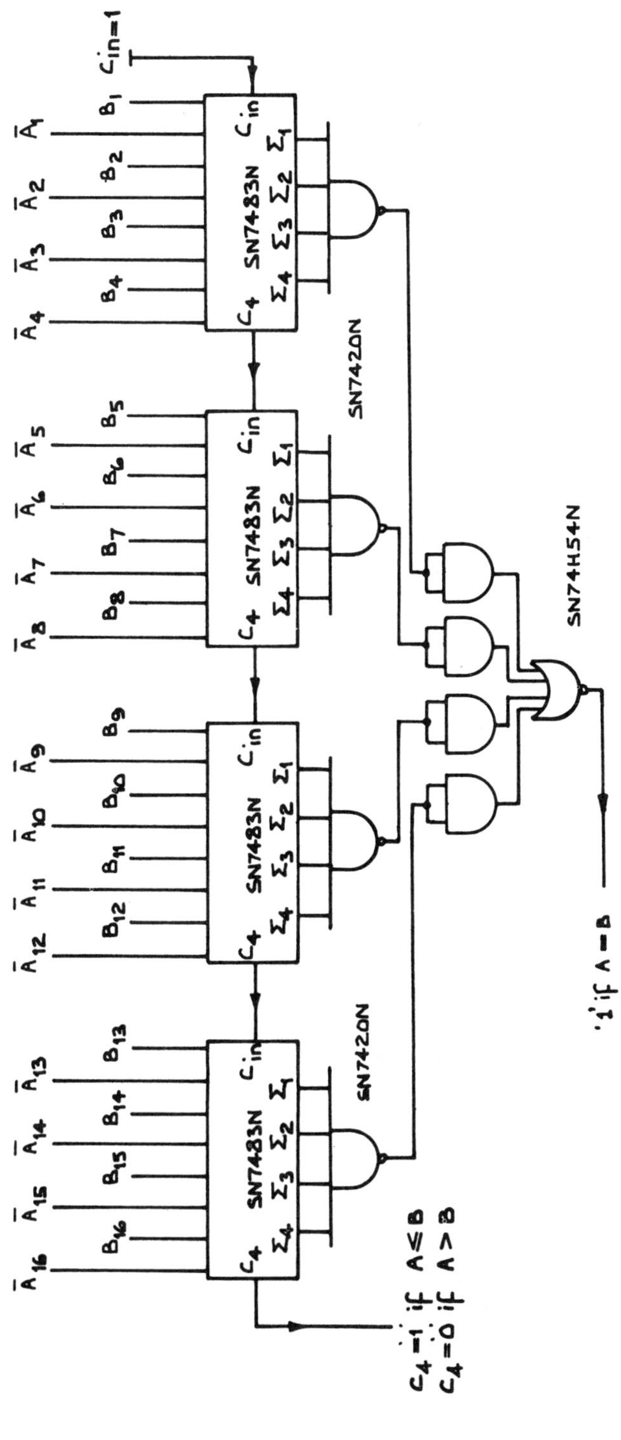

FIGURE 7 16—Bit Comparator

SERIAL GRAY TO BINARY CONVERTER

The logic required to convert Gray code to binary is easily determined if the Gray code is available in parallel form. The conversion from serial Gray to binary is not so easily accomplished.

However, some forms of Gray code, i.e. reflected binary, have properties which simplify the conversion. Figure 8 shows the most common of these codes.

	G_4	G_3	G_2	G_1
0	0	0	0	0
1	0	0	0	1
2	0	0	1	1
3	0	0	1	0
4	0	1	1	0
5	0	1	1	1
6	0	1	0	1
7	0	1	0	0
8	1	1	0	0
9	1	1	0	1
10	1	1	1	1
11	1	1	1	0
12	1	0	1	0
13	1	0	1	1
14	1	0	0	1
15	1	0	0	0

FIGURE 8. Reflected Binary Gray Code

If the most significant digit is disregarded, numbers 8 to 15 will be found to be a 'mirror image' of numbers 0 to 7.

It can be shown that if R_j is the jth bit in the reflected code and B_j is the corresponding bit in the equivalent binary number, then

$B_n = R_n$ **where n is the most significant digit**

$$B_j = B_{j+1}\cdot\overline{R}_j + \overline{B}_{j+1}\cdot R_j$$

(These are the equations commonly used for parallel conversion).

The second equation may be interpreted as a difference equation defining the operation of a bistable which is carrying out the serial Gray to serial binary conversion.

The logic diagram, truth table and wiring diagram for the Serial Gray to Binary Converter are shown in Figures 9A, 9B and 9C.

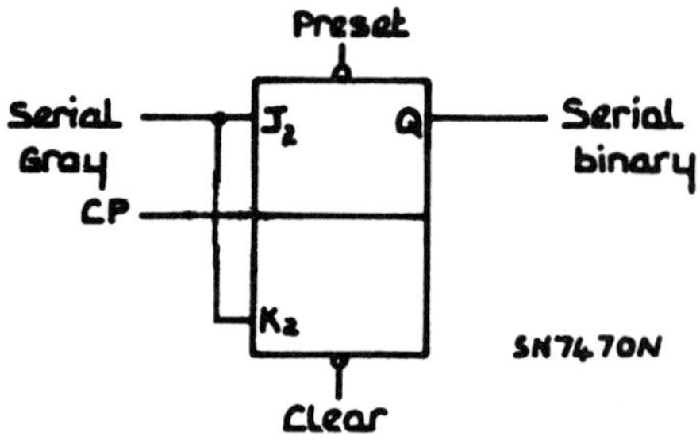

FIGURE 9A Logic Diagram

t_n		t_{n+1}
J	K	Q
0	0	Q_n
0	1	0
1	0	1
1	1	$\overline{Q}_n$

FIGURE 9B Truth Table

An SN7470N single-phase J–K flip-flop is used to perform the conversion. The J* and K* inputs are at 0, the J_1 and K_1 inputs are at 1 and the serial Gray code is applied to the J_2 and K_2 inputs, the most significant digit first. The bistable must be cleared to zero before the conversion commences.

Example

Gray 9 = 1101. Reference to the Truth Table will show that the Q output for input 1101 is 1001, i.e. binary 9.

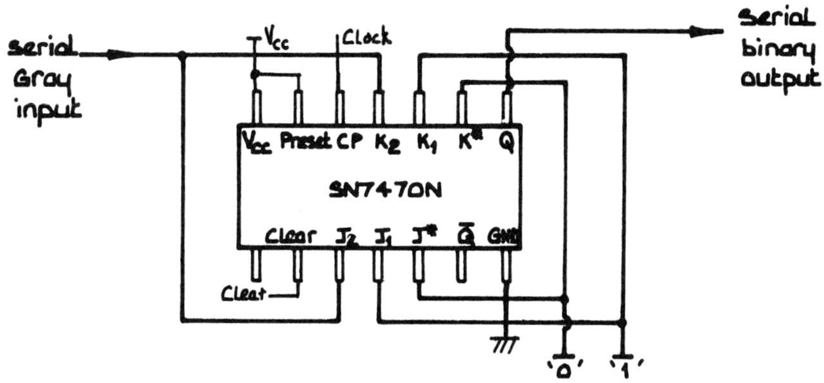

FIGURE 9C Serial Gray to Binary Converter

The same principle may be applied to a parallel Gray to serial binary converter. Figure 10 illustrates the circuit configuration.

The shift register of SN7474N D-type flip-flops is cleared and then the parallel Gray input is transferred into the register. The number is then shifted, most significant digit first, into an SN7470N J–K flip-flop which converts to serial binary at the Q output.

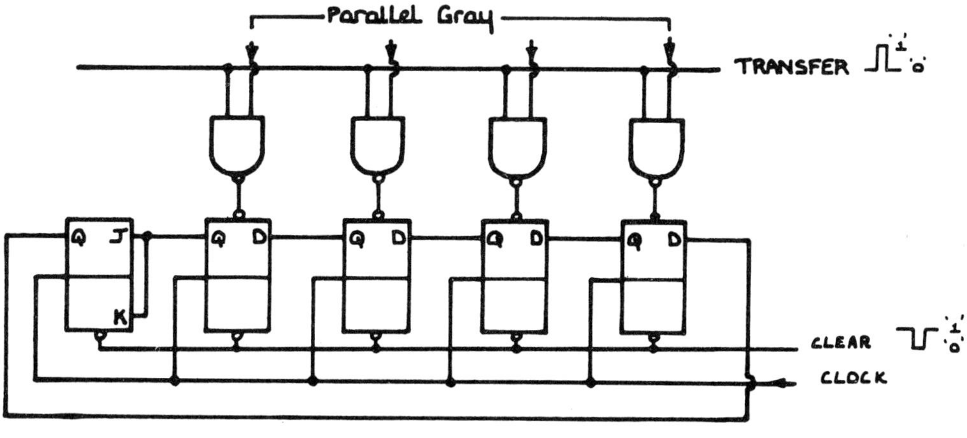

FIGURE 10 Parallel Gray to Serial Binary Converter

DECADE COUNTER WITH UNCONDITIONAL PRESET FACILITY

This section describes a high speed, asynchronous, BCD decade counter with strobed information entry that is unconditional on the state of the count.

The basic counter comprises four D-type flip-flops, the first dividing by two and the remaining three connected to divide by five.

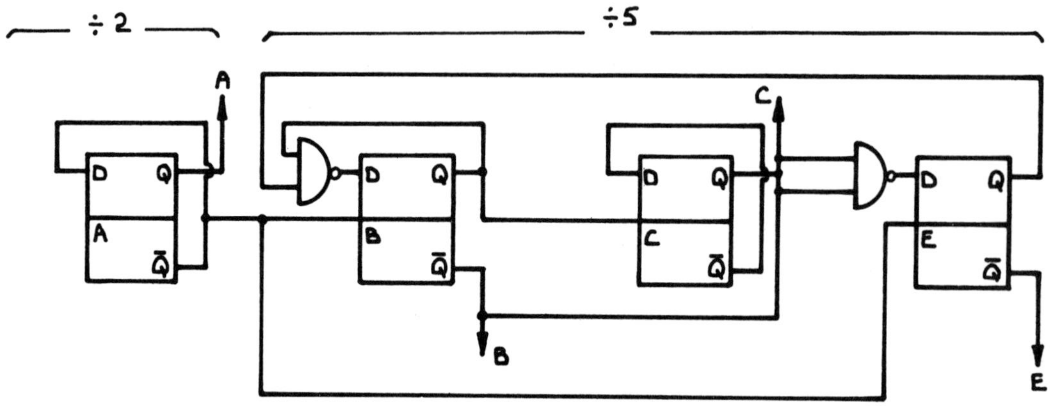

FIGURE 11

Considering first the divide-by-five section, bistables B.C and E, the required truth table (Figure 12) is as follows:–

C.P.	B_1	C_2	E_4
0	0	0	0
1	1	0	0
2	0	1	0
3	1	1	0
4	0	0	1
0	0	0	0

FIGURE 12

Stage B divides by two except when state 001 is reached. Output B_1 then remains at a '0'. A D-type flip-flop is used with the $\overline{Q}$ output fed back to the input as shown in Figure 13A. Since the operation of this unit must be modified when the E_4 output is a '1', the $\overline{E}_4$ output should be fed back via an AND gate so that as $\overline{E}_4$ goes to '0' the D input of B_1 goes to '0' and its output becomes a '0' on the following clock pulse. However, since only NAND gates were available in the Texas Instruments range this type is used and the B_1 and $\overline{B}_1$ outputs are interchanged as shown in Figure 13B. The B_1 output therefore appears at the Q output of the flip-flop.

Stage C operates as a straight divide-by-two stage, clocked when B_1 goes from '1' to '0', i.e. clocked by $\overline{B}_1$.

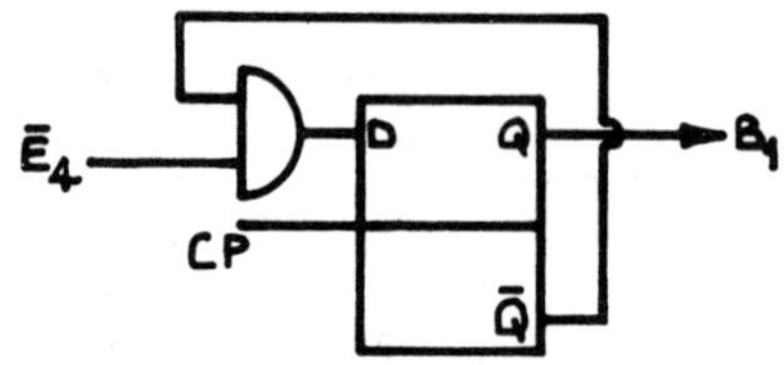

FIGURE 11A

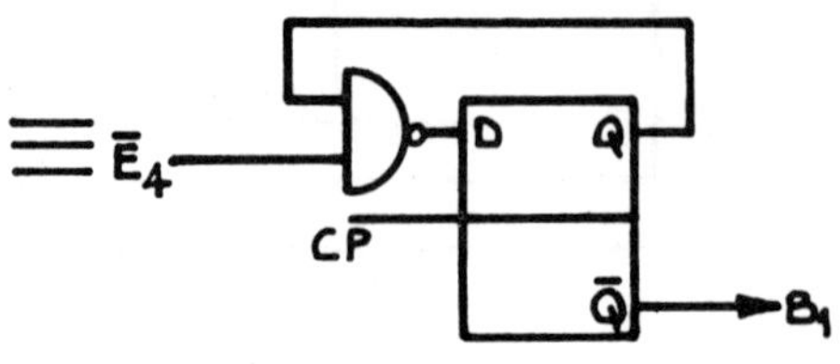

FIGURE 11B

Stage E is clocked from the input since E_4 = '1' for one clock period only, i.e. the period following the state 110. This stage could have its D input steered by B.C. However, since NAND gates are used the Q output must be used as E_4 and the D input must be steered by B.C.

Stage A, which divides by two, is added to produce a decade counter.

When considering the preset and clear arrangement it must be borne in mind that in D-type flip-flops a '0' on the preset input sets the Q output to a '1' and a '0' on the clear input sets the Q output to a '0'. The circuit shown in Figure 13 allows information to be entered into the counter upon application of a clock pulse without first clearing it. A clear line is provided to allow all outputs to be set to zero if required. The circuit operates as follows:

A '1' at Input A results in a '0' at the output of gate 1A upon application of a strobe pulse. The A output is therefore set to a '1'. A '0' at Input A gives a 1 at the output of gate 1A. This is ANDed with the strobe pulse and inverted by 2A to provide a '0' at the clear input of A. The A output is therefore set to a '0'.

Information is entered into bistable C in the same way as A. A '1' at Input B results in a '0' at the output of gate 1B so that the output is set to a '1'.A '0' at Input B gives a 1 at the input of 2B resulting in a '0' at the present input of B. The B output is therefore set to a '0'.

Information is entered into bistable E in the same way as B.

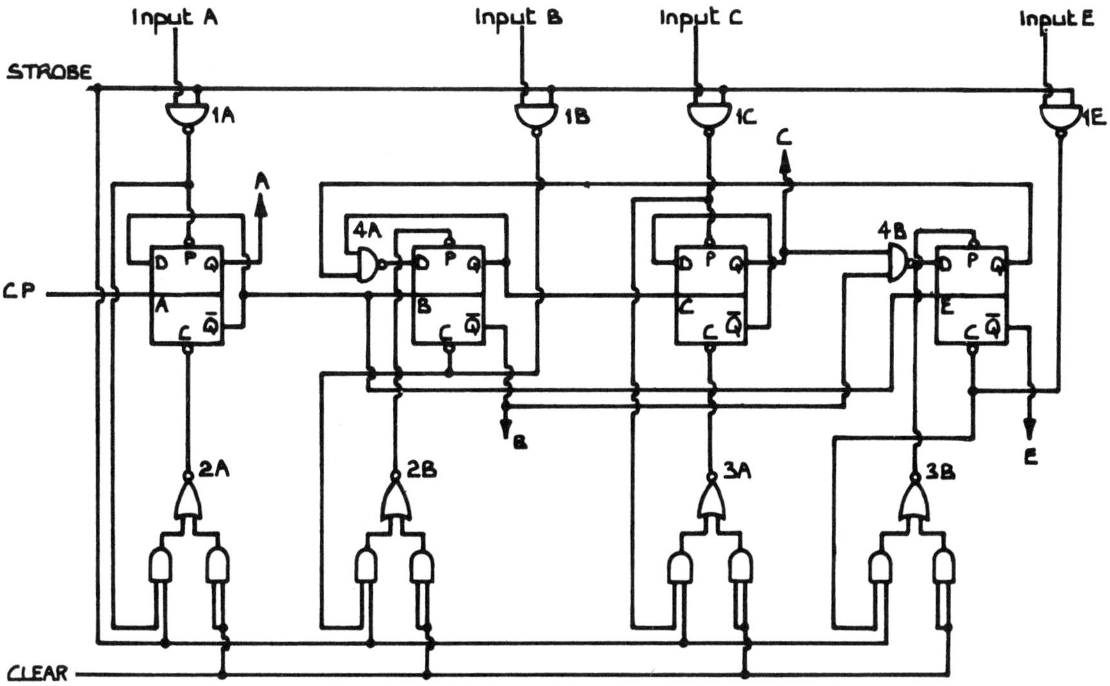

FIGURE 13

Components

Gates 1A, 1B, 1C, 1D		SN7400N Quad Two-input Gate
Gates 4A, 4B	½ x	SN7400N
Bistables A, B, C, E	2 x	SN7474N Dual D-type Flip-flops
Gates 2A, 2B, 3A, 3B	2 x	SN7451N Dual AND-OR INVERT Gates

DIVIDING BY NUMBERS OTHER THAN TEN WITH THE SN7490N

This section describes the use of the SN7490N decade counter for division ratios other than ten. By using appropriate external connections, the counter can be made to divide by a considerable number of indexes, many of which still give the output in true binary coded decimal (BCD).

No external interconnections are needed, of course, if either a divide by two or a divide by five is required. Flip-flop A is used as a binary element for the divide by two function and independently the BD input is used to obtain binary divide by five at the B, C and D outputs.

The counter has gated reset lines so that count inputs can be inhibited and all outputs returned to a logical zero or to a BCD count of nine. The count/reset tables for the SN7490N are shown below in Figure 14 (An X indicates that either a logical 1 or logical 0 may be rpesent).

Count	Output BCBA	Reset/Count R_{01}	R_{02}	R_{91}	R_{92}	Output DCBA
0	0000	1	1	0	X	0000
1	0001	1	1	X	0	0000
2	0010	X	0	1	1	1001
3	0011	0	X	1	1	1001
4	0100	1	1	1	1	1001
5	0101	X	0	X	0	Count
6	0110	0	X	0	X	Count
7	0111	0	X	X	0	Count
8	1000	X	0	0	X	Count
9	1001					

FIGURE 14

This shows that to reset to zero (all '0's) both the R_{01} and R_{02} must be logical '1's, i.e. $R_0 = R_{01}.R_{02}$, and to reset to nine, both $R_{9.1}$ and $R_{9.2}$ must be logical '1's, i.e. $R_{9.1}.R_{9.2}$.

By using the R_0 or R_9 facility, a single SN7490N can be modified to divide by any number from three to nine inclusive. Since the counter is monolithic, all four bistables require approximately the same time for setting or re-setting. This means that if any one bistable has been reset, the remaining three will have also been reset. Sufficient propagation delay exists internally to ensure that any other counter which is reset from the same gate inputs will be reset correctly.

To modify the count to a BCD divide by nine, the counter can be reset to 0000 as soon as the state 1001 has been reached, i.e. R_0 = A.D. The count sequence for such a counter is shown in Figure 15A and the logic diagram in Figure 15B.

Count	DCBA	
0	0000	
1	0001	
2	0010	
3	0011	
4	0100	
5	0101	
6	0110	
7	0111	
8	1000	
9T	1001	where 9T is a transition state
	0000	

FIGURE 15A

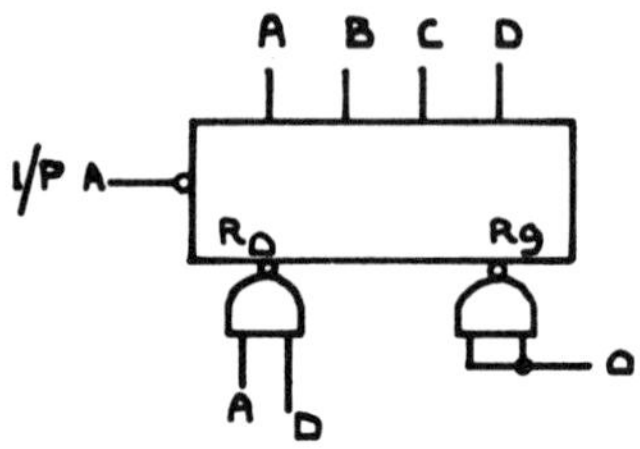

FIGURE 15B

For the divide by nine shown, outputs B, C and D can be used. However, care must be taken with the A output since a transitional state occurs before the counter is reset (see Figure 16).

Applying the same principle the counter can be modified to any other numbers from three to eight inclusive. BCD outputs are obtained with all numbers except seven, e.g.

	BCD ÷ 8	÷ 7	BCD ÷ 6	BCD ÷ 4	BCD ÷ 3
	DCBA	DCBA	DCBA	DCB	DCB
0	0000	0000	0000	000	000
1	0001	0001	0001	001	001
2	0010	0010	0010	010	010
3	0011	0011	0011	011	011
4	0100	0100	0100	T 100	000
5	0101	0101	0101	000	
6	0110	T 0110	T 0111		
7	0111	1001	0000		
8 T	1000	0000			
	R_0=D, R_9=0	R_0=0, R_9=CB	R_0=CB, R_9=0	R_0=D, R_9=0	R_0=CB, R_9=0

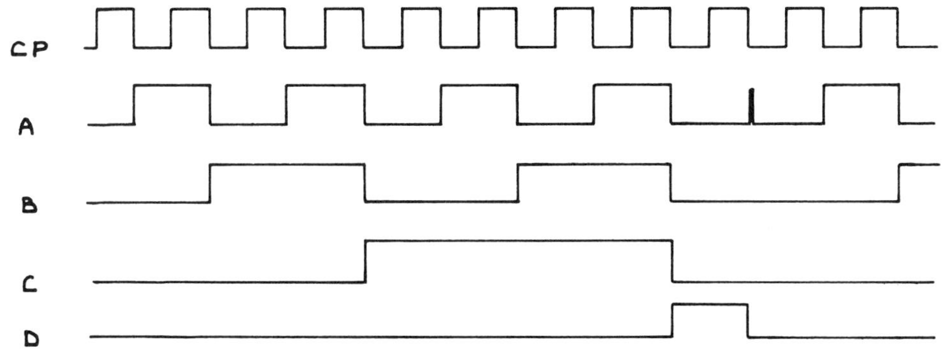

FIGURE 16

Division ratios greater than ten can be performed with two decades in two ways:–

1. A desired division ratio can be achieved by using the R_0 gates, if it can be recognised from no more than any two outputs from the two decades. A BCD count is obtained. For example, BCD divide by 84:–

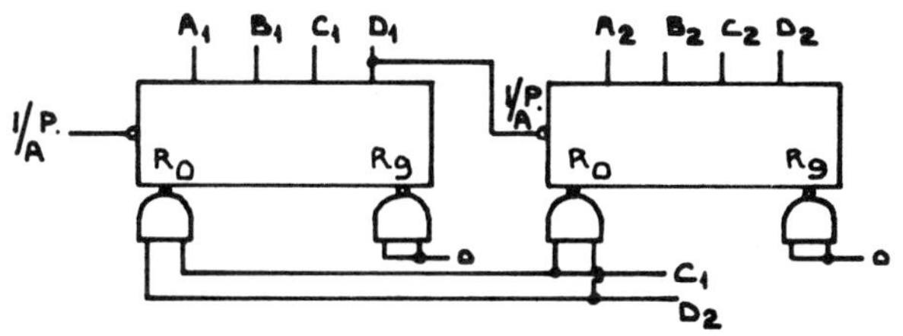

   The two decade counters count normally until state eighty-four is reached, i.e. 0100, 1000. This is detected by the R_0 gates and the counters reset to zero. Typical division ratios obtained by this method are 14, 18, 20, 24, 28, 42, 44, 48, 81, 82, etc.

2. If the desired division ratio can be split into its lowest common multiples and these can be obtained by any of the forementioned methods, then the desired ratio can again be obtained although the count will not be BCD, e.g. 54 can be produced by a divide by nine and a divide by six.

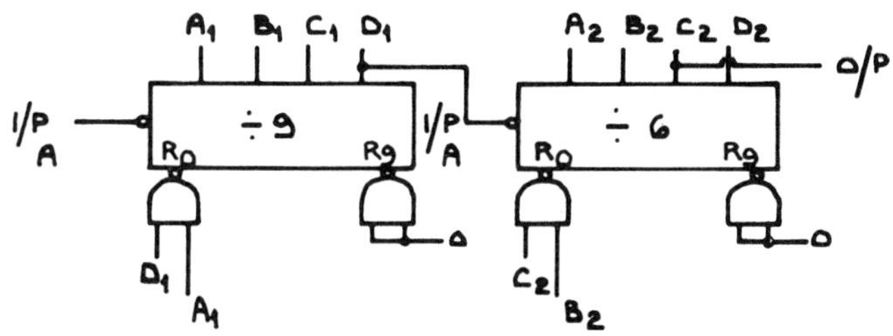

* All the circuits in the above sections cannot be guaranteed to operate successfully at temperature extremes due to differential propagation delays.

TWENTY-FOUR HOUR CLOCK USING ONLY SN7490 DECADE COUNTERS

This twenty-four hour clock uses only SN7490N decade counters for its various divider stages. Division ratios by numbers other than twn are accomplished by a reset method as described in the previous section.

The block diagram for the clock is shown in Figure 17, and the connection diagram in Figure 19.

If the SN74141N Nixie tube decoder/drivers are to be used for decoding, the divide by six and divide by twenty-four stages must count in binary coded decimal (BCD). The count sequence of the SN7490N is modified as shown in Figures 18A and 18B.

a) BCD coded divide by six

P. (I/P)	A B C D
1.	1 0 0 0
2	0 1 0 0
3	1 1 0 0
4	0 0 1 0
5	1 0 1 0
6 T	0 1 1 0
6	0 0 0 0

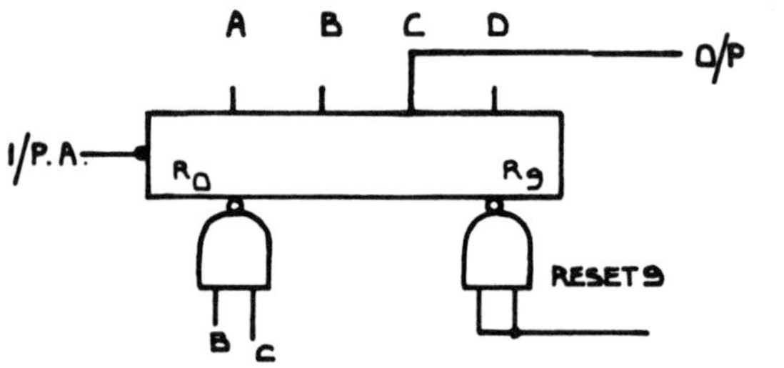

FIGURE 18A

The normal divide by ten sequence of the SN7490N is modified by resetting the counter to zero when it has reached a particular state, (0110 for divide by six.) The resetting is performed by the two input R_0 gates, i.e. R_0 = B.C.

b) BCD coded divide by twenty-four

P.	I/P	Units A_1	B_1	C_1	D_1	Tens A_2	B_2	C_2	D_2
1		1	0	0	0	0	0	0	0
2		0	1	0	0	0	0	0	0
3		1	1	0	0	0	0	0	0
4		0	0	1	0	0	0	0	0
:		:	:	:	:	:	:	:	:
22		0	1	0	0	0	1	0	0
23		1	1	0	0	0	1	0	0
24	T	0	1	0	0	0	1	0	0
24		0	0	0	0	0	0	0	0

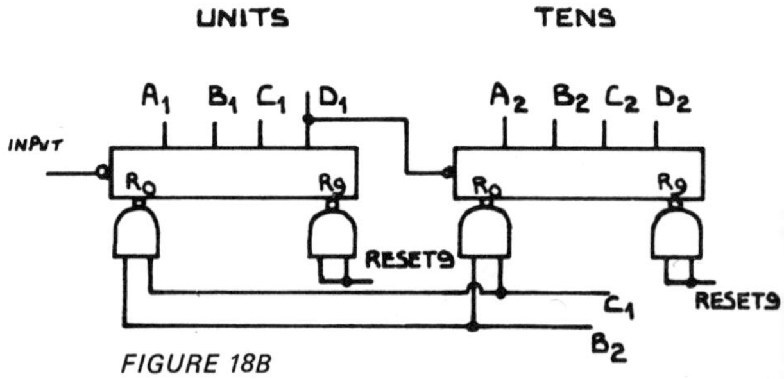

FIGURE 18B

The twenty-four state (0100, 0100) is detected by the R_0 gates and both decades reset to zero.

The 100kHz output from the crystal oscillator is counted down by counters SN1 – SN5 (divide by tens) and SN6, SN7 (divide by 60) to give one pulse per minute. The following stages SN8 and SN9 form a divide by 60 whose outputs are decoded by units SN12 and SN13 to display minutes and tens of minutes. The hours output from SN9 is counted by SN10 and SN11 which form a divide by twenty-four. These outputs are now decoded by SN14 and SN15 which drive the hours and tens of hours display tubes.

A manual reset switch is incorporated to allow all stages to be set to nine. The next input pulse will then reset the clock to zero.

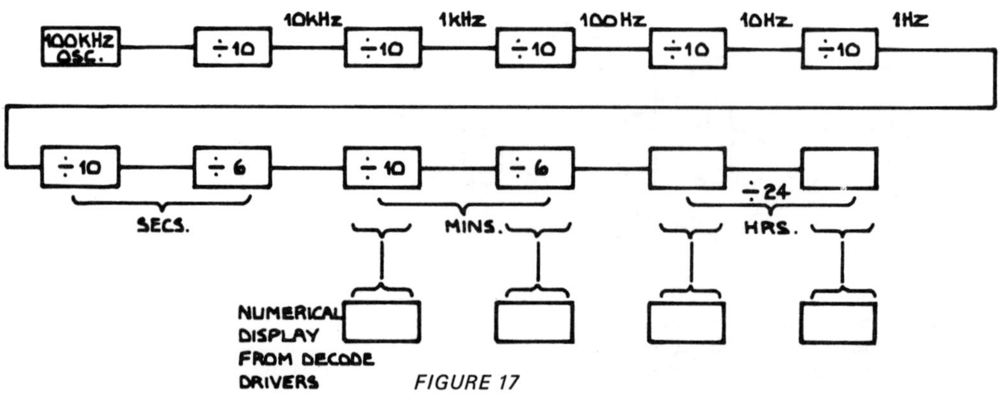

FIGURE 17

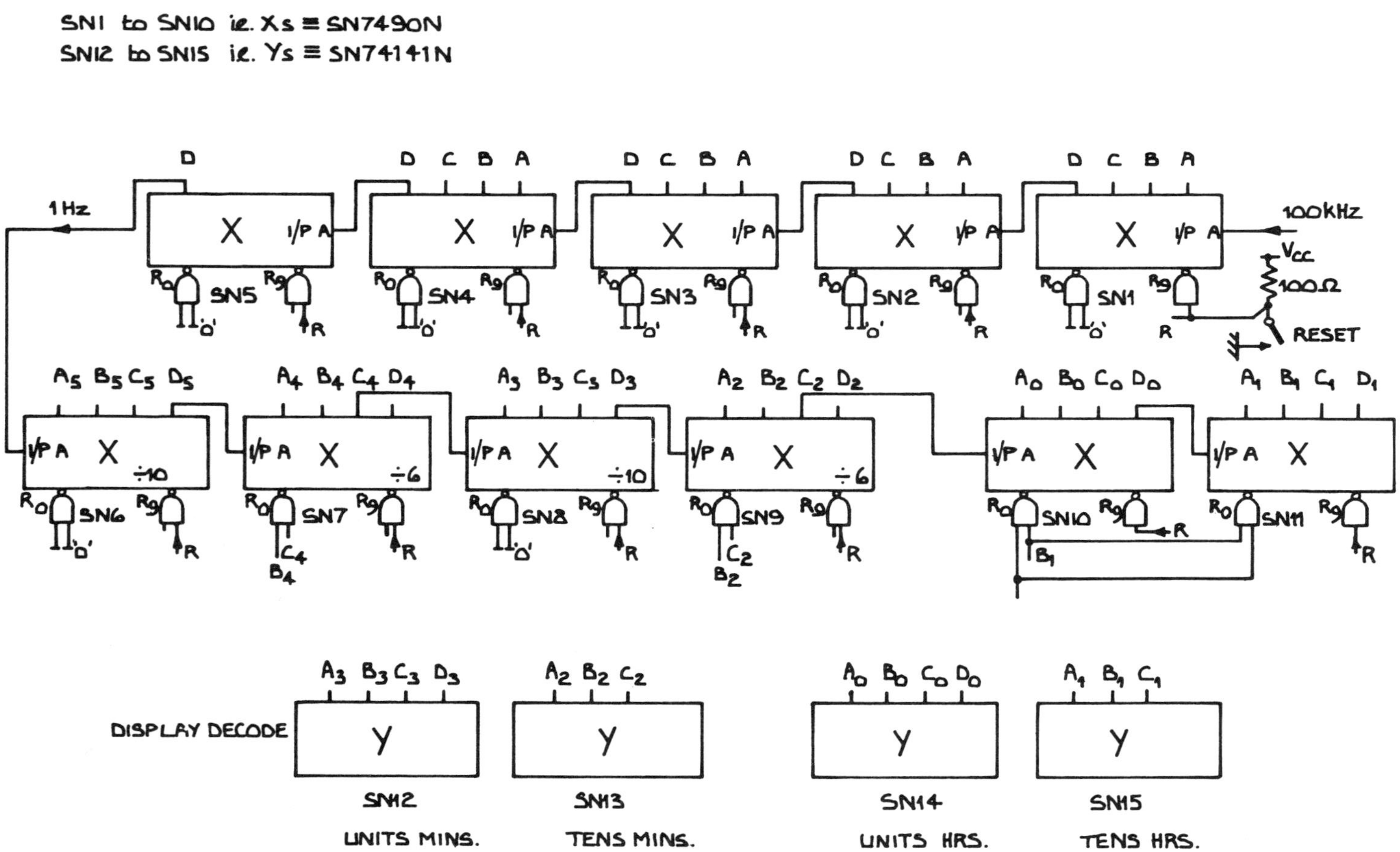
SN1 to SN10 ie. Xs ≡ SN7490N
SN12 to SN15 ie. Ys ≡ SN74141N
1 Hz
100 KHz
Vcc
100 Ω
RESET
SN1
SN2
SN3
SN4
SN5
SN6
SN7
SN8
SN9
SN10
SN11
÷10
÷6
DISPLAY DECODE
SN12
UNITS MINS.
SN13
TENS MINS.
SN14
UNITS HRS.
SN15
TENS HRS.

FIGURE 19

A DIGITAL PHASE METER

The phase meter described in here uses digital techniques to give the phase (measured as a time difference) between two input signals of the same frequency.

The two input signals are shaped in low hysteresis Schmitt trigger circuits using integrated circuit amplifiers type SN72702 (See Figure 20). The outputs of the Schmitt circuits are buffered with SN7400N two input gates, G1 and G2, used as inverters. Clamp diodes are included to prevent the outputs of the Schmitts going much above the supply voltage to the gates and damaging the input base-emitter junctions of the gates.

The outputs of the Schmitts are in antiphase due to the input configurations, so if ϕ_1 and ϕ_2 are the input signals, the output of the gate G3 will be $\phi_1 . \overline{\phi_2}$ as shown in Figure 21, i.e. the output goes to a zero when the leading input goes positive and returns to a '1' when the lagging input goes negative.

A 1MHz clock is gated with a 'measure phase' signal in gate G5. The outputs from gates G3 and G5 are NORed in gate G4 the output of gate G4 going to a '1' when all the inputs to gates G3 and G5 are at '1'. Binary coded decimal counters are driven from the output of gate G4.

On the application of a positive going step to the 'measure phase' input, the output of G7 goes to a '1' assuming that the input to G7 from flip-flop F1 is at a zero. (G6 and G7 form a bistable which remembers which input was last positive). This puts a '1' on the D-input of flip-flop F1. On the first negative excursion of ϕ_2, the '1' is transferred to the Q output of F1, thus opening gate G5. Clock pulses are fed to the counter chain when $\phi_1.\overline{\phi_2}$ is at a '1'. Since the clock rate is 1MHz, the counter chain will show the time difference between the two input signals in microseconds.

Since the Q output of flip-flop F1 is now at a '1', the bistable formed by G6 and G7 changes over putting a zero on the D-input of F1. On the next negative going excursion of ϕ_2 the zero is transferred from the D-input to the Q which unlocks the G6, G7 bistable ready for the next 'measure phase' pulse.

Flip-flop F2 is used to sense the sign of the time difference shown in the counters. If ϕ_1 leads ϕ_2, then the D-input of F2 will be at a '1' when ϕ_1 goes to a '1' and clocks the flip-flop. If ϕ_1 lags ϕ_2, then the D-input will be at a zero when ϕ_1 goes to a '1'. Under these conditions, the Q output will be at zero when ϕ_1 lags ϕ_2.

The potentiometers at the inputs to the Schmitts should be set to give a 1 : 1 mark space ratio at the output for an input signal of the order of 10mV.

Because of the fast risetimes present, care must be taken to ensure adequate supply decoupling. Figure 21 shows operating waveforms.

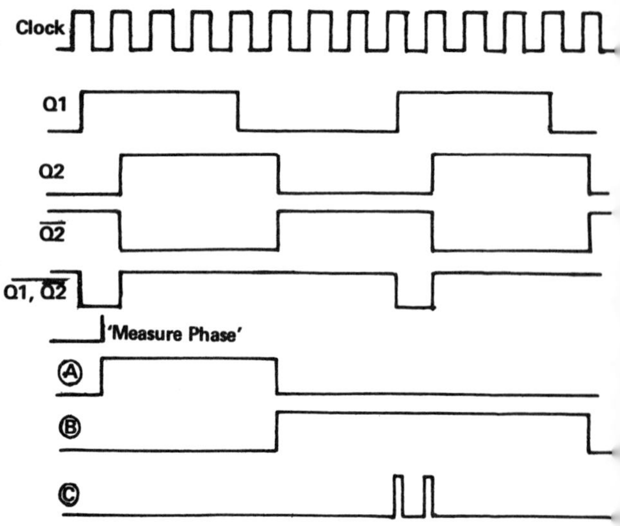

FIGURE 21

Performance

Hysteresis of Schmitt triggers = 5mV
Minimum slewing rate of signals at input = 13V
Risetime at output of Schmitts = 30ns.
Fall time at output of Schmitts = 70ns.
Maximum logic error = $\pm 1/2\mu$s.

Components

2	–	SN7270N	1	–	0.1 μF capacitors
1	–	SN7400N	1	–	50 μF capacitors
1	–	SN7402N	1	–	500μF capacitors
1	–	SN7474N	2	–	50Ω resistors
5	–	SN7490N	2	–	470Ω resistors
7	–	1S44	2	–	520Ω resistors
4	–	10pF capacitors	2	–	470kΩ resistors
1	–	25nF capacitors	2	–	10kΩ trimpots

Conclusions

The number of B C D counters required will depend on the lowest frequency and maximum phase difference to be measured. To be able to measure any phase difference of a 5Hz signal with a 1MHz clock, a five decade counter would be required.

The phase difference could be read out of the counters directly in degrees, say, if the oscillator was run at 180 times the frequency of the input signal.

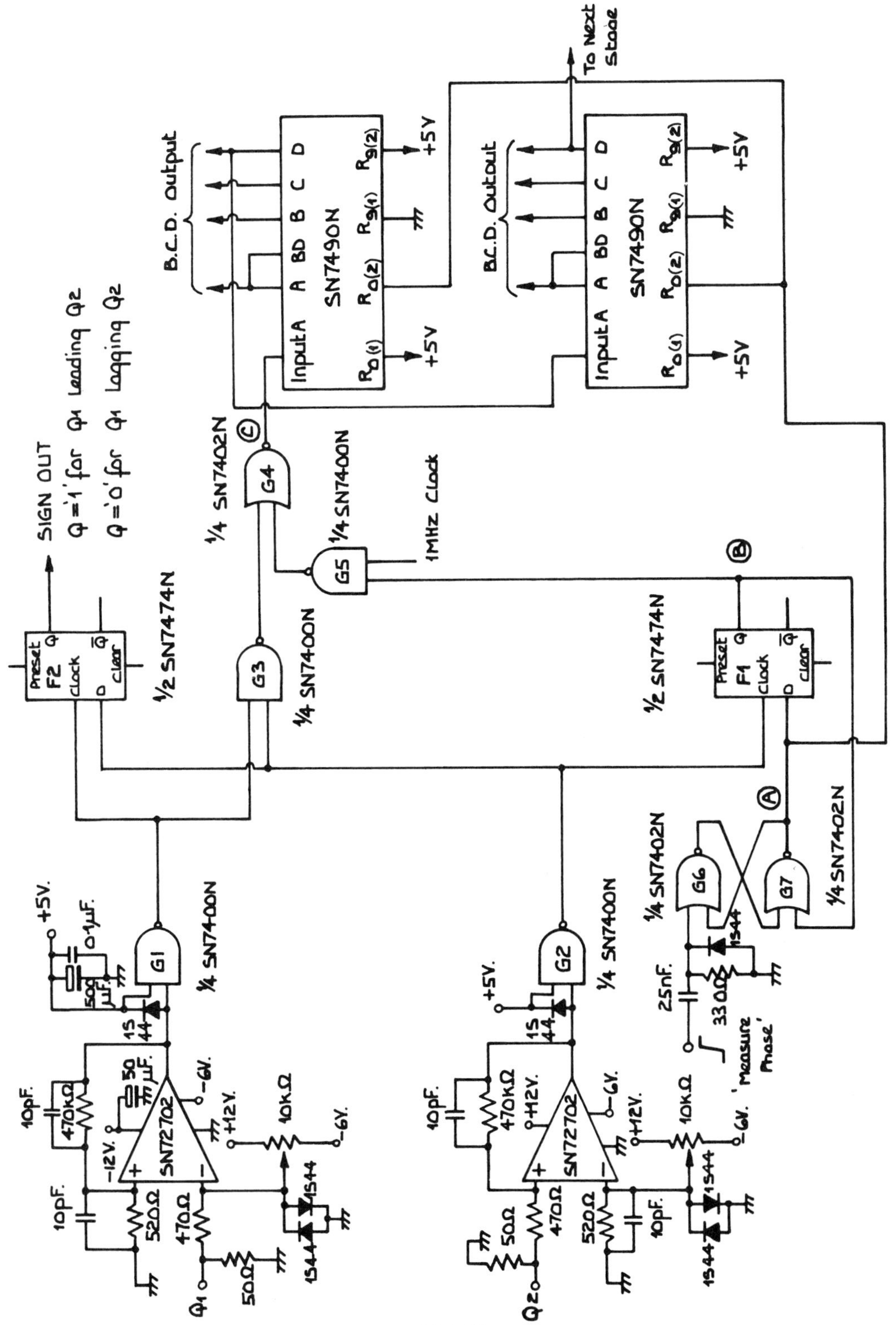

SIGN OUT
Q='1' for Q1 Leading Q2
Q='0' for Q1 Lagging Q2
B.C.D. Output
B.C.D. Output
To Next Stage
SN7490N
SN7490N
1/2 SN7474N
1/2 SN7474N
1/4 SN7400N
1/4 SN7402N
1MHz Clock
1/4 SN7402N
SN72702
SN72702
'Measure Phase'

FIGURE 20

XVI TTL I.C.S IN INDUSTRIAL NOISE ENVIRONMENTS

by Bob Parsons

Noise in electrical systems can be a difficult and elusive problem to define. This chapter has been written to act as a guide to users of TTL. Noise may be loosely defined as undesired electrical signals present in any electrical system. By careful design most noise problems can be alleviated before a system is built. Similarly, by systematic investigation, problems can be removed from pre-production equipment.

Noise considerations in digital systems may be summarised as shown in Figure 1.

Each topic shown cannot be considered in isolation, but must take into consideration the influence of other factors, for example, internal noise only becomes important if it is significant compared with data levels.

INTERNAL NOISE GENERATION

High frequency signal components due to fast transition times at the outputs of logic elements are the main source of internally generated noise. These components are still present when data rates are low, demanding that high frequency techniques should be used no matter what the pulse repetition rate.

As device logic levels change, circuit currents flowing also change. These changes are due to:–

(i) Different currents required to maintain the new logic level.

(ii) Transients due to lines charging and discharging.

(iii) Conduction overlap of the transistors in a TTL output stage.

Figure 2 shows the stray circuit elements associated with these transients.

Capacitance C_S is the capacitance to ground and surroundings of the output connection from gate G1 to the input of gate G2. L_S and L_g are the self inductances of the ground and supply lines respectively.

When the gate output changes from a logical '0' to a logical '1' capacitance C_S must be charged through inductance L_S from the supply. The charging current I_S being C_S. V/t_r where V is the voltage difference between the logical '1' and '0' levels and t_r is the voltage rise time at the gate output. For standard TTL with typical rise and fall times of 7ns or less, interconnection charging currents can be greater than 8mA per output. This together with conduction

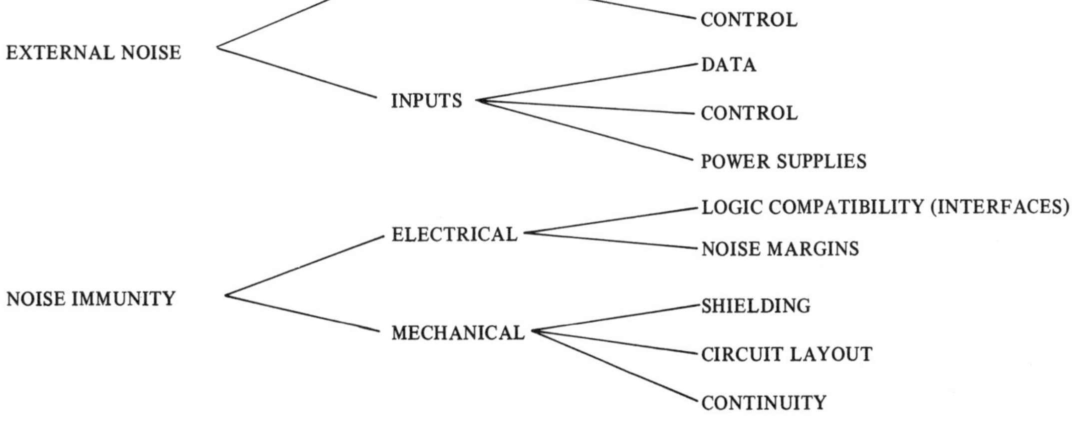

FIGURE 1

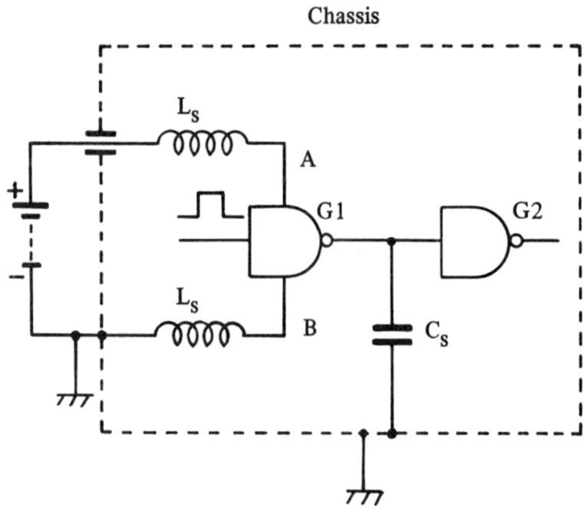

FIGURE 2

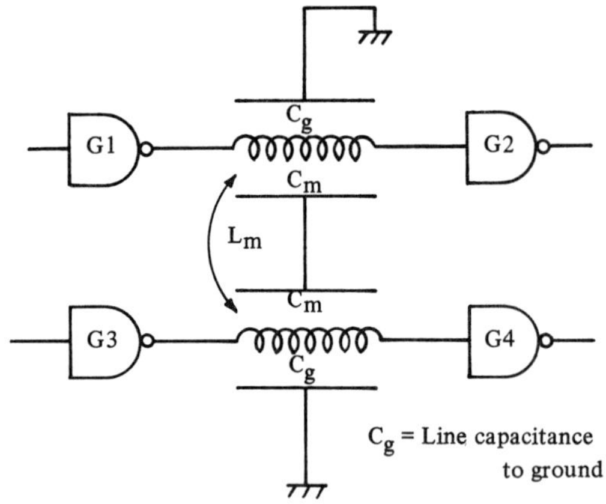

FIGURE 3

overlap in the output stage indicates that care should be taken with system layout so as to minimise supply line impedance and interconnection inductance.

In a practical system the following steps should be taken to minimise the generation of internal noise.

(i) The power supply distribution system should be of low impedance. This can be achieved by two methods:

  (a) A low impedance transmission line formed by a stripline above a ground plane – or line.

  (b) Medium impedance line with decoupling to ground at regular intervals by discrete R.F capacitors, e.g. disc ceramics. Decoupling should be carried out every 8 to 10 packages. Use 0.01μF to 0.1μF or 200pF per totem pole output, whichever is the greater value. These should be distributed throughout the circuit, not lumped together. Both ends of a long ground bus-bar should be returned to a common point.

(ii) Gates that drive lines must be decoupled at the package. The decoupling capacitor and transmission line ground should be commoned as near to the driving device ground as is practical.

(iii) Power gates formed by paralleling gates should utilise a single package with its decoupling capacitor.

Noise that has gained entry to equipment as well as that which is internally generated can couple into adjacent signal and power lines. This can take place through common impedances or as shown in Figure 3.

Here two lines carrying signals, or signal and noise, are in close proximity to one another. Cross coupling exists between these two lines due to mutual capacitance and inductance C_m and L_m. In order to analyse this configuration, it is convenient to consider C_m and L_m forming a mutual impedance Z_m between the two lines. if the line impedances are Z_o, the gate output impedances are Z_1, the logic swing at the output of gate G1 is V_s, and the coupled component of V_s at the input of G4 is V_{in}, then, as shown in the Appendix:

$$V_{in} = V_s/(1.5 + Z_m/Z_o)(1 + Z_1/Z_o) \quad \ldots\ldots \quad 1$$

The following practical conclusions can be drawn from equation 1.

1. To minimise cross talk between two lines their mutual impedance should be as high as possible and the line impedance as low as possible.
2. Such lines should be coaxial cables with low impedances Z_o and very little external field giving a high Z_m.

In practice the line impedance cannot be too low or the logic swing at the receiving gate will be restricted and line reflections will be significant compared with logic levels.

Stripline conductors above a ground plane can have controlled impedances giving excellent crosstalk properties.

Single wires, whilst being inexpensive and convenient, have poor cross-talk and reflection characteristics limiting their usable length to 25cm or less with standard TTL.

For lines less than 5m in length twisted pairs with Z_o between 50Ω and 150Ω are satisfactory provided that the sending gate is restricted to line driving. (This is explained in more detail in the section on driving transmission lines).

EXTERNAL NOISE

In any digital system provision must be made for the transmission of input and output data and control signals between the logic system and its peripherals. The current and voltage levels chosen for transmission are generally incompatible with TTL as they are dictated by considerations such as line impedance, noise environment, contact resistance, etc.

The influence of external noise on a TTL logic system is determined mainly by the type of interface circuits used to transfer signals into the logic system. These circuits cannot be designed effectively unless the input requirements of a TTL gate are understood.

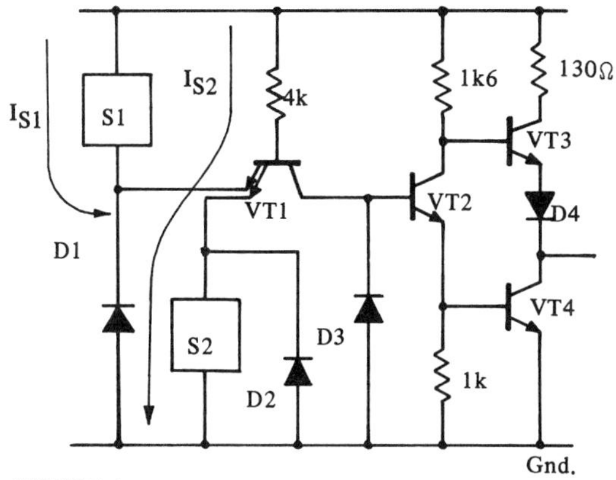

FIGURE 4

Figure 4 shows the circuit diagram of a standard TTL gate, where I_{S1} is the current to be 'sourced' and I_{S2} to be 'sunk' by the driver circuits S1 and S2.

Transistor VT1 is a multiemitter transistor with a 4kΩ base resistor to the supply voltage V_{CC}. If any input emitter is grounded base current flows out of the grounded emitter saturating transistor VT1 and Turning 'off' transistor VT2 allowing its collector to rise towards V_{CC}. The output of emitter follower transistor VT3 is, therefore, a logical '1'. If all input emitters of VT1 are high, i.e. logical '1', then its base collector diode is forward biassed and transistors VT2 and VT4 are saturated. The collector of VT4 is therefore at $V_{CE(sat)}$, i.e. logical '0'. Since the input transistor is now operating in an inverse mode with the collector to base diode forward biassed, an inverse emitter current, dependent upon inverse gain and base current, must flow into each emitter. The input voltage at which VT1 changes from normal to inverse mode of operation, i.e. the gate threshold voltage, is determined by its collector potential. This is defined by the V_{BE}s of VT2 and VT4, giving a typical threshold of 1.4V.

A typical transfer characteristic of the circuit of Figure 4 is shown in Figure 5.

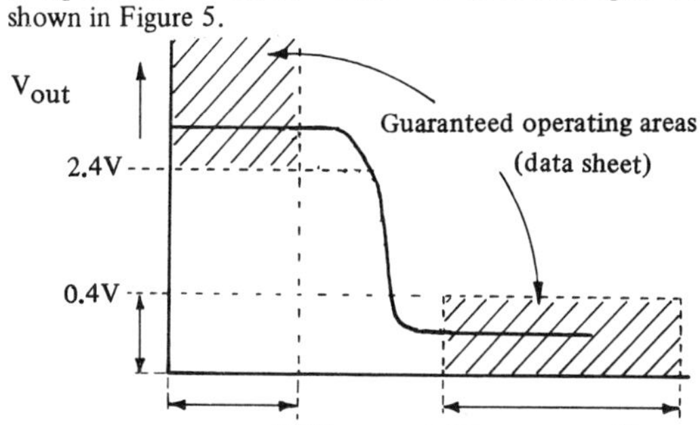

FIGURE 5

To obtain maximum noise immunity logical '0' and '1' input levels should be far removed from the gate threshold. This means that any interface circuit must be able to sink the base current of VT1 when in the logical '0' state and source the inverse emitter current when in the logical '1' state in order to maintain maximum noise immunity.

Interface Circuits

An interface circuit must perform the following operations:–

(i) Transform input logic levels to those of TTL. This must be achieved without loss of noise margin.

(ii) Filter out unwanted noise whose frequency components lie outside the signal frequency range. Noise that occurs within the signal frequency spectrum, i.e., comparable with data bit rate, can only be removed by logical means.

(iii) The interface circuit must take into consideration the transfer characteristics of the circuit that it is driving. The basic TTL gate is a saturating amplifier with a range of input voltages around its threshold where the gate will act as a high gain (> 55dB power gain at 10MHz.) high frequency linear amplifier.

The rise and fall times of data edges into TTL required for oscillation-free switching, are dependent upon a number of factors. These include the driving source impedance, gate loading, layout and supply decoupling. In practice, interface circuits should have rise and fall times (10 – 90%) <50ns between logical '0' and '1' levels.

Further problems arise if waveforms with slow edges are used to clock bistables such as the SN7474 dual D type. The clock noise immunity is degraded and the data hold time has to be increased to ensure correct data transfer.

Interface Filters

Suitable interface circuits consist of a low pass filter followed by an impedance converter. The low pass filter can be single or multiple section CR or LC. The impedance converter has to satisfy the conditions outlined above. An ideal converter is the SN7413N Schmitt trigger. Logically this device performs the same logic function as the SN7420 dual 4 input NAND gate but the upper and lower input thresholds are 1.7V and 0.9V typically.

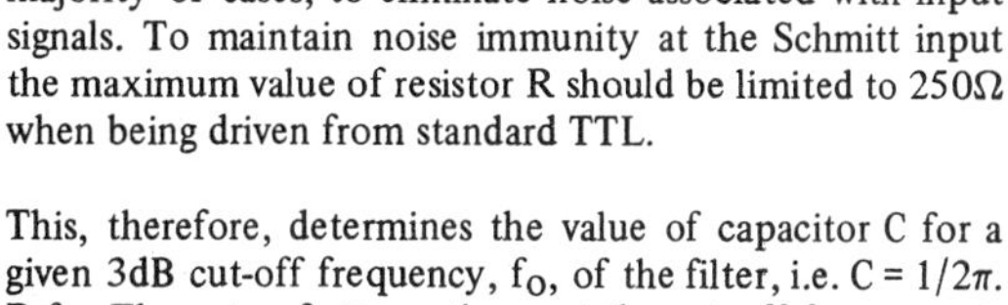

FIGURE 6

The filter shown in Figure 6 is all that is needed, in the majority of cases, to eliminate noise associated with input signals. To maintain noise immunity at the Schmitt input the maximum value of resistor R should be limited to 250Ω when being driven from standard TTL.

This, therefore, determines the value of capacitor C for a given 3dB cut-off frequency, f_o, of the filter, i.e. $C = 1/2\pi . R.f_o$. The rate of attenuation past the cut-off frequency is 6dB/octave. For larger rates of attenuation and faster response times multipole Chebyshev filters can be used. Since the input to the Schmitt is always positive with respect to ground, capacitor C may be a large value electrolytic.

Low pass filter, Chebyshev

These filters should be used when digital equipment is required to operate under conditions of extreme electrical noise. The Chebyshev filter is characterised by an amplitude frequency response that ripples in the pass band and falls off rapidly past the cut-off frequency f_o. The general form of this type of filter is shown in Figure 7.

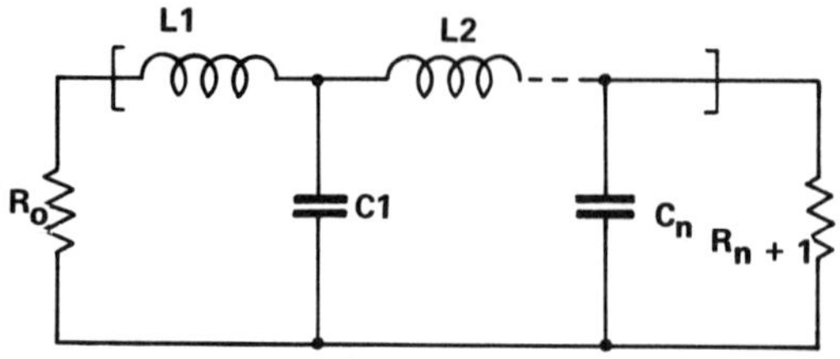

FIGURE 7

The amplitude frequency response of such a filter is shown in Figure 8.

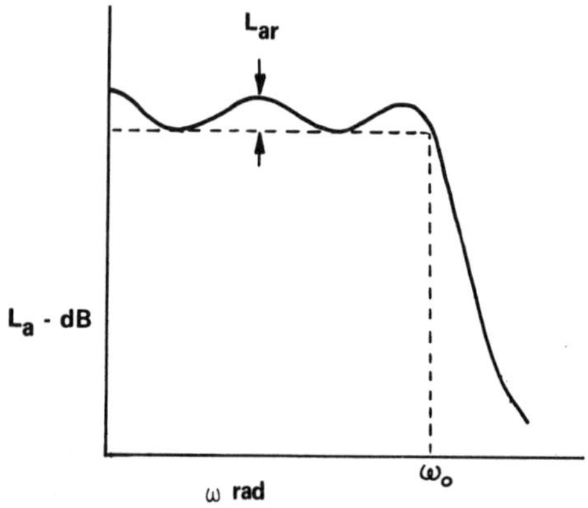

FIGURE 8

This circuit form is well suited to mechanical construction since the capacitive elements can be lead throughs. The required D.C. operating conditions for the Schmitt trigger are easily satisfied since the filter has almost zero impedance at D.C.

A compromise has to be made between the number of elements and the rate of attenuation past the cut-off frequency. In practice, a four element filter that has an attenuation of 60dB at $4\omega_o$ is usually satisfactory. The load impedance $R_n + 1$ is determined by the input impedance of the Schmitt trigger, which in turn determines the filter source impedance R_o. For this reason it is more practical to use a different filter design which will match a low source impedance to the input impedance of the Schmitt trigger (4kΩ approx.)

Such a design is shown in Figure 9 and component values are tabulated in Figure 9A for various cut-off frequencies.

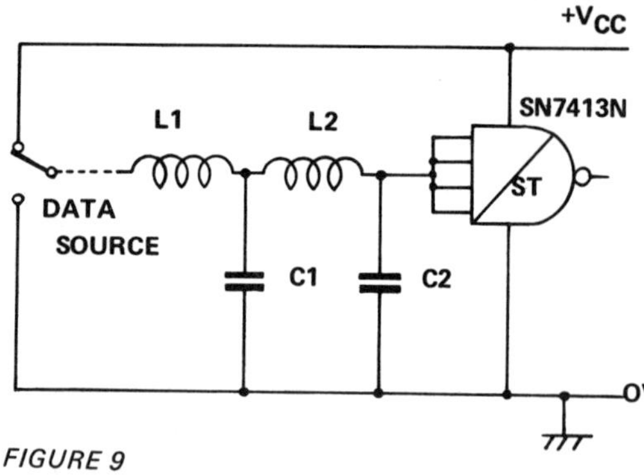

FIGURE 9

Element values for 4 element filter with L_{ar} = 1dB and L_a = −60dB at 4 ω_o.

f_o(kHz)	L1(mH)	C1(pF)	L2(mH)	C2(pF)
100	9.0	420	8.2	760
200	4.5	220	4.1	380
500	1.8	84	1.6	150
1000	0.9	42	0.8	76

FIGURE 9A

Since the inductive and capacitive components are $\propto 1/f_o$. element values for any cut off frequency f_o may be calculated from the tables given. Additional components may be added to the circuit of Figure 9 to enable it to handle very large voltage surges at the filter input as shown in Figure 10.

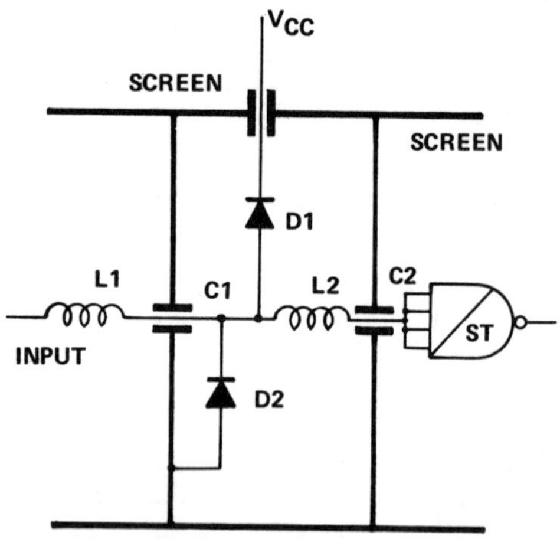

Constructional Form

FIGURE 10

Diodes D1 and D2 limit the input swing into the Schmitt trigger, they should be situated after inductor L1 so that the rate of rise of voltage across diodes D1 and D2 is limited.

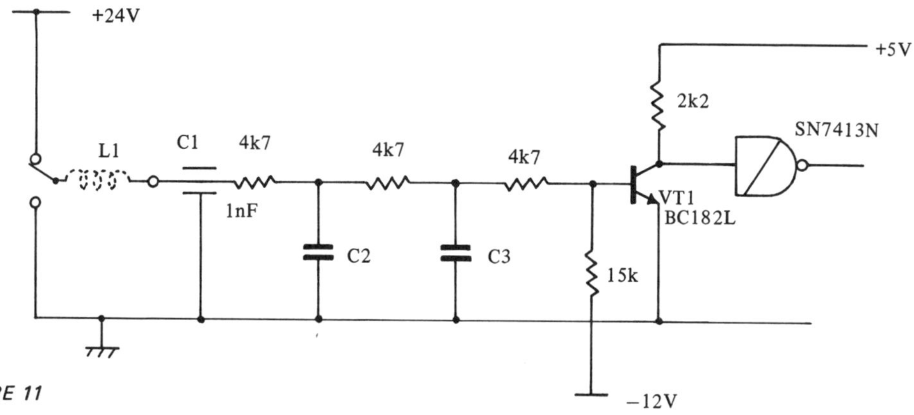

FIGURE 11

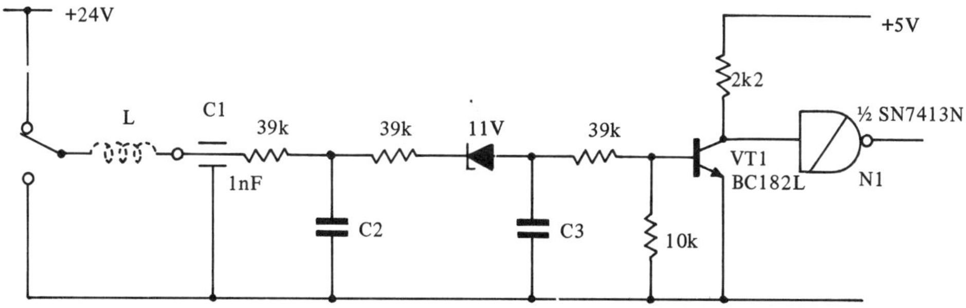

FIGURE 12

The previous filter circuits have logical threshold voltages that are the same as TTL. In many systems it is advantageous to have higher thresholds in order to overcome voltage drops down lines and contact resistance etc. Such circuits are shown in Figures 11 and 12.

Both circuits consist of CR low pass filters followed by a transistor buffer and a Schmitt trigger. Capacitor C1 is a lead through ceramic. This, together with L (lead inductance), forms a low pass filter removing high frequency components that would otherwise, due to layout, be very difficult to remove. The logic threshold for the circuit of Figure 11 is arranged to be 12V midway between the input logic levels, 0 and 24V. The values of C1, C2 and C3 should be chosen to give the required frequency response.

Both circuits have a series resistor between capacitor C3 and transistor VT1. This limits the rate of discharge of C3 through the base emitter diode of transistor VT1.

Optical Interfaces

In many industrial systems it is possible for very large transient common mode voltages to exist between equipments. In cases such as these an isolated coupler such as reed relays, transformers or optical devices must be used in order to transmit data. An alternative is to attenuate the signal and common mode offset until the common mode voltage is sufficiently low to be acceptable. The signal is then detected by a high gain differential comparator. This system does, however, have poor noise rejection unless common mode paths are carefully balanced.

An economical method giving a high degree if isolation (>1kV) and reasonable speed is the optical coupler, as shown in Figure 13.

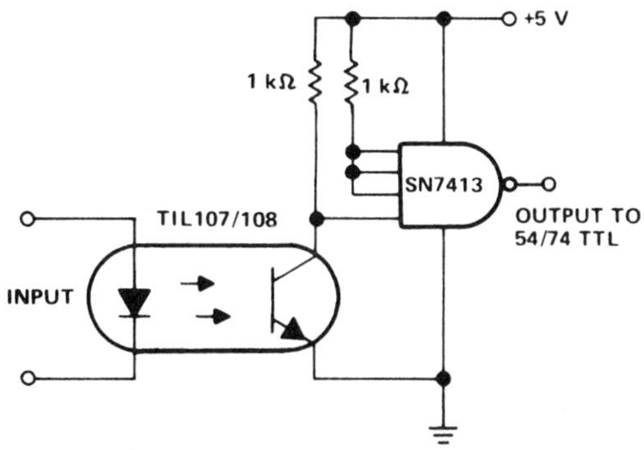

FIGURE 13

The optical isolator consists of an infrared emitter and receiver diode electrically isolated from each other, but maintained in the same package. The emitter diode forward current is in the order of 15mA to obtain a receiver current of 2mA (TIL108). A suitable buffer between the diode output and TTL input must be provided which in the simplest form would be a TTL Schmitt trigger. A maximum bit rate of approximately 60kHz can be obtained with the circuit of Figure 13. If higher bit rates are required then the TIL 107/8 should be followed by a wideband operational amplifier such as the SN7511 operated in the virtual earth mode.

DRIVING TRANSMISSION LINES

When the interconnections used to transfer digital information become long enough so that line propagation delay is equal to or greater than the pulse transition times, the effects of reflections must be considered. These reflections occur because most TTL interconnections are not terminated in their characteristic impedance. Reflections cause reduced noise margins, excessive delays, ringing, and overshoot. Some method must be used to analyse these reflections. Basic transmission line equations are applicable but unwieldy, since neither the gate's input nor output impedance is constant. Transmission-line characteristics of TTL interconnections can best be analysed by a simple graphic technique using Bergeron Diagrams.

Figure 15 shows a typical TTL interconnection. Open wires will usually have characteristic impedances of 150 to 200Ω depending upon the amount of capacitive coupling to ground. To evaluate a logical '1' to '0' transition, Figure 16 is used. The line $-1/Z_O(Z_O = 200\Omega)$ which represents the tranmission line is superimposed on the output characteristic curves. Since evaluation of a logical '1' to '0' transition is desired, the $-1/Z_O$ line starts at the point of intersection of the impedance curves of the input and output for a logical '1' condition; this point is labelled t_S. The slope $-1/Z_O$ then proceeds towards the logical '0' output curve. At time t_O, the driver output voltage is determined by the intersection of $-1/Z_O$ and the output 0 curve ($\simeq +0.25$V). The transmission-line slope now becomes $+1/Z_O$ and is

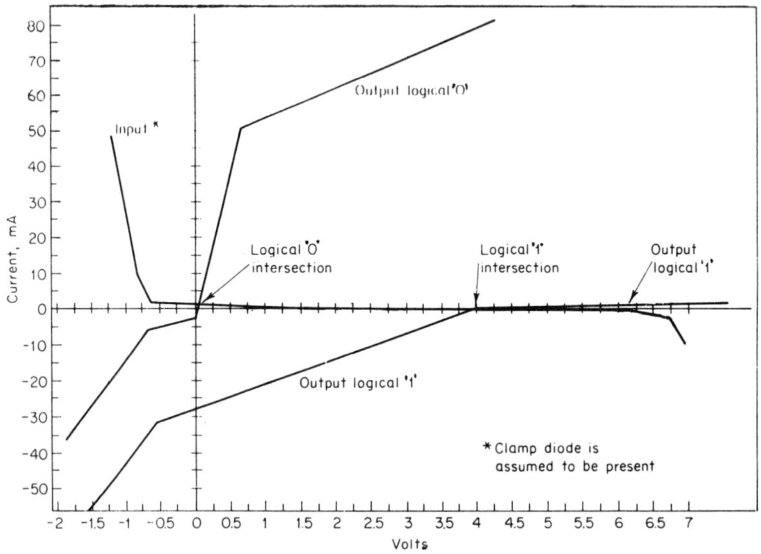

FIGURE 14 Output and input characteristics of a typical Series SN54/74 device at $T_A = 25°C$, $V_{CC} = 5.0V$.

Figure 14 shows plots of a gate input and both (logical '0' and '1') conditions of the output for a typical SN54/74 device. The output curves are plotted in the normal way with positive slopes, but the input characteristic is inverted, since it will be at the receiving end of a transmission line. The logical '0' and '1' intersections are indicated on the plot. These points are the steady-state values which will be observed on a lossless transmission line.

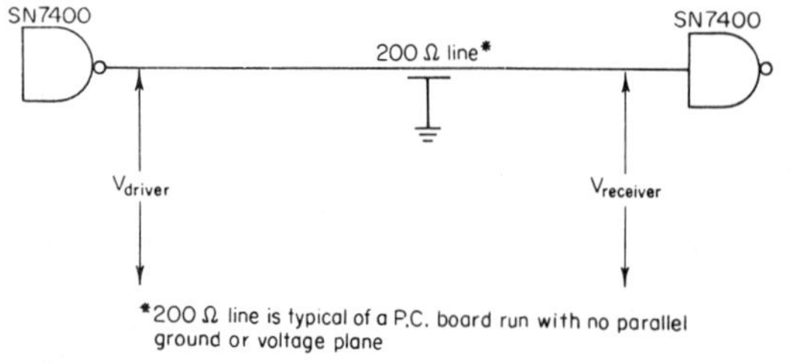

FIGURE 15

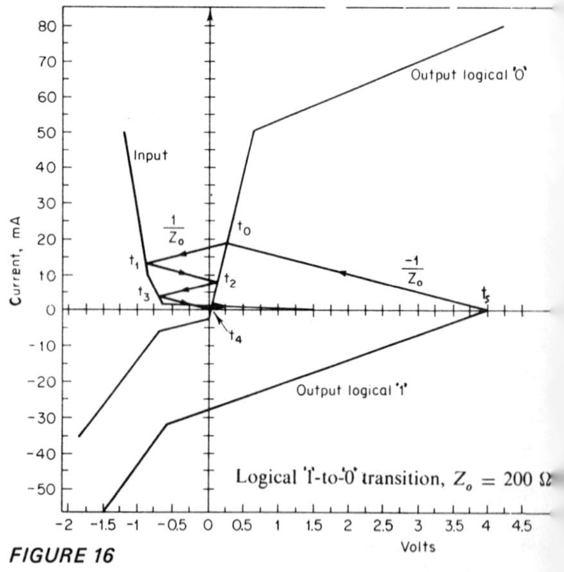

FIGURE 16

drawn toward the input curve. At time $t_1 (t_{n+1} - t_n =$ time delay of line), the receiving gate sees -0.75V. Now the line slope changes back to $-1/Z_o$, and the graph continues until the intersection of the input and output curves for a logical '0' is approached. Figure 17 plots driver and receiver voltages versus time in this case.

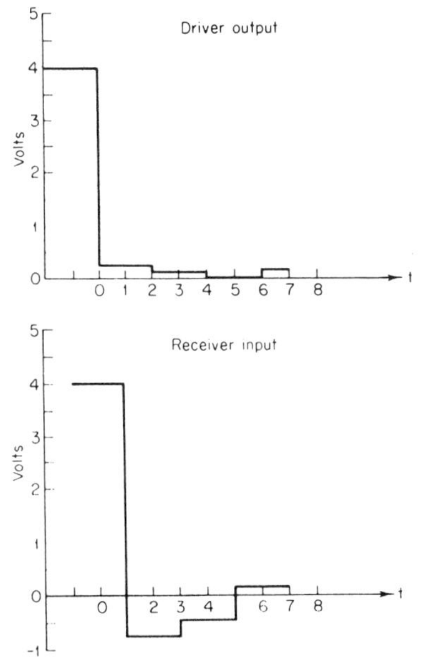

$Z_o = 200\ \Omega$ - logical '1'-to-'0' transition.

FIGURE 17

A logical '0' to '1' transition is treated in the same manner (Figure 18). The line $-1/Z_o$ now starts at the intersection for a logical '0', t_s. At time t_o the driver's output rises to +2.4V and at time t_1, the receiving gate's input goes to + 5.0V. Reflections then occur on the line because both ends are now terminated in a high impedance. This ringing decreases in amplitude as a steady-state condition is approached, see the insert of Figure 18. Both the output and input voltages are plotted in Figure 19 for this case.

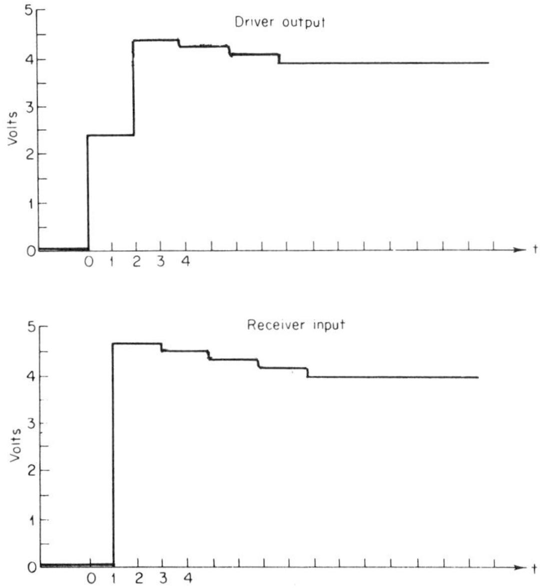

$Z_o = 200\ \Omega$-logical '0'-to-'1' transition.

FIGURE 19

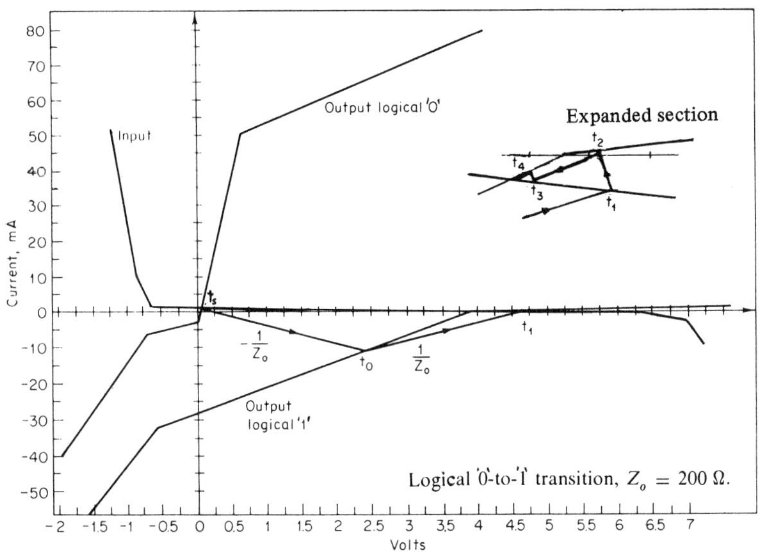

Logical '0'-to-'1' transition, $Z_o = 200\ \Omega$.

FIGURE 18

In Figures 20 to 27 a similar analysis is applied to 50Ω and 100Ω transmission lines. Notice that a decrease in line impedance does not significantly change the logical '1' to '0' transition characteristics. This is due to the low driver output impedance in the '0' state and the presence of an integral clamping diode on the input of the receiving gate.

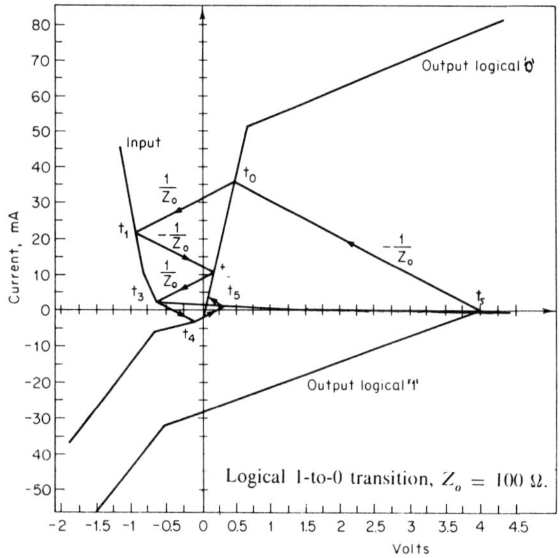

FIGURE 20

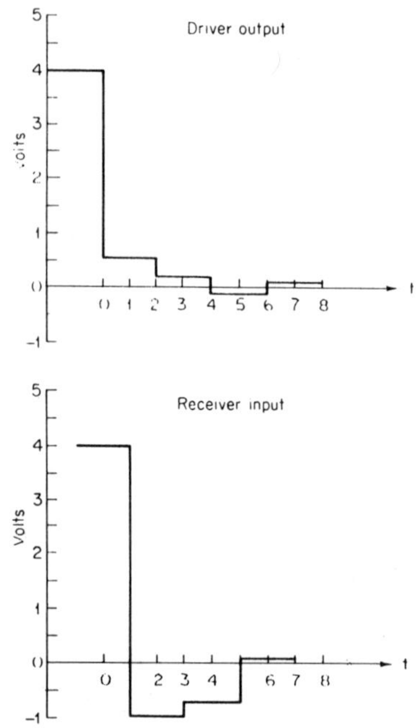

Z_o = 100Ω – logical '1' to '0' transition

FIGURE 21

Line impedance does, however, significantly alter the '0' to '1' transition characteristics. As the line impedance increases, overshoot at the receiving end also increases. This overshoot should be limited to no more than +5.5V to avoid emitter to emitter breakdown of the receiving gate. It is advisable to connect all emitters of a receiving gate together if an overshoot >5.5V is expected, i.e. use one receiving gate per line.

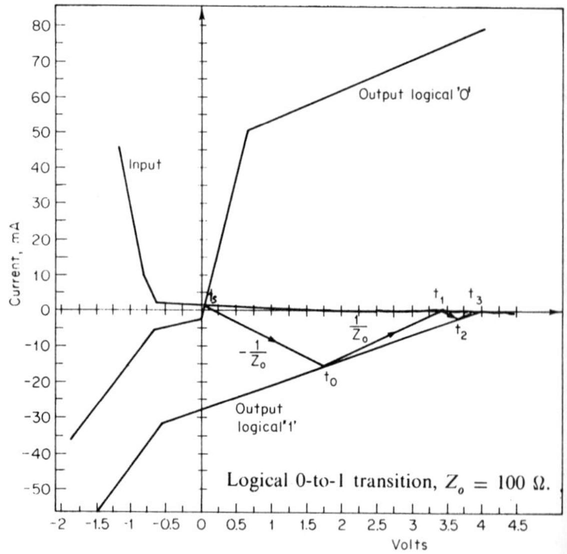

FIGURE 22

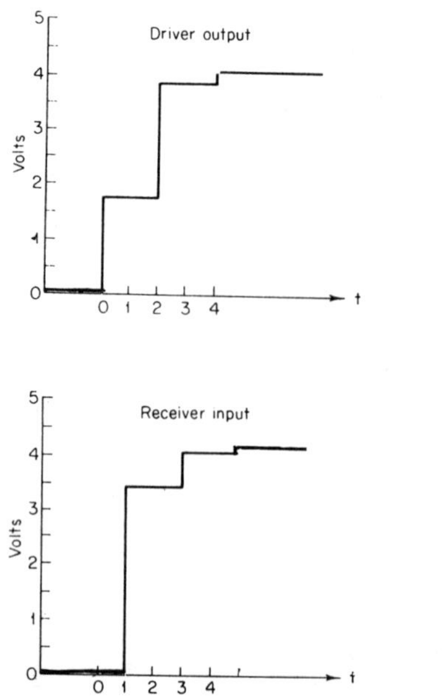

Z_o = 100Ω – logical '0' to '1' transition

FIGURE 23

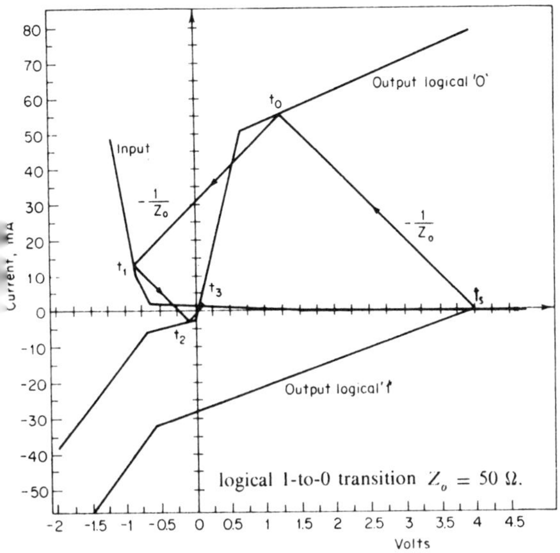

FIGURE 24

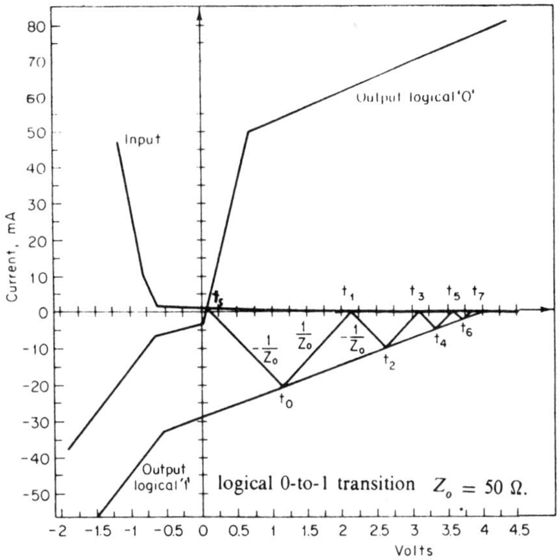

FIGURE 26

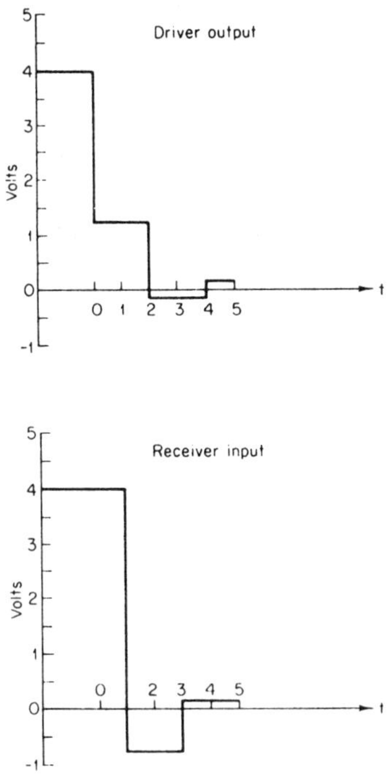

Z_o = 50Ω – logical '1' to '0' transition

FIGURE 25

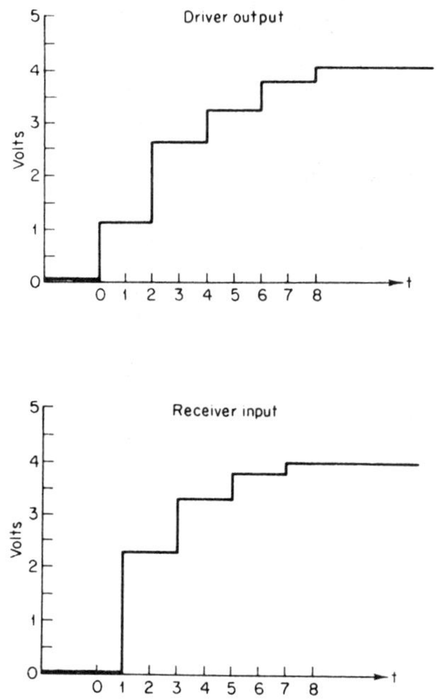

Z_o = 50Ω – logical '0' to '1' transition

FIGURE 27

Low impedance lines significantly affect noise margins of gate B placed as shown in Figure 28. Note that receiver B will see only +1.25V for two times delays of the transmission line. (Figure 25). Its output can therefore not be guaranteed until time t_2, or two transmission line delays.

The oscilloscope photographs of Figure 29 show the effectiveness of the graphical techniques. In most cases, the calculated and experimental values of voltage steps agree very well. The ringing that appears for the open wire is not immediately obvious, but this is because the input and output curves in this region lie practically along the positive horizontal axis. At the scale used for the graphical analysis, it is difficult to predict beyond the first few reflections. However this method of analysis does provide useful information. It should, of course, be remembered that the characteristics have been approximated by straight line sections. Accurate plots can be easily obtained by means of a curve tracer such as the Tektronix Type 576.

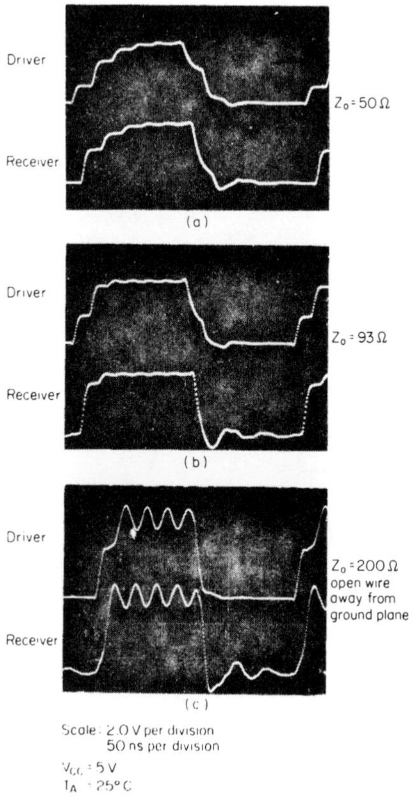

FIGURE 29

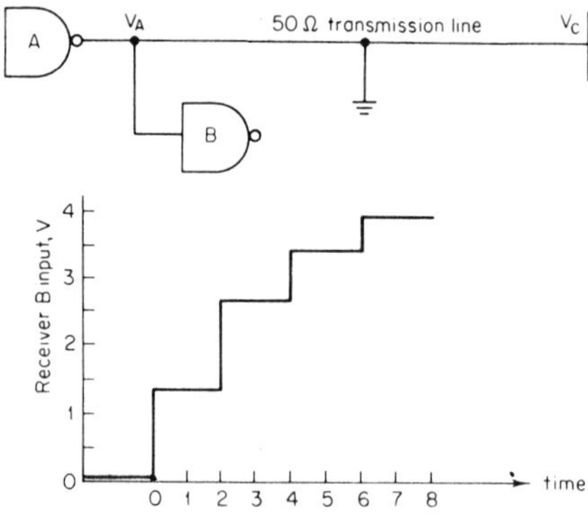

FIGURE 28

DESIGN RULE SUMMARY

The following rules have been established for the minimization of transmission-line effects in TTL systems.

1. Use direct wire interconnections that have no specific ground return for lengths up to about 10in. only. A ground plane is always desirable.

2. Direct wire interconnections must be routed close to a ground plane if longer than 10 in. and should never be longer than 20 in.

3. When using coaxial or twisted-pair cables, design around approximately 100Ω characteristic impedance. Coaxial cable of ≃ 100Ω impedance is recommended. For twisted pair. No. 26 or No. 28 S.W.G. wire with thin insulation twisted about 30 turns per foot works well. Higher impedances increase cross talk, and lower impedances are difficult to drive.

4. Ensure that transmission-line ground returns are carried through at both transmitting and receiving ends.

5. Connect reverse termination at driver output to prevent negative overshoot.

6. Decouple line-driving and line-receiving gates as close to the package V_{CC} and ground pins as practical, with a 0.1μF capacitor.

7. Gates used as line drivers should be used for that purpose only. Gate inputs connected directly to a line-driving output could receive erroneous inputs due to line reflections, long delay times introduced, or excessive loading on the driving gate.

8. Gates used as line receivers should have all inputs tied together to the line. Other logic inputs to the receiving gate should be avoided, and a single gate should be used as the termination of a line.

9. Any device that derives internal feedback from its outputs should not be used for line driving. Such devices are bistables, shift registers and monostables.

GENERAL RULES

Further rules have been established for general use of TTL in systems:

1. Power supply. Ripple <5% and regulation <5%. RF filter the primary supply.

2. Decouple every 5 – 10 packages with low inductance disk ceramics of value 0.01 to 0.1μF.

3. A ground plane is desirable when the printed circuit board contains a large number of packages. If no ground plane is used the ground line should be as wide as possible and the area around the board periphery used as a distribution system.

4. Data rise and fall times into TTL devices should be <50ns except in the case of Schmitt triggers.

5. Unused inputs of ALL TTL devices should be tied to guaranteed logic levels, e.g. inputs of NAND/AND gates may be tied to V_{CC} where V_{CC} is guaranteed to ALWAYS be <5.5V. Otherwise tie to V_{CC} through a series resistor, several inputs may share a common resistor.

6. Increased fan out can be obtained by paralleling devices in the same package.

7. Gate expanders should lie as close as possible to the gate being expanded. This avoids capacitive loading and noise pick-up.

8. Input data to master slave JK flip flops should not be changed when the clock is high.

9. If a ground plane is not used the ground distribution should be arranged as an interlinking mesh.

10. Gates that drive back wiring via an edge connector should be mounted near the connector ground in order to provide a low impedance return for line currents.

APPENDIX

Deviation of Crosstalk between lines

Gates 1 and 3 have output impedances Z_1. For simplicity, consider the inputs of gates 2 and 4 to be open circuits. The lines, whose transition times are longer than the rise time of the pulse edges, both have a characteristic impedance Z_o, and a mutual coupling impedance Z_m, as in Figure 30. Let the logic swing be V_s.

Now consider Figure 31. The swing V_s produces a step V_L which travels down the line. It is coupled by the impedance Z_m into the other line. The voltage induced travels to the gate input where, on being reflected by the open circuit, it is doubled. Therefore, the voltage at the gate $V_{in} = 2V_1$.

$$V_1 = (V_L.Z_o/2)/(Z_m + Z_o.3/2) \quad V_L = V_s.Z_o/(Z_1 + Z_o)$$

Therefore:–

$$V_{in}/V_s = 1/(1.5 + Z_m/Z_o).(1 + Z_1/Z_o)$$

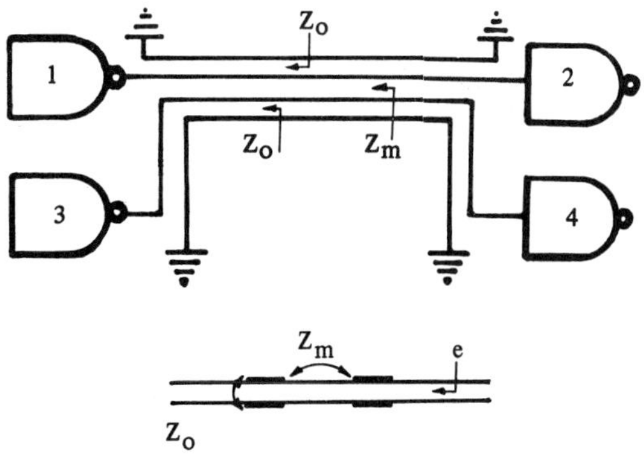

FIGURE 30

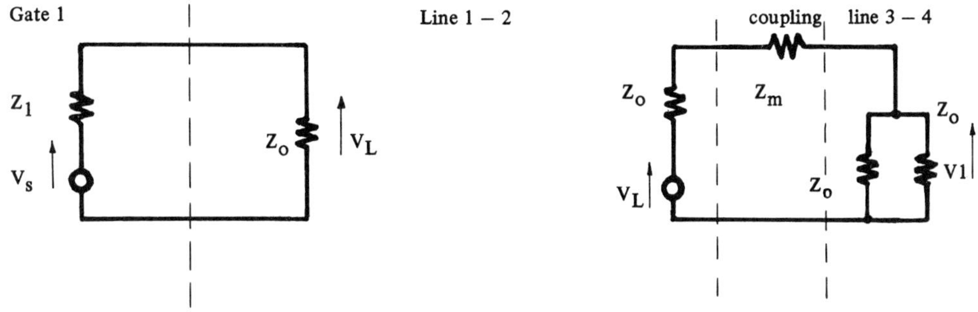

FIGURE 31

Index